Mαthematics
of Planet Earth

A Primer

Advanced Textbooks in Mathematics

ISSN: 2059-769X

Published

The Wigner Transform
by Maurice de Gosson

Periods and Special Functions in Transcendence
by Paula B Tretkoff

Mathematics of Planet Earth: A Primer
by Darryl Holm, Colin Cotter, Davoud Cheraghi, Ben Calderhead, Jochen Bröcker, Tobias Kuna, Ted Shepherd, Hilary Weller and Beatrice Pelloni edited by Dan Crisan

Forthcoming

Conformal Maps and Geometry
by Dmitry Belyaev

Published by

World Scientific Publishing Europe Ltd.

57 Shelton Street, Covent Garden, London WC2H 9HE

Head office: 5 Toh Tuck Link, Singapore 596224

USA office: 27 Warren Street, Suite 401-402, Hackensack, NJ 07601

Library of Congress Cataloging-in-Publication Data
Names: Holm, Darryl D., author. | Crisan, Dan, editor.
Title: Mathematics of planet earth : a primer / by Darryl Holm (Imperial College London, UK),
    Colin Cotter (Imperial College London, UK), Davoud Cheraghi (Imperial College London, UK),
    Ben Calderhead (Imperial College London, UK), Jochen Bröcker (University of Reading, UK),
    Tobias Kuna (University of Reading, UK), Ted Shepherd (University of Reading, UK),
    Hilary Weller (University of Reading, UK), Beatrice Pelloni (Heriot-Watt University, UK) ;
    edited by Dan Crisan (Imperial College London, UK).
Description: New Jersey : World Scientific, [2017] | Series: Advanced textbooks in mathematics |
    Includes bibliographical references and index.
Identifiers: LCCN 2017024278| ISBN 9781786343826 (hc : alk. paper) |
    ISBN 9781786343833 (pbk : alk. paper)
Subjects: LCSH: Mathematics--Textbooks.
Classification: LCC QA37.3 .H65 2017 | DDC 510--dc23
LC record available at https://lccn.loc.gov/2017024278

British Library Cataloguing-in-Publication Data
A catalogue record for this book is available from the British Library.

Desk Editors: V. Vishnu Mohan/Mary Simpson

Typeset by Stallion Press
Email: enquiries@stallionpress.com

Advanced Textbooks in Mathematics

Mαthematics
of Planet Earth
A Primer

**Ben Calderhead, Davoud Cheraghi,
Colin Cotter, Darryl Holm**

Imperial College London, UK

Beatrice Pelloni

Heriot-Watt University, UK

**Jochen Bröcker, Tobias Kuna,
Ted Shepherd, Hilary Weller**

University of Reading, UK

Edited by Dan Crisan
Foreword by Christiane Rousseau

NEW JERSEY · LONDON · SINGAPORE · BEIJING · SHANGHAI · HONG KONG · TAIPEI · CHENNAI · TOKYO

Foreword

It has been eight years since work on Mathematics of Planet Earth (MPE) started.... The idea for MPE came naturally to me. I am a dreamer, I like to explore the world around me with my mathematical eyes, and I was concerned with the challenges our planet was facing. There was more than just climate change and sustainability. I could imagine how kids could be interested in learning about the movements of our planet in the solar system, or discovering how a female Danish mathematician, Inge Lehmann, had discovered the inner core of the Earth in 1936. During the years 2010–2012, there was a fever in the world, with institutes and learned societies preparing dozens of workshops, long-term programs, public lectures and outreach activities, which took place during the year 2013, under the title Mathematics of Planet Earth 2013, or MPE2013. So much has been done in one year, and yet... so much more remains to be done.

Working for MPE has intensified my concern about global change and the challenge of sustainability. I had the privilege to learn so much and to meet so many scientists concerned with the state of the planet and the urgency for action: enriching meetings, but not at all reassuring. Some years ago, I made the calculation of what the rise to sea level would be if all the glaciers of Antarctica and Greenland were to melt (without counting thermic expansion of the oceans). When I found an answer of the order of 70–80 m, I looked for the mistake in my calculations but, unfortunately, there was none. This is where I started noticing that the experts never speak of the potential sea rise to come, but only of the potential sea rise "up to 2100". In 2014, during the general assembly of the International Council of Science (ICSU), I learnt from a lecture of Nancy Bertler that the West-Antarctica ice sheet is under intense observation: a potential collapse could bring a sea rise of 4–8 m before 2100.... On similar matters, it is only the 2013 IPCC report that takes into account the methane — a much stronger greenhouse gas than CO_2 — to be released by the melting of the permafrost in the Northern regions. But we learnt very recently that the level of methane in the atmosphere has significantly increased in the last few years, and that this methane does not yet come from the melting of the permafrost... Another ingredient that is missing in our models... The 2015 Paris Agreement was at least one good news. But governments change and may not respect the commitment of their countries... We discover new problems faster than we find solutions.

MPE2013 has enlightened the role that mathematical sciences have to play in addressing the planetary challenges, and the need to train a new generation of scientists to work on these problems. MPE2013 was also an unprecedented cooperation around the world. At the end of 2013, it turned into MPE. Indeed, the work is already starting. The most difficult ground work is ahead of us. Steps have been taken forward around the world: the new Doctoral Training Centre in Mathematics of Planet Earth, a joint enterprise of Imperial College London and the University of Reading, is an inspiring example where collaboration allows training researchers with both a high and a broad expertise. But there is also a need for action outside these high-level training centres. For a mathematician or a student, it is easy to hear a few colloquium talks on MPE topics and realise that there are indeed interesting mathematical problems. This is great, but far from sufficient: we need research in these areas. The problems of global change are very complex because so many elements are intertwined: oceans, atmosphere, soil, ice, vegetation. The equations are nonlinear and allow for feedback phenomena, tipping points and chaotic dynamics. How can one start contributing to the study or solutions of these problems? If we want a significant number of students and mathematicians to make the jump and orient their research on these issues, then there is a need for shortest paths. This is what this book intends to do around the themes of weather and climate.

The different chapters of the book cover a broad spectrum of subjects related to climate studies. The first chapter concentrates on the modelling of climate. It highlights the physical laws that underpin the climate models: energy balance, conservation of angular momentum, etc. It then shows the large-scale phenomena: Hadley circulation in the tropics, wind-driven ocean circulation, etc. Emphasis is put on the origin of uncertainties in predictions coming from various sources: feedback mechanisms, percentage of emitted CO_2 remaining in the atmosphere, type of clouds, etc. The very pedagogical PDE chapter provides a short path to the main equations of climate modelling and show the link between PDE and ODEs. The chapter on data and probability gives a short route to the modern probabilistic and statistical methods to treat the climate data and test the models. In particular, it highlights the importance of Bayesian statistics when only a small amount of data is available and yet, we need to adjust the models to these data. The chapter on dynamical systems presents the mathematics of chaotic systems, and the conditions leading to chaotic behaviour. The chapter on numerical methods focuses on the simulation of models for weather prediction. It explains the desired characteristics of numerical schemes: convergence, stability, consistency, order of accuracy, etc., the differences between explicit and implicit schemes, and focuses on the simulation of the one-dimensional linearised shallow-water equations.

This book is a great addition to the literature on Mathematics of Planet Earth, and it will interest a wide range of students, young researchers and established scientists.

Christiane Rousseau
Université de Montréal, Canada

Introduction

We live in exciting times! Mathematicians, as well as meteorologists, physicists, computer scientists, statisticians, are working together to shape up a new area of Mathematics: Mathematics of Planet Earth (MPE). Similar to Mathematical Biology and Mathematical Finance, Mathematics of Planet Earth is defined not through its subject matter, but through its area of application. Its applications are directed towards the planetary issues that we face today: climate change, quantification of uncertainty, move to an economy of sustainability, preservation of biodiversity, natural hazards, financial and social systems, adaptation to change and many others. Such issues give rise to an abundance of challenging scientific problems with a strong multidisciplinary characteristics.

The scientific community has responded strongly to these demands. A massive volume of research papers are routinely published in general circulation journals (e.g., *Discrete and Continuous Dynamical Systems, Journal of Computational and Applied Mathematics, Journal of Nonlinear Science, Mathematics Today, Nonlinearity, Physica D*) as well as specialist ones (e.g., *Dynamics and Statistics of the Climate System, Environmental and Ecological Statistics, International Journal on Geomathematics, Journal of Atmospheric Sciences, Stochastic Environmental Research and Risk Assessment, Mathematical Geosciences*). A large number of books related to the Mathematics of Planet Earth have been published by all major publishers including the American Mathematical Society, Cambridge University Press, Institute of Physics Publishing, Springer Verlag, the Society of Industrial and Applied Mathematics, Princeton University Press and World Scientific.

The activity of the researchers in this area is supported by the new Society of Industrial and Applied Mathematics Activity Group on

Mathematics of Planet Earth (SIAG/MPE).[1] The group provides a forum for mathematicians and computational scientists to study Planet Earth, its life-supporting capacity and the impact of human activities. Activities of the SIAG include the biennial SIAM Conference on Mathematics of Planet Earth[2] and minisymposia at SIAM Annual Meetings and other conferences. In addition, there is a thriving network linking researchers across the globe to develop the Mathematics of Planet Earth: The Mathematics and Climate Research Network (MCRN) is a virtual organisation of active researchers in mathematics and the geosciences.

The future of any discipline, particularly a burgeoning one, relies on its capacity to continuously attract graduate students and young researchers. This is precisely the aim of the Centre for Doctoral Training in the Mathematics of Planet Earth (MPECDT). The Centre started its activities in 2013, the year of the Mathematics of Planet Earth,[3] following an award from the UK's Engineering and Physical Sciences Research Council. The MPECDT is providing a bespoke graduate training programme using the expertise in the Mathematics and Meteorology departments at two leading UK universities, Imperial College London and the University of Reading, in partnership with other UK and international scientific institutions, including the UK Met Office, the European Centre for Medium Range Weather Forecasts (ECMWF), the German Weather Service (DWD), the National Centre for Atmospheric Science (NCAS), the National Centre for Earth Observation (NCEO) and the National Physical Laboratory (NPL), etc.

The material incorporated in this textbook is largely based on five bespoke courses given to students enrolled at the MPECDT. For the last three years, we have welcomed three cohorts of PhD students who have studied these courses. All contributors to the textbook have been directly involved in teaching the corresponding courses offered to MPECDT students in their first year of study as part of a Master of Research in Mathematics. Since there are very few other similar resources, we have decided to disseminate our course material to other interested graduate students and young researchers. Indeed, with the notable exception of Kaper and Engler (2013), we are not aware of a similar textbook on the Mathematics of Planet Earth.

---

[1] http://www.siam.org/activity/mpe.
[2] See www.siam.org/meetings/mpe16/ for the 2016 edition of the Conference.
[3] www.mpe2013.org.

It would be futile to attempt a description of the many applications of the Mathematics of Planet Earth. Their length and breadth is already too wide to address in a single volume. Instead, this textbook aims to provide the starting point (hence the choice of *a Primer* in the title) for acquiring the mathematics background required for pursuing research in the Mathematics of Planet Earth with particular emphasis on applications related to atmosphere and oceans, including numerical weather prediction, climate change and oceans modelling.

By its very nature, Mathematics of Planet of Earth comes at the confluence of several areas of "traditional" Mathematics: partial differential equations, dynamical systems, probability and statistics, and numerical analysis. Naturally, the textbook includes a chapter dedicated to each of these four areas. In the following, we give a succinct description topics covered in each chapter.

To illustrate the application area and motivate the choice of the material incorporated in the textbook, we begin a chapter that describes the atmosphere and ocean circulation. The goal of Chapter 1 is to show how many of the key observed features of the general circulation of the atmosphere and oceans may be explained from the fundamental physical constraints that are represented mathematically in the governing equations. Most of the emphasis is on aspects of climate, meaning the time-averaged behaviour of the system (including higher-order moments such as correlations of fluctuations), but some discussion is also provided of atmospheric weather systems, meaning the day-to-day variations of the state of the atmosphere. The observed features are explained through the mathematical model of the relevant physical processes with connections to the partial differential equations of geophysical fluid dynamics. Crucial aspects of the general circulation are covered, including the wind-driven ocean circulation, stationary Rossby waves and atmospheric transient disturbances.

Chapter 2 introduces the notion of a partial differential equation (PDE), and the basic definitions and results concerning PDEs. We describe the governing equations for an ideal fluid and give the fundamental properties, before analysing in greater detail the case of one-dimensional flows. We cover an example of a three-dimensional system of PDEs governing large-scale atmospheric flows, of particular mathematical interest. In the second part of the chapter, we cover the various representations and approximations of ideal shallow water dynamics in a rotating frame. These rotating shallow water (RSW) equations possess a slow–fast decomposition in which they reduce approximately to quasigeostrophic (QG) motion (conservation of energy and potential vorticity) plus nearly decoupled equations for gravity waves in an asymptotic expansion in small Rossby number. The solution properties of the RSW equations are discussed, and some alternative representations of the RSW equations which highlight the slow + fast interactions are given. Finally, the RSW equations and the thermal RSW equations are derived as Euler–Poincare equations from Hamilton's principle in the Eulerian fluid representation.

Chapter 3 covers the basics of probability theory and Bayesian computation. We begin with the most basic concepts including probability spaces, integration, distributions and independence. Some elementary statistics notions are also incorporated in this chapter, as well as conditional expectation, convergence in distribution, and the Central Limit Theorem. The second part of the chapter addresses computational aspects of statistical inference. We cover the concept of Bayesian inference and the class of probabilistic methods known as Monte Carlo integration, as well as some simple numerical simulation methods for arbitrary random variables. Finally, we introduce the Markov Chain Monte Carlo method.

Chapter 4 offers some a basic introduction to dynamical system theory, touching on topics such as entropy, ergodicity and transfer operators. A good understanding of these topics is essential to understand the qualitative mechanism which drives the power of stochastic predictions over deterministic predictions. The theory of dynamical systems studies the time evolution of systems in general terms. Their evolution is given by a law as for example a recursion relation, a (partial) differential equation, an integral equation or even a random evolution. The theory of dynamical systems takes a global and more qualitative viewpoint trying to work out properties which are genuine and have features independent of the details of the considered dynamics; in other words, a more qualitative than quantitative approach. This chapter covers properties which are generalisable to large complex systems, like the Earth's climate system itself. Two paradigmatic toy examples, rotation and expanding maps are studied in detail. The concept of chaos is introduced and related to symbolic dynamics, structural stability and topological entropy. As an introduction to probabilistic methods, basic ergodic theory is presented, and for expanding, hence chaotic, systems further properties are derived such as decay of correlation, fluctuations and linear response.

Chapter 5 gives a taster of some numerical methods used in atmosphere and ocean models and their numerical analysis. The full equations have advective and faster wave terms, and so we start by considering separately numerical methods for the advection and second-order wave terms. We introduce finite difference methods as well as the concepts of stability, consistency, numerical dispersion and convergence. Von Neumann stability analysis is described and used to predict numerical dispersion. In the second part of the chapter, we introduce iterative methods for solving the massive linear systems that arise in operational weather, ocean and climate. We explain how these systems arise and describe the main workhorse for large-scale linear solvers, the conjugate gradient solver. An analysis of the convergence motivates the introduction of preconditioners. Finally, we recall some classical iterative methods in the context of preconditioning and study their properties.

The authors of the book would like to thank the students of the MPECDT who were the first to sample the material contained in this textbook over the last three years. We benefited greatly from their useful comments and constructive feedback. Special thanks are due to Alastair Gregory for his continuous assistance in assembling the manuscript. Last but not least, we would like to thank World Scientific for the opportunity to help us disseminate our course material to other interested graduate students and young researchers. We owe special thanks to Laurent Chaminade and Mary Simpson for their patience and diligence in helping us complete the project.

Dan Crisan
Director MPECDT
Imperial College London, UK

March 2017

Contents

Beatrice Pelloni and Darryl Holm

Chapter 1

The General Circulation
of the Atmosphere and Oceans

Theodore G. Shepherd

University of Reading, UK

1.1 Introduction

The goal of this chapter is to show how many of the key observed features
of the general circulation of the atmosphere and oceans may be explained
from the fundamental physical constraints that are represented mathematically in the governing equations. Most of the emphasis is on aspects of
climate, meaning the time-averaged behaviour of the system (including
higher-order moments such as correlations of fluctuations), but some discussion is also provided of atmospheric weather systems, meaning the day-
to-day variations of the state of the atmosphere. The observed features are
explained through the simplest possible mathematical model of the relevant
physical processes, and there are many connections to the PDE models of
geophysical fluid dynamics discussed in Chapter 2.

Analytical solutions of the governing equations are restricted to a few
very special cases, and even the simplest dynamical models can generally only be solved numerically (Chapter 5). Also, comparisons between
models and observations, as well as the predictability of weather, raise
issues of probability and representativeness (Chapter 3), whilst system-
level (or emergent) behaviour requires understanding of dynamical systems
(Chapter 4). None of that is touched upon in this chapter.

It is rather remarkable how such crucial aspects of the general circulation as tropical rain belts, deserts, storm tracks, coastal upwelling regions
(important for fisheries), and western oceanic boundary currents — also

westerly and easterly surface winds, which used to be enormously important in the days of sailing ships — can be explained in a qualitative way
from basic physical principles. This is the basis of the concept of using a
hierarchy of models to represent the atmosphere and oceans. By including
more detailed mathematical representations of the relevant physical processes, the model solutions can be expected to become more quantitatively
accurate. Nevertheless, the essential causal relationships are represented in
the simplest models, and these are often the basis of diagnostic theories that
are used to understand the workings of the more complex models, such as
those used for operational weather and climate prediction.

Some thoughts are also provided at the end of this chapter concerning what can be said about climate change. Here, the difference between
qualitative and quantitative understanding becomes crucial.

1.2   Zonally Integrated View of Atmospheric and Oceanic Circulation

The atmosphere and oceans obey the laws of fluid dynamics, represented by
the Navier–Stokes equations which express conservation of energy, momentum, and mass (see Chapter 2, Sec. 2.2 for the incompressible case, and
Chapter 5, Sec. 5.1 for the compressible case). For the atmosphere, these
are complemented by a conservation law for moisture. The steady forms
of these conservation laws represent physical balances, and much can be
deduced about the circulation from them. It is convenient to start by considering the zonally (or longitudinally) integrated balances.

1.2.1   *Energy balance*

Because the Earth is approximately spherical, incident solar radiation is
approximately proportional to the cosine of latitude, modulated by the
seasonal cycle. Hence, most solar radiation is absorbed in the tropics. The
outgoing longwave radiation (OLR), which depends on atmospheric temperature according to the Stefan–Boltzmann (blackbody radiation) law, is
also greatest in the tropics because that is where temperatures are highest,
but the equator-to-pole contrast in OLR is weaker than that in the absorbed
solar radiation (ASR). Overall, the net OLR must balance the ASR, otherwise the atmospheric temperature adjusts to make it so. As a result, there
is a net heating of the climate system in the tropics (ASR > OLR), where
temperatures are relatively high, and a net cooling in the higher latitudes

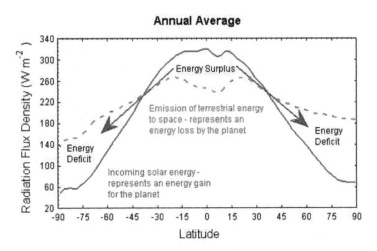

Figure 1.1. Energy balance of the climate system as a function of latitude. The emission of terrestrial radiation is the OLR, and the incoming solar energy is the ASR. Reproduced with permission from Suomi virtual museum, University of Wisconsin-Madison (http://profhorn.meteor.wisc. edu/wxwise/museum/a2main.html).

(OLR > ASR), where temperatures are relatively low (Fig. 1.1). Conservation of energy implies a downgradient transfer of energy by the climate system from the tropics to higher latitudes. Indeed the second law of thermodynamics implies that it must be this way.

Observations show that most of this poleward energy transport is accomplished by the atmosphere. The oceanic heat transport is important regionally, and for the time evolution of climate (since the heat capacity of the oceans is much greater than that of the atmosphere), but it can be ignored for a basic understanding of the energy balance of the climate system. Thus, the rest of Sec. 1.2 focuses on the atmosphere.

There is also a vertical transfer of energy within the atmosphere, because most of the ASR is absorbed at the Earth's surface, whilst the longwave emission to space occurs from the atmosphere itself (Fig. 1.2). This is because the atmosphere is largely transparent to solar radiation but opaque to longwave radiation. The latter property is what gives rise to the greenhouse effect (Sec. 1.6). Because the atmosphere is heated from below and cooled from above, it undergoes convective instability (much as soup heated on a stove), which leads to turbulent heat transport. In contrast, the ocean is heated from above so it is not subject to spontaneous instability; its motion is mainly forced mechanically, by the atmosphere (Sec. 1.3).

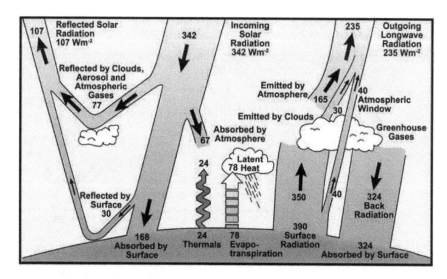

Figure 1.2. The energy balance of the climate system. Most of the solar radiation that is not reflected is absorbed at Earth's surface. But most of the OLR is emitted from the atmosphere itself. This gives rise to the greenhouse effect. Reproduced from Intergovernmental Panel on Climate Change (IPCC) Fourth Assessment Report, 2007.

1.2.2 *Energy equation*

The atmosphere is normally treated as an ideal gas, for which the 1st law of thermodynamics takes the form

$$dQ = c_p dT - \frac{1}{\rho} dp. \tag{1.1}$$

Here, dQ is the diabatic heat input, c_p is the heat capacity at constant pressure, T is temperature, ρ is density, and p is pressure. This applies to a fluid parcel, so leads to the Lagrangian equation

$$\rho c_p \frac{DT}{Dt} = \frac{Dp}{Dt} + \rho \frac{DQ}{Dt}, \tag{1.2}$$

where $\frac{D}{Dt} = \frac{\partial}{\partial t} + \vec{v}\cdot\nabla$ is the Lagrangian or material derivative, following a fluid parcel (see Chapter 2, Definition 2.7), where $\vec{v}$ is velocity. It is conventional in meteorology to use the symbol Q for the diabatic heating rate, i.e., for $\rho \frac{DQ}{Dt}$ above, which leads to

$$\rho c_p \frac{DT}{Dt} = \frac{\partial p}{\partial t} + \vec{v}\cdot\nabla p + Q. \tag{1.3}$$

The second term on the right-hand side is the so-called pressure-work term, which is related to kinetic energy and thus follows from Newton's second law (in a rotating frame)

$$\frac{D\vec{v}}{Dt} + 2\vec{\Omega} \times \vec{v} = -\frac{1}{\rho}\nabla p + \vec{g} + \vec{F}. \tag{1.4}$$

Here, $\vec{\Omega}$ is the angular rotation of the Earth, $\vec{g} = -g\hat{z}$ is the gravitational acceleration, $\hat{z}$ is the unit vector in the vertical direction, and $\vec{F}$ is friction. Taking the scalar product of this equation with the velocity, the Coriolis term drops out and we obtain

$$\frac{D}{Dt}\left(\frac{1}{2}|\vec{v}|^2\right) = -\frac{1}{\rho}\vec{v}\cdot\nabla p - wg + \vec{v}\cdot\vec{F}, \tag{1.5}$$

where w is the vertical component of velocity. The pressure-work term in this equation can be cancelled with that in the thermodynamic equation above, whilst $w = \frac{Dz}{Dt}$ where z is the vertical spatial coordinate. This leads to

$$\rho\frac{D}{Dt}(K + \Phi + c_p T) = \frac{\partial p}{\partial t} + Q + \rho(\vec{v}\cdot\vec{F}), \tag{1.6}$$

where $K = \frac{1}{2}|\vec{v}|^2$ is the kinetic energy, $\Phi = gz$ (known as the geopotential) is the gravitational potential energy, and $c_p T$ is the internal energy (all per unit mass).

In practice, the kinetic energy is negligible (approximately $10^2\,\mathrm{m}^2/\mathrm{s}^2$) compared with the other two forms of energy (approximately $10^5\,\mathrm{m}^2/\mathrm{s}^2$), and the frictional loss term is similarly negligible. Furthermore, a significant component of Q is the latent heat release from condensation of water vapour (as occurs in rising air, as the temperature and pressure decrease; see the Appendix), which is really a conversion between different forms of energy. Thus, we may write

$$\rho\frac{D}{Dt}(\Phi + c_p T + Lq) = \frac{\partial p}{\partial t} + \tilde{Q}, \tag{1.7}$$

where L is the latent heat of evaporation, q is the specific humidity, and $\tilde{Q}$ is the residual diabatic heating from radiation and surface fluxes. The quantity in brackets is known as the moist static energy. This equation provides a local representation of the atmospheric energy budget that is widely used in theoretical studies. It is important to note that $\vec{v}$ is crucial for energy transport, but not for energy itself.

1.2.3 Fluxes

The mass continuity equation $\frac{\partial \rho}{\partial t} + \nabla \cdot (\rho \vec{v}) = 0$ leads to the integral relation

$$\frac{d}{dt} \int_V \rho dV = \int_V \frac{\partial \rho}{\partial t} dV = -\int_V \nabla \cdot (\rho \vec{v}) dV = -\int_A \rho \vec{v} \cdot \hat{n} dA \qquad (1.8)$$

over a fixed volume V, where A is the surface bounding V and $\hat{n}$ is the outward normal. The last step uses the Gauss divergence theorem (Chapter 2, Sec. 2.1.3). This equation expresses the fact that changes in the mass within V are related to the fluxes of mass into or out of V.

Consider an arbitrary scalar H satisfying $\frac{DH}{Dt} = \frac{\partial H}{\partial t} + \vec{v} \cdot \nabla H = S$. (In the case of energy, for example, H is the moist static energy and $\rho S = \tilde{Q}$.) In order to write a budget for H, it is necessary to re-cast the advection term $\vec{v} \cdot \nabla H$ in flux form.

$$\frac{\partial}{\partial t}(\rho H) = \rho \frac{\partial H}{\partial t} + H \frac{\partial \rho}{\partial t}$$

$$= -\rho \vec{v} \cdot \nabla H + \rho S - H \nabla \cdot (\rho \vec{v}) = -\nabla \cdot (\rho H \vec{v}) + \rho S. \qquad (1.9)$$

Therefore,

$$\frac{d}{dt} \int_V \rho H dV = \int_V \frac{\partial(\rho H)}{\partial t} dV = -\int_A \rho H \vec{v} \cdot \hat{n} dA + \int_V \rho S dV. \qquad (1.10)$$

In the steady limit, this implies a balance between the flux of H into or out of V, and the sources or sinks within V:

$$\int_A \rho H \vec{v} \cdot \hat{n} dA = \int_V \rho S dV. \qquad (1.11)$$

An application of this relationship is that the northward flux of H across a latitude circle must equal the loss (via S) in the polar cap bounded by that latitude circle:

$$\int_0^\infty \int_0^{2\pi} \rho H v a \cos(\phi) d\lambda dz \bigg|_{\phi=\phi_1} = -\int_{\phi \geq \phi_1} \rho S dV. \qquad (1.12)$$

Here, ϕ is latitude, λ is longitude (in radians), a is the radius of the Earth, and the scalar v is the meridional (northward) component of velocity, which is oppositely directed to the outward normal of the northern hemisphere polar cap (hence the minus sign).

Although the northward flux of H, $\rho H v$, appears to be a cubic quantity, the vertical integral in altitude z can be replaced by an integral over pressure p, using the hydrostatic relation $\rho g dz = -dp$. This eliminates the factor of

ρ and renders the flux of H quadratic, Hv. It is then useful to decompose the zonal average as follows:

$$\frac{1}{2\pi} \int_0^{2\pi} Hv d\lambda = \overline{Hv} = \overline{H}\,\overline{v} + \overline{H'v'}, \tag{1.13}$$

where $v' = v - \overline{v}$ is the departure from the zonal average, since $\overline{\overline{H}v'} = 0 = \overline{H'\overline{v}}$. The physical interpretation of these terms is that $\overline{H}\,\overline{v}$ is the northward flux of H by the zonal-mean meridional circulation, and $\overline{H'v'}$ is the northward flux of H by the eddies (departures from the zonal mean). These considerations apply to any scalar H.

Observations show that the poleward energy transport in the tropics is dominated by the mean component, through the Hadley circulation, whilst in the extratropics it is dominated by the eddy component (Fig. 1.3). The latter can be viewed as downgradient heat transport by macroturbulence, generated by baroclinic instability (Sec. 1.5), although there is also a contribution from stationary eddies (Sec. 1.4).

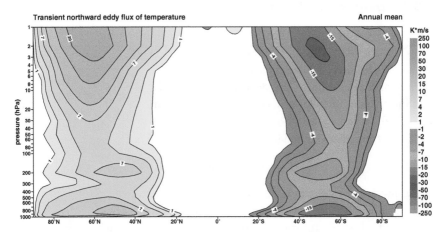

Figure 1.3. Northward heat flux by time varying (transient) eddies (departures from zonal mean flow), which is primarily associated with extratropical storms driven by baroclinic instability. The flux is poleward in both hemispheres, providing downgradient energy transport by atmospheric macroturbulence. Within the tropics, the poleward energy flux is provided by the Hadley circulation, which is a zonal mean feature and thus not visible here. It is conventional in meteorology to use pressure as the vertical coordinate, as this provides a mass weighting. 1000 hPa is approximately the average surface pressure, and 250 hPa is approximately 10 km altitude. Note that the northern hemisphere is placed on the left. Reproduced from ERA-40 Atlas, European Centre for Medium-range Weather Forecasts (http://www.ecmwf.int/s/ERA-40_Atlas/docs/).

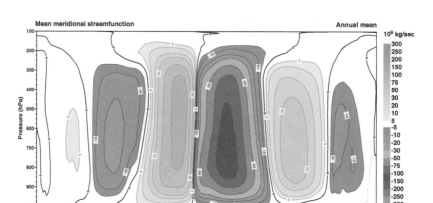

Figure 1.4.   Streamfunction of the zonally averaged mass flux in the latitude–altitude plane. The streamfunction shows ascent in the deep tropics, and descent in the subtropics of both hemispheres. There is also a weaker reverse circulation at higher latitudes, but this is strongly compensated by eddy fluxes and has no physical significance on its own. Reproduced from ERA-40 Atlas.

1.2.4   *Hadley circulation*

The zonally averaged tropical circulation consists of rising motion in the deep tropics, driven by the latent heat release associated with (moist) convective instability, which then moves poleward before descending to the surface and returning to the deep tropics (Fig. 1.4). This meridional overturning circulation is known as the Hadley circulation. Its latitudinal extent is limited by conservation of (axial) angular momentum.

The atmospheric angular momentum (per unit mass)

$$M = a\cos(\phi)\big(\Omega a\cos(\phi) + u\big) \tag{1.14}$$

is the product of the moment arm (distance from Earth's axis of rotation) and the absolute zonal velocity, the latter consisting of the velocity associated with Earth's rotation as well as the zonal component of velocity in the rotating frame, u. M is conserved following an air parcel, apart from torques. Now imagine a parcel of air (actually a zonal ring of air) rising at the equator ($\phi = 0$) with $u = 0$. As the air moves poleward aloft, then if M is conserved,

$$a\cos(\phi)\big(\Omega a\cos(\phi) + u\big) = M(\phi)$$

$$= M(0) = a^2\Omega \implies u(\phi) = \frac{\Omega a\sin^2(\phi)}{\cos(\phi)}. \tag{1.15}$$

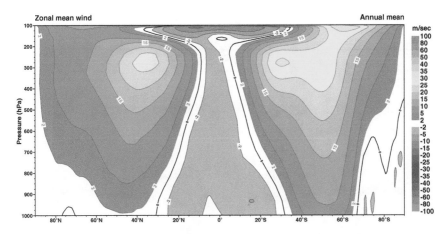

Figure 1.5. Zonal mean zonal wind. Maxima are found in the subtropical upper troposphere corresponding to the poleward limit of the Hadley circulation, which through conservation of angular momentum acts to increase the zonal flow as it moves away from the equator. The secondary jet evident in the Southern Hemisphere is considered to be eddy-driven (see Sec. 1.5). Reproduced from ERA-40 Atlas.

The latter expression increases away from the equator, and diverges as $\phi \to \pm\frac{\pi}{2}$ (the poles). A physical upper limit on u is provided by the equator-to-pole temperature difference together with thermal-wind balance (see the Appendix), and the assumption that $u \approx 0$ at the Earth's surface, because of friction. Using the above relation, this implies a limit on the poleward extent of the Hadley circulation. For Earth parameters, the limit is approximately 30 degrees latitude, which is remarkably close to the observed extent (Fig. 1.4) (Held and Hou, 1980). In reality, angular momentum is not conserved because of damping by eddy momentum fluxes in the upper troposphere (Sec. 1.5), but this simple theory gives a reasonable first approximation.

The poleward limit of the Hadley circulation has important implications for climate. The maximum zonal wind speed in the upper troposphere occurs at this latitude, a feature known as the subtropical jet (Fig. 1.5). There is a strong seasonal cycle to the Hadley cell; during the solstice seasons, the ascending branch is located in the summer hemisphere and the main cell extends across the equator. Thus, the subtropical jet is by far the strongest in the winter season. At the limit of the Hadley circulation, the air is forced to descend. Since descending air warms adiabatically, moisture cannot condense and this suppresses rainfall, leading to the subtropical band of deserts found on Earth (Fig. 1.6).

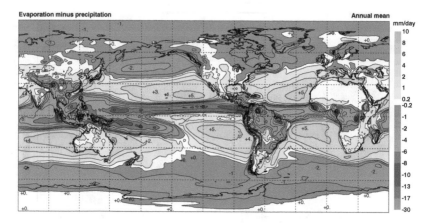

Figure 1.6.   Evaporation minus precipitation. There is net evaporation in the subtropical oceanic regions, corresponding to the descending branch of the Hadley circulation. Over land these regions can be deserts, which have neither precipitation nor evaporation. There is net precipitation over the inter-tropical convergence zones of the deep tropics, representing the tropical rain belts, and over latitudes poleward of roughly 45 degrees latitude. This implies a moisture flux from the subtropics to both lower and to higher latitudes. Reproduced from ERA-40 Atlas.

1.2.5   Moisture budget

Turbulent motions generically lead to downgradient transport of tracers. Thus, in the extratropics, where meridional fluxes are dominated by the eddies, we find a downgradient and hence poleward flux of moisture. In the tropics, however, the fluxes are dominated by the mean motion associated with the Hadley circulation. The moisture flux is determined by the lower branch of the Hadley circulation, since that is where most of the moisture resides (the upper troposphere is comparatively dry), and thus is directed equatorward. Hence, the tropical moisture transport is upgradient (Fig. 1.7), and provides the moisture needed for the tropical rain belts associated with the rising branch of the Hadley circulation (Fig. 1.6). Upgradient transport is possible since the motion is coherent. Although thermodynamic constraints require the energy flux to be downgradient (Sec. 1.2.1), moisture represents only a minor component of the moist static energy, so there is no contradiction.

1.2.6   Angular momentum budget

From Newton's second law (see above), it can be shown that

$$\frac{DM}{Dt} = -\frac{1}{\rho}\frac{\partial p}{\partial \lambda} + \frac{a\cos(\phi)}{\rho}\frac{\partial \tau_x}{\partial z}, \tag{1.16}$$

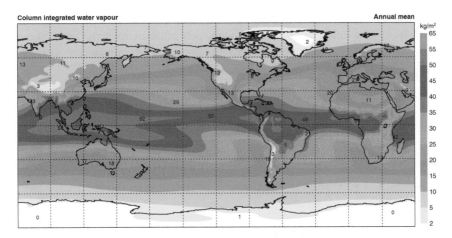

Figure 1.7.   Total water vapour in the atmosphere. The largest values are found over the tropical rain belts. Whilst the moisture flux to high latitudes is downgradient, the moisture flux to tropical latitudes is upgradient. Reproduced from ERA-40 Atlas.

where τ_x is the zonal component of the frictional stress. The right-hand side of this equation represents the torques that can change the atmospheric angular momentum.

Taking an average over longitude λ, the first term on the right-hand side vanishes unless the integral (which is taken at constant z) is interrupted by topography. In the presence of a single mountain, the integral will yield a contribution proportional to the difference in pressure between the two sides of the mountain, known as the pressure or mountain torque. For example, if the pressure is higher on the western side of the mountain than on the eastern side (which is generally true if the flow is eastward), then the atmosphere acts to accelerate the Earth, and M is transferred from the atmosphere to the Earth.

Integrating the above equation (after weighting by ρ) in z, the second term on the right-hand side leads to a surface torque $-a\cos(\phi)\tau_{xs}$, as there is no contribution from the top of the atmosphere. This is known as the friction torque. Thus, the only torques acting on the atmosphere are those associated with interaction with the lower surface, either the solid Earth or the ocean. The interactions with the solid Earth are manifest in small, but discernible, variations in the length of day. The interactions with the ocean are the origin of the wind-driven ocean circulation (Sec. 1.3).

We now apply the flux considerations from Sec. 1.2.3 with $H = M$. Taking the time average and integrating over the sphere, the net torque must vanish: $\int \rho S dV$. Since the mountain torque is generally of the same sign as the friction torque (e.g., an eastward surface flow is generally retarded by both terms), it follows that the surface winds must be westward over some parts of the Earth, and eastward over other parts. At any given latitude, the northward flux of angular momentum is

$$\int_0^\infty \int_0^{2\pi} \rho M v a \cos(\phi) d\lambda dz = \int_0^\infty \int_0^{2\pi} \rho \big(\Omega a \cos(\phi) + u\big) v a^2 \cos^2(\phi) d\lambda dz.$$
(1.17)

In steady state, mass conservation implies no net northward mass flux: $\int_0^\infty \int_0^{2\pi} \rho v a \cos(\phi) d\lambda dz = 0$. Hence, only the relative component of angular momentum, i.e., that associated with u, contributes to the northward flux:

$$\int_0^\infty \int_0^{2\pi} \rho u v a^2 \cos^2(\phi) d\lambda dz.$$
(1.18)

Apart from the density factor, this is again a quadratic quantity and can be decomposed into mean and eddy fluxes. In the tropics, the mean flux $\bar{u}\,\bar{v}$ dominates to give a poleward angular momentum flux from the Hadley circulation. In the extratropics, the eddy flux $\overline{u'v'}$ dominates, and its sense depends on the nature of the eddies. As we will see later (Sec. 1.5), this flux is also poleward, corresponding to propagation of Rossby waves primarily from the extratropics into the tropics. This is not a given; under different conditions (e.g., other planets, or palaeoclimates) it could be the other way around. But on present-day Earth, the eddy momentum flux is out-of-the tropics. As a result, we have westward surface flow in the tropics (known as easterlies), providing a source of atmospheric angular momentum, and eastward surface flow in the extratropics (known as westerlies), providing a sink of atmospheric angular momentum (Fig. 1.5).

1.3   The Wind-Driven Ocean Circulation

1.3.1   *Ekman transport*

In the surface layer of both atmosphere and ocean, where friction is important, the steady horizontal momentum balance departs from geostrophic balance (the small-Rossby-number, inviscid limit of Newton's second law; see Chapter 2, Sec. 2.5):

$$f\hat{z} \times \vec{u} = -\frac{1}{\rho}\nabla p + \frac{1}{\rho}\frac{\partial \vec{\tau}}{\partial z},$$
(1.19)

where $\vec{u}$ is horizontal velocity and $f = 2\Omega \sin(\phi)$ is the Coriolis parameter. But geostrophic velocity is defined by $f\hat{z} \times \vec{u_g} = -\frac{1}{\rho}\nabla p$. Hence, we may write

$$f\hat{z} \times \vec{u}_{ag} = \frac{1}{\rho}\frac{\partial \vec{\tau}}{\partial z}, \tag{1.20}$$

where $\vec{u}_{ag} = \vec{u} - \vec{u_g}$. Taking the cross product of $\hat{z}$ with this equation, and using the identity $\hat{z} \times \hat{z} \times \vec{u}_{ag} = -\vec{u}_{ag}$ since $\vec{u}_{ag} \cdot \hat{z} = 0$, yields

$$-f\vec{u}_{ag} = \frac{1}{\rho}\frac{\partial}{\partial z}(\hat{z} \times \vec{\tau}). \tag{1.21}$$

The physical interpretation is that friction induces a departure from geostrophic balance, such that the flow converges within a low-pressure system; there is then a unique three-way vector balance between the pressure gradient force, the Coriolis force, and the frictional drag.

Integrating this relation in the vertical yields

$$\int_{z_1}^{z_2} \rho\vec{u}_{ag} dz = -\left[\frac{1}{f}\hat{z} \times \vec{\tau}\right]_{z_1}^{z_2}. \tag{1.22}$$

Typically, the stress is confined to the boundary layer, maximising in magnitude at the surface $z = 0$ and decaying to zero at the top of the atmospheric boundary layer, $z = z_{b,atm}$, and at the bottom of the oceanic boundary layer, $z = -z_{b,oc}$. The same stress $\vec{\tau}_s$ applies to both atmosphere and ocean at the surface where they meet. This equation may then be applied to give the so-called Ekman transport for both atmosphere and ocean

$$M_{Ek,atm} = \int_0^{z_{b,atm}} \rho\vec{u}_{ag} dz = \frac{1}{f}\hat{z} \times \vec{\tau}_s,$$

$$M_{Ek,oc} = \int_{-z_{b,oc}}^0 \rho\vec{u}_{ag} dz = -\frac{1}{f}\hat{z} \times \vec{\tau}_s = -M_{Ek,atm}. \tag{1.23}$$

The fact that the atmospheric and oceanic Ekman transport are equal and opposite is a consequence of Newton's 3rd law, and reflects the overall conservation of angular momentum; friction leads to an exchange of angular momentum between atmosphere and ocean.

Ekman transport has implications for vertical velocity through conservation of mass, which under steady conditions gives the constraint

$\rho \nabla \cdot \overrightarrow{u} + \frac{\partial(\rho w)}{\partial z} = 0$. Hence,

$$[\rho w]_{z_1}^{z_2} = -\int_{z_1}^{z_2} \rho \nabla \cdot \overrightarrow{u} \, dz = -\int_{z_1}^{z_2} \rho \nabla \cdot \overrightarrow{u}_{ag} dz$$

$$= \nabla \cdot \left[\frac{1}{f} \hat{z} \times \overrightarrow{\tau} \right]_{z_1}^{z_2} = -\hat{z} \cdot \nabla \times \left[\frac{\overrightarrow{\tau}}{f} \right]_{z_1}^{z_2} \tag{1.24}$$

using the fact that $\nabla \cdot \overrightarrow{u}_g = 0$ to a first approximation (assuming the length scale characterising the flow is much smaller than a). Now use the boundary condition $w = 0$ at $z = 0$, which applies to both atmosphere and ocean. Then,

$$\rho w_{Ek,atm} = \rho w(z = z_{b,atm}) = \hat{z} \cdot \nabla \times \frac{\overrightarrow{\tau}_s}{f},$$

$$\rho w_{Ek,oc} = \rho w(z = -z_{b,oc}) = \hat{z} \cdot \nabla \times \frac{\overrightarrow{\tau}_s}{f}. \tag{1.25}$$

This implies that the vertical pumping into or out of the boundary layer (known as the Ekman layer) is in the same sense for both atmosphere and ocean.

As an example, consider the case of a tropical cyclone. This is a low-pressure atmospheric vortex, with converging Ekman transport. In the atmosphere, the Ekman pumping is upward, leading to latent heat release and intensification of the cyclone. In the ocean, the Ekman pumping is similarly upward, bringing cold water from depth up to the surface (and weakening the cyclone). Indeed, tropical cyclones are observed to leave tracks of cold surface waters in their wake as they move (Price, 1981).

1.3.2　*Regions of systematic oceanic upwelling*

The atmosphere is characterised by persistent subtropical high pressure systems over the oceans (Fig. 1.8). At eastern boundaries of the oceans, the atmospheric surface flow is therefore equatorward, which implies that the Ekman transport is offshore and the Ekman pumping upwards. This coastal upwelling accounts for the cold surface waters found along the western coasts of continents (Fig. 1.9), which are good for fishing.

A special case is the Antarctic ocean. Here, the atmospheric flow is circumpolar, which induces Ekman upwelling around the entire Antarctic coastline. This is a key component of the oceanic thermohaline circulation.

At the equator, the change in sign of the Coriolis parameter means that a coastline is not required to generate systematic upwelling. The westward atmospheric flow in the tropics induces equatorward convergence in

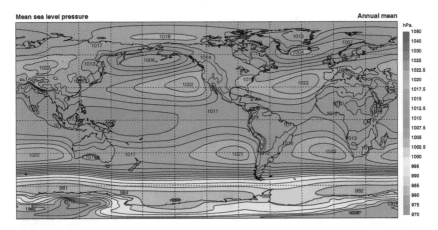

Figure 1.8. Atmospheric surface pressure shows persistent anticyclones over the mid-latitude oceans. Reproduced from ERA-40 Atlas.

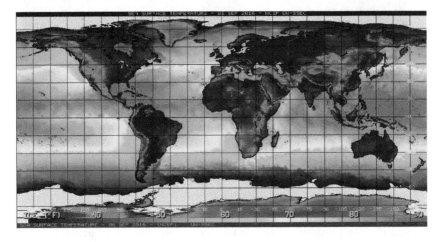

Figure 1.9. Observed sea-surface temperatures on a particular date (September 6, 2016). The winds associated with the subtropical anticyclones drive oceanic upwelling, leading to cold surface waters, along the western coasts of the continents. The equatorial cold tongue is also apparent in the eastern Pacific. Reproduced from SSEC, University of Wisconsin-Madison (http://www.ssec.wisc.edu/data/sst/). Provided courtesy of University of Wisconsin-Madison Space Science and Engineering Center.

both hemispheres (the lower branch of the Hadley circulation), and upward Ekman pumping over the equator (feeding the ascending branch of the Hadley circulation). Associated with this is poleward surface flow in the ocean, which is responsible for some of the oceanic heat transport, and

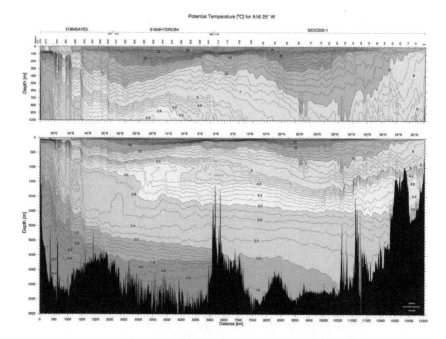

Figure 1.10. The potential temperature along a north–south section through the Atlantic Ocean shows the effects of equatorial upwelling bringing cold waters close to the surface, reflected in the bowing of the contours. Note the different vertical scale in the two panels; the top panel shows only the top 1000 m. Reproduced from the WOCE Atlantic Ocean Atlas (https://doi.org/10.21976/C6RP4Z) (Koltermann, K.P., V.V. Gouretski and K. Jancke. Hydrographic Atlas of the World Ocean Circulation Experiment (WOCE). Volume 3: Atlantic Ocean (eds. M. Sparrow, P. Chapman and J. Gould). International WOCE Project Office, Southampton, UK, 2011, ISBN 090417557X).

upward Ekman pumping in the equatorial ocean (Figs. 1.10 and 1.11). The latter leads to the so-called "cold tongue" of sea surface temperatures over the equator, most apparent in the tropical Pacific Ocean (Fig. 1.9). This feature is important for climate, as it tends to suppress atmospheric convection and cause the atmospheric inter-tropical convergence zone in the Pacific, and associated rain belts, to split in two (Fig. 1.6).

1.3.3 *Western boundary currents*

We now return to the Ekman balance and consider what happens to the flow below the Ekman layer. Integrating the modified geostrophic balance

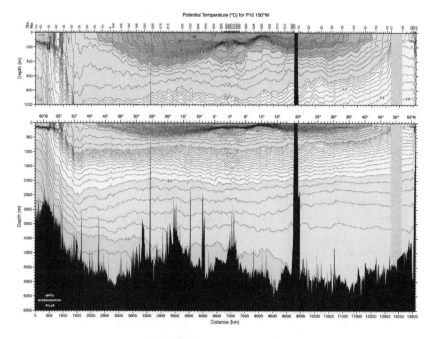

Figure 1.11. The same as in Fig. 1.10, but for the Pacific Ocean. A different colour scale is used for the Pacific, because of the overall colder waters. Reproduced from the WOCE Pacific Ocean Atlas (https://doi.org/10.21976/C6WC77) (Talley, L.D., 2007. Hydrographic Atlas of the World Ocean Circulation Experiment (WOCE). Volume 2: Pacific Ocean (eds. M. Sparrow, P. Chapman and J. Gould), International WOCE Project Office, Southampton, UK, ISBN 0-904175-54-5.).

discussed earlier over the depth of the ocean (down to $z = -H$), yields

$$-fV = -\frac{\partial P}{\partial x} + \tau_{s,x}, \qquad fU = -\frac{\partial P}{\partial y} + \tau_{s,y}, \qquad (1.26)$$

where $U = \int_{-H}^{0} \rho u\, dz$, $V = \int_{-H}^{0} \rho v\, dz$, $P = \int_{-H}^{0} p\, dz$. Here, for simplicity we use Cartesian coordinates, with x the zonal coordinate and y the meridional coordinate, and also assume the abyssal flow is stagnant so that there is no bottom friction. Taking the curl to eliminate the pressure gradient term leaves

$$f\left(\frac{\partial U}{\partial x} + \frac{\partial V}{\partial y}\right) + \frac{df}{dy}V = \frac{\partial \tau_{s,y}}{\partial x} - \frac{\partial \tau_{s,x}}{\partial y}. \qquad (1.27)$$

But,

$$\frac{\partial U}{\partial x} + \frac{\partial V}{\partial y} = \int_{-H}^{0} \rho\left(\frac{\partial u}{\partial x} + \frac{\partial v}{\partial y}\right) dz = -\int_{-H}^{0} \frac{\partial(\rho w)}{\partial z} dz = -[\rho w]_{-H}^{0} = 0. \qquad (1.28)$$

Hence,

$$\beta V = \hat{z} \cdot (\nabla \times \vec{\tau}_s), \qquad (1.29)$$

where $\beta = \frac{df}{dy}$. This is known as the Sverdrup relation.

The combination of atmospheric mid-latitude surface westerlies and tropical surface easterlies provides an anti-cyclonic surface stress on the ocean, and hence an equatorward net mass flux. This has a natural physical interpretation in terms of the vorticity balance: an anti-cyclonic torque reduces the absolute vorticity of a fluid parcel, which in the time average must be achieved by a drift toward the equator (where the vorticity associated with the Earth's rotation is zero).

However, if such a flow existed at all longitudes, it would violate mass conservation. This raises the question of where the poleward return flow occurs. Whenever a mathematical analysis violates a physical constraint, it implies that some essential physics has been omitted. In this case, the relevant physics is frictional stress along the lateral boundaries of the ocean (i.e., the coastlines), where the ocean is in motion but is also in contact with the solid Earth.

If the poleward flow occurs along the western boundary of the ocean, then the meridional frictional stress weakens toward the east, $\frac{\partial \tau_y}{\partial x} < 0$. This provides a cyclonic torque on the ocean, which is what is needed to balance the anti-cyclonic torque from the wind stress over the open ocean. Thus, a steady balance is possible. On the other hand, if the poleward flow occurs along the eastern boundary of the ocean, no such balance is possible. We thus conclude that the poleward return flow within the subtropical wind-driven circulation occurs via western boundary currents. The same conclusion can also be reached by consideration of the initial-value problem. Western boundary currents are a key feature of the oceanic circulation, most notably the Gulf Stream in the Atlantic and the Kuroshio in the Pacific (Fig. 1.12).

1.4 Stationary Rossby Waves in the Atmosphere

1.4.1 *Theory*

The atmosphere exhibits significant persistent zonal asymmetries, which are important for regional aspects of climate. Zonal asymmetries obey Rossby-wave dynamics, thus can be understood from the simplest model equation including Rossby waves. This is the barotropic (i.e., vertically uniform)

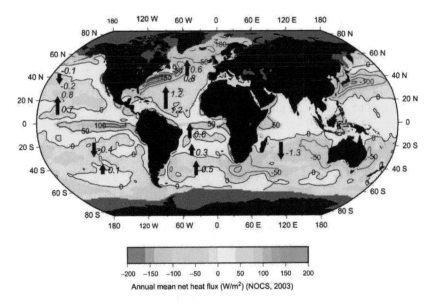

Figure 1.12. The net heat flux between atmosphere and ocean shows a strong heat flux into the atmosphere from the warm waters of the Gulf Stream and Kuroshio. The enhanced heat flux into the ocean in the equatorial cold tongue in both Pacific and Atlantic is also apparent. Reproduced with permission from Copyright © 2011, Elsevier. All rights reserved. (Talley, L. D., Pickard, G. L., Emery, W. J., and Swift, J. H. (2011). *Descriptive Physical Oceanography* (Academic Press)); derived from National Oceanographic Centre Southampton data.

vorticity equation on the mid-latitude β-plane, in the presence of a uniform zonal flow $\overline{u}$:

$$\left(\frac{\partial}{\partial t} + \overline{u}\frac{\partial}{\partial x}\right)\xi + \beta v = 0, \tag{1.30}$$

where $\xi = \frac{\partial v}{\partial x} - \frac{\partial u}{\partial y} = \nabla^2\psi$ is the vorticity and $u = -\frac{\partial \psi}{\partial y}$, $v = \frac{\partial \psi}{\partial x}$. The equation follows from the Lagrangian conservation of the absolute vorticity $f + \xi$ (see Chapter 2, Sec. 2.7.4). Since the equation is homogeneous in x, y and t, we may seek wave solutions by writing $\psi = \text{Re}\big(\psi_0 e^{i(kx+ly-\omega t)}\big)$, which yields

$$(-i\omega + i\overline{u}k)(-k^2\psi_0 - l^2\psi_0) + ik\beta\psi_0 = 0 \implies \omega = \overline{u}k - \frac{\beta k}{\kappa^2},$$

$$\kappa^2 = k^2 + l^2. \tag{1.31}$$

This represents the dispersion relation for barotropic Rossby waves, in the non-divergent regime where the length scale is much smaller than the external Rossby deformation radius ($\mathcal{F} \ll 1$, in the notation of Chapter 2, Sec. 2.7.4). Note that this regime is valid for the atmosphere, but not for the ocean. The dispersion relation specifies the frequency ω of a Rossby wave with zonal wavenumber k and meridional wavenumber l. The $\bar{u}k$ term is the familiar Doppler shift, arising from advection by the mean flow.

Note that the zonal phase speed of Rossby waves, $c_x = \frac{\omega}{k} = \bar{u} - \frac{\beta}{\kappa^2}$, is always westward (i.e., negative) relative to the zonal flow, but can be either westward or eastward relative to Earth, depending on the zonal flow. This westward relative propagation can be understood by considering the velocity anomaly induced by a vorticity anomaly ξ, and the effect this has on $\frac{\partial \xi}{\partial t}$ through βv. Note that Rossby waves are dispersive, since the zonal phase speed depends on wavenumber: longer waves (smaller κ) travel more rapidly, relative to the mean flow.

To understand persistent zonal asymmetries in the atmosphere, we need to consider stationary waves, with $c_x = 0$. These are potentially important since many forcing mechanisms are fixed in space and thus can only force stationary waves: examples include mountains, thermal forcing from land-sea contrasts, and thermal forcing from sea-surface temperature anomalies such as El Niño, which evolve slowly enough to be considered stationary from an atmospheric perspective. Stationary waves are seen to exist only in westerly flow, $\bar{u} > 0$, and must have a total wavenumber $\kappa = \kappa_s = \sqrt{\frac{\beta}{\bar{u}}}$ ($\bar{u} > 0$), which depends on the zonal flow speed.

In the tropics and the summertime stratosphere, the easterly zonal flow implies there are no stationary Rossby waves, and any such waves propagating into these regions are evanescent. In the mid-latitude troposphere, with weak westerlies ($\bar{u} \approx 20\,\text{m/s}$), the stationary wavelength $\lambda_s = \frac{2\pi}{\kappa_s} \approx 7000\,\text{km}$. Since the circumference of the Earth at 45 degrees latitude is approximately $28,000\,\text{km}$, this implies a non-dimensional zonal wavenumber of $k = 4$. In the mid-latitude winter stratosphere, with strong westerlies ($\bar{u} \approx 50\,\text{m/s}$), λ_s is longer and $k = 1$ or 2. This is because longer waves have a larger relative phase speed which can offset a stronger zonal flow.

1.4.2 *Some principal stationary wave sources*

In the tropics, horizontal temperature gradients are weak (since f is small) so the steady thermodynamic equation reduces to a balance between vertical motion (providing adiabatic heating/cooling) and diabatic heating/cooling Q:

$$\rho w \frac{\partial \theta}{\partial z} \propto Q. \tag{1.32}$$

Here, θ is potential temperature (see the Appendix). Thus, convection leads to latent heat release, providing the heating Q to drive deep ascent, $\rho w > 0$. It follows that $\frac{\partial(\rho w)}{\partial z} > 0$ in the lower troposphere and $\frac{\partial(\rho w)}{\partial z} < 0$ in the upper troposphere. Now considering the vorticity equation including a stretching component (see Chapter 2, Sec. 2.10.1 for the equivalent in the rotating shallow water equations),

$$\frac{\partial \xi}{\partial t} + \beta v = -f\delta = f\frac{1}{\rho}\frac{\partial(\rho w)}{\partial z}, \tag{1.33}$$

where $\delta = \frac{\partial u}{\partial x} + \frac{\partial v}{\partial y}$ is the horizontal divergence, this deep ascent leads to horizontal convergence, vortex stretching and a cyclonic tendency in the lower troposphere, and to horizontal divergence, vortex squashing and an anti-cyclonic tendency in the upper troposphere. In the steady limit, if the diabatic forcing is sustained, a Sverdrup balance results with $\frac{f}{\rho}\frac{\partial(\rho w)}{\partial z} = \beta v$. Thus, over the convective ascent region, $v < 0$ in the upper troposphere and $v > 0$ in the lower troposphere (Fig. 1.13). Consideration of how this steady balance is achieved from the initial-value problem leads to the conclusion that an anti-cyclone is spun up in the upper troposphere to the west of the ascent region, and a cyclone in the lower troposphere. Thus, the local response is baroclinic: i.e., opposite-signed in the upper and lower troposphere.

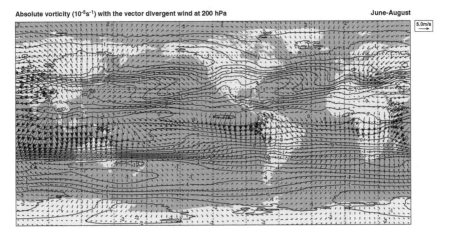

Absolute vorticity (10⁻⁵s⁻¹) with the vector divergent wind at 200 hPa   June-August

Figure 1.13.   The upper tropospheric divergent flow (arrows), here for boreal summer, shows a Sverdrop balance in the tropics just north of the equator, above the regions of most intense convection. Reproduced from ERA-40 Atlas.

But in the upper troposphere, the westerly zonal flow supports propagating barotropic stationary Rossby waves, forced by the anti-cyclonic tendency. These waves provide teleconnections (long-distance causal relationships) between tropical sea-surface temperature anomalies (which induce convective heating anomalies) and the midlatitudes, which is why El Niño affects weather in midlatitudes.

In the midlatitudes, to provide a realistic model, we need to include the effects of a zonal flow in the vorticity balance, as with the Rossby-wave model itself. However, the stretching term $-f\delta$ needs to be included to provide a forcing. For stationary waves, this leads to

$$\bar{u}\frac{\partial \xi}{\partial x} + \beta v = -f\delta. \tag{1.34}$$

One important source of vortex stretching is flow over mountains. As air moves over a mountain, there is horizontal divergence and an anti-cyclonic tendency on the upstream side of the mountain as air parcels are compressed in the vertical, and horizontal convergence and a cyclonic tendency on the downstream side as air parcels are extended in the vertical. Now there are three terms in the balance and which of the two terms on the left-hand side dominates depends on the length scale of the mountain, because $\bar{u}\frac{\partial \xi}{\partial x}$ contains two more spatial derivatives than does βv. For sufficiently long waves, the latter term dominates meaning that the dominant balance is Sverdrup and a cyclonic circulation develops over the mountain. For sufficiently short waves, the dominant balance is instead $\bar{u}\frac{\partial \xi}{\partial x} = -f\delta$ and an anti-cyclonic circulation develops over the mountain. In practice, it is a bit of both, but for continental-scale mountains in the atmosphere, the latter balance tends to dominate. An idealised example is shown in Chapter 5.

Propagation of stationary waves away from their source regions can be studied using ray-tracing theory, using the geometrical optics/WKB approximation (Hoskins and Karoly, 1981).

1.5 Transient Disturbances in the Atmosphere

1.5.1 *Extratropical storms*

Extratropical storms, or cyclones, are responsible for the day-to-day variations known as weather. They are characterised in many different ways, both Lagrangian (following the storms) and Eulerian (fixed in space). The storms are concentrated in preferred locations known as storm tracks. They

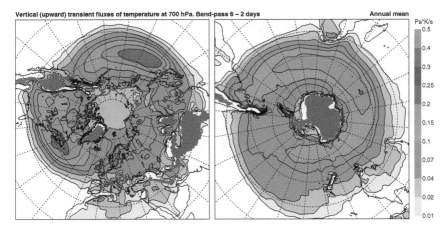

Figure 1.14. Vertical heat fluxes in the lower troposphere (approximately 3 km altitude) from eddies with timescales of $2-6$ days, representing extratropical storms. In the northern hemisphere they are most intense over the western part of the ocean basins. Reproduced from ERA-40 Atlas.

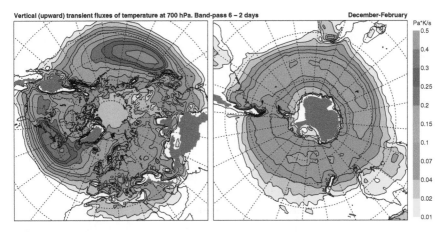

Figure 1.15. The same as in Fig. 1.14, but just for boreal winter. By comparing the two figures, it is evident that in both hemispheres, the heat fluxes are stronger in the winter season and weaker in the summer season. Reproduced from ERA-40 Atlas.

are strongest over oceans and weakest over land; concentrated in the mid-latitudes; strongest in the winter season and weakest in summer; and with strongest amplification in the presence of strong meridional temperature gradients, over the warm western ocean boundary currents (Figs. 1.14 and 1.15).

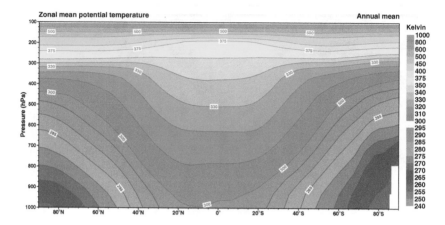

Figure 1.16. Distribution of potential temperature, a quantity that accounts for the effect of atmospheric compressibility on temperature (see the Appendix). Temperatures are highest in the tropics and lowest at the poles, with the maximum horizontal gradient in midlatitudes. This baroclinicity provides the potential energy that drives extratropical storms. Reproduced from ERA-40 Atlas.

The principal explanation for extratropical storms is the background meridional temperature gradient (Fig. 1.16), hence explaining why they are concentrated in midlatitudes, and strongest in the winter season when the equator-to-pole temperature difference is the greatest. Other factors affecting storm magnitude include moisture, static stability and surface friction (hence explaining why they are strongest over the oceans, and mainly amplify over the warm western boundary currents; see Fig. 1.12).

The role of the background meridional temperature gradient is formulated in the mathematical theory of baroclinic instability (Pedlosky, 1987). The source of energy for the instability is the gravitational potential energy associated with the equatorward temperature gradient, though its release is strongly constrained by rotation through the Coriolis force. In the presence of rotation, the background isentropic surfaces in the atmosphere may be sloping since the pressure gradient force can be balanced by the Coriolis force; they slope poleward and upward (Fig. 1.16), with potential temperature θ increasing as one moves upward (since static stability is positive) or equatorward (since the atmosphere is warmer in the tropics). To a first approximation, air parcels conserve θ. If an air parcel moves poleward and upward, but at a shallower angle than the background isentropic surfaces, it finds itself in an environment of smaller θ and thus experiences positive buoyancy, reinforcing the rising motion. Similarly, an equatorward and downward displacement at a shallower angle than the background isentropic surfaces leads to negative buoyancy, reinforcing the sinking motion. Motions in this so-called "wedge of instability" can thus experience the positive feedback needed for instability. At the same time, they transport heat downgradient (Fig. 1.3).

This picture is over-simplified because baroclinic instability is not really a parcel instability; instead, it is a wave instability which is constrained nonlocally (Thorpe *et al.*, 1989). In any case, the linear theories of baroclinic instability are not directly applicable to the atmosphere, which is closer to a state of macroturbulence (Held, 1999). Nevertheless the "Eady growth rate"

$$\frac{1}{N}\left|\frac{\partial \overline{T}}{\partial y}\right| \propto \frac{f}{N}\frac{\partial \overline{u}}{\partial z} \tag{1.35}$$

(N is the Brunt–Väisälä frequency, which characterises static stability) which emerges in these linear instability theories has proven to be a useful index of baroclinicity and hence of storm growth.

Individual storms grow and decay, with a characteristic timescale of a week or so. Idealised numerical simulations show that the growth stage follows the conceptual picture captured in the classic Norwegian frontal model, inferred from observations in the early part of the 20th century, whilst the decay stage occurs barotropically (Simmons and Hoskins, 1978). This "baroclinic lifecycle" is now the accepted theoretical model for extratropical storms.

1.5.2 *Predictability*

The explosive growth of weather systems through the process of dynamical instability places limits on deterministic predictability of weather, even with perfect knowledge of the governing equations (which of course we do not really have). Even perfect observations will be limited in their spatial resolution, so the initial condition of a weather forecast can never be perfectly known. Consider the effect of an initial error in small (hence unobserved) length scales; through multiscale interactions (via the advective nonlinearity), it can be expected to influence the larger scales of motion through an upscale cascade of error. The timescale on which this occurs can be estimated heuristically as follows. Let τ_L be the time for error on horizontal length scale L to introduce error on length scale $2L$. Then, the predictability time at scale L, if the initial error is at scale $(1/2)^N L$, is

$$T_N = \tau_{L/2} + \tau_{L/4} + \tau_{L/8} + \cdots + \tau_{L/(2^N)} = \sum_{n=1}^{N} \tau_{\left(\frac{1}{2}\right)^N L}. \qquad (1.36)$$

Now, what is τ_L? Dimensional analysis suggests $\tau_L \sim \left(k^3 E(k)\right)^{-0.5}$ where k is the horizontal wavenumber, $E(k)$ is the spectral density of kinetic energy, and $L = \frac{1}{k}$.

For two-dimensional turbulence, which is arguably relevant to large-scale atmospheric dynamics (because of rotation and stratification), we expect $E(k) \sim k^{-3}$ and thus $\tau_L \sim$ const., independent of spatial scale. It follows that $T_N \to \infty$ as $N \to \infty$, hence there is no predictability limit. This may seem counter-intuitive in light of the butterfly effect in chaos, but for low-order (i.e., finite-dimensional) chaos (see Chapter 4), with bounded Liapunov exponents — and τ_L^{-1} is effectively a Liapunov exponent — error growth is only exponential in time. This means that for any given lead time, the forecast error can be made arbitrarily small provided the initial error is kept sufficiently small. Thus, low-order chaos is not inherently unpredictable for a given lead time.

But in the atmosphere, the small-scale dynamics is three-dimensional and it is this behaviour that determines the predictability limit. On small scales, we expect $E(k) \sim k^{-5/3}$ and then $\tau_L \sim \left(k^{4/3}\right)^{-0.5} \sim k^{-2/3} \sim L^{2/3}$. Thus, the Liapunov exponent τ_L^{-1} increases with decreasing L, and we have not low-order but rather multi-scale chaos. This heuristic argument gives physically sensible results: for example, for extratropical cyclones, with $L \sim 3000\,\mathrm{km}$, the error growth timescale $\tau_L \sim 2$ days; whereas for thunderstorms, with $L \sim 10\,\mathrm{km}$, the error growth timescale is a factor of $300^{2/3} \sim 45$ shorter, namely $\tau_L \sim 1\,\mathrm{h}$ (Fig. 1.17).

This scale-dependence of the error growth timescale has profound implications, because

$$\lim_{N \to \infty} T_N \sim \sum_{n=1}^{\infty} \left(\frac{L}{2^N}\right)^{2/3} < \infty, \qquad (1.37)$$

which implies finite predictability. Thus, imagine that the atmosphere is observed perfectly down to a horizontal spatial resolution of $10\,\mathrm{km}$. Even if the observational network were improved to reach a spatial resolution of $1\,\mathrm{km}$, this would only extend predictability of the large scales by an hour or so — clearly a losing battle. The predictability limit of the atmosphere

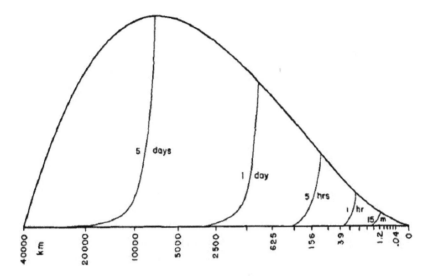

Figure 1.17. Schematic of the contamination of the energy spectrum by error, with an upscale cascade of error from small to large horizontal length scales. The timescale shortens dramatically as the length scale becomes smaller (Lorenz, 1969).

is estimated to be about 2 weeks (Lorenz, 1969). This multi-scale chaos is the real butterfly effect.

1.5.3   *Tropical dynamics*

In the tropics, the dynamics of weather systems are quite different from the extratropics, because of the small Coriolis parameter. One implication is that horizontal temperature gradients are necessarily small, so the dominant instability is not baroclinic but rather (moist) convective. A second implication is that vortex stretching is weak, so nonlinearity tends to be weak.

Indeed, large-scale tropical disturbances are remarkably linear, in comparison with their extratropical counterparts. A unique phenomenon of the tropics are equatorially trapped waves, notably the equatorial Kelvin wave. It is akin to a surface water wave (see Chapter 2, Sec. 2.11) in that its restoring force is gravity, but the role of Earth's rotation manifests itself through the β-effect, which is not zero at the equator and which means that the wave can only propagate eastward.

A major feature of tropical dynamics is the Madden–Julian oscillation, a complex eastward-propagating phenomenon with a period of around 40–50 days (Zhang, 2005). It is generally understood as a convectively coupled equatorial Kelvin wave.

Outside the deep tropics, vortex stretching becomes non-zero and this allows the formation of tropical cyclones (known as hurricanes when they occur over the Atlantic Ocean). They propagate poleward but eventually get ripped apart by the horizontal shear associated with the mid-latitude jet stream.

1.5.4   *Eddy momentum fluxes*

The baroclinic eddies generated in the midlatitude storm tracks can lead to eddy-driven jets through their momentum fluxes. This is in fact a general property of rotating, stratified fluids, and is believed to be the explanation for the atmospheric jets observed on giant planets such as Jupiter. The phenomenon is most easily understood through the barotropic non-divergent model previously used for unforced atmospheric Rossby waves. In this case, the zonal-mean zonal flow equation simplifies to

$$\frac{\partial \overline{u}}{\partial t} = -\frac{\partial}{\partial y}\left(\overline{u'v'}\right), \tag{1.38}$$

since the zonal-mean meridional velocity $\overline{v}$ vanishes. The right-hand side of this equation represents the eddy momentum flux convergence, which arises from the advective nonlinearity. To determine the nature of the eddy fluxes, we need to consider the dynamics of the eddies, which follows from the eddy component of the barotropic vorticity equation

$$\frac{\partial \xi'}{\partial t} + \overline{u}\frac{\partial \xi'}{\partial x} + \hat{\beta}v' = 0, \quad \hat{\beta} = \beta - \frac{\partial^2 \overline{u}}{\partial y^2}. \tag{1.39}$$

This leads to

$$\frac{\partial}{\partial t}\left(\frac{1}{2}\frac{\overline{\xi'^2}}{\hat{\beta}}\right) + \overline{v'\xi'} = 0. \tag{1.40}$$

With some algebra, and recalling that $u' = -\frac{\partial \psi}{\partial y}$, $v' = \frac{\partial \psi}{\partial x}$, $\xi' = \nabla^2 \psi$, it can be shown that $\overline{v'\xi'} = -\partial\left(\overline{u'v'}\right)/\partial y$, which shows that convergence of poleward momentum flux corresponds to a poleward vorticity flux, and divergence to an equatorward flux.

Now, let

$$A = \frac{1}{2}\frac{\overline{\xi'^2}}{\hat{\beta}}, \qquad \vec{F} = -\overline{u'v'}\hat{y}. \qquad (1.41)$$

These are known in meteorology as the Eliassen–Palm (E–P) wave activity and flux, and arise from the Hamiltonian structure of the dynamics (see Chapter 2 for further discussion about Hamiltonian representations of fluid dynamics) (Shepherd, 2003). (They generalise to the baroclinic equations, so the theory outlined here generalises as well.) The earlier equation can then be written in the conservation-law form

$$\frac{\partial A}{\partial t} + \nabla \cdot \vec{F} = 0, \qquad (1.42)$$

where here $\nabla = \hat{y}\frac{\partial}{\partial y}$. More generally, we may allow for generation and dissipation of eddies through

$$\frac{\partial A}{\partial t} + \nabla \cdot \vec{F} = D. \qquad (1.43)$$

So long as $\hat{\beta} > 0$, which is generally the case in the free atmosphere, then $A \geqslant 0$ and one can sensibly talk about the "amount" of wave activity. The advantage of this formulation is that other measures of wave activity, such as the eddy kinetic energy, do not satisfy such a conservation law. Moreover in the WKB limit, the wave activity flux $\vec{F} = \vec{c}_g A$, where $\vec{c}_g$ is the group velocity (Vanneste and Shepherd, 1999).

We now consider locally plane waves with x and y wavenumbers given by k and l. With some algebra, it can be shown that $-\overline{u'v'} = \frac{1}{2}kl|\psi_0|^2$. But for Rossby waves, $c_{gy} = \frac{\partial \omega}{\partial l} = 2\hat{\beta}kl/\kappa^2$. It follows that

$$\text{sgn}(c_{gy}) = \text{sgn}(-\overline{u'v'}), \quad \hat{\beta} > 0. \qquad (1.44)$$

Therefore, meridional propagation of wave activity implies an oppositely signed meridional momentum flux. So if jets are a source of waves (because the vertical zonal wind shear implies baroclinicity, leading to baroclinic instability), then they are self-maintaining. Indeed, the atmosphere exhibits mid-latitude jets, notably during the summer season, that are distinct from the subtropical jets associated with the poleward limit of the Hadley circulation, and are understood to be eddy driven (Fig. 1.5).

This theory also provides the explanation for tropical easterlies, discussed in Secs. 1.2.6 and 1.3. The extratropical storm tracks provide a source of eddies which propagate into the tropics, and induce a poleward momentum flux out of the tropics.

1.6   Climate Change

1.6.1   *Global-mean warming*

Authoritative assessments of the scientific basis of climate change can be found in the periodic reports of the Intergovernmental Panel on Climate Change (IPCC). An early report which succinctly summarises the key issues is the Charney report of the US National Academy of Sciences, published in 1979.[1] It was remarkably prescient and is an excellent primer.

Carbon dioxide (CO_2) is a naturally occurring greenhouse gas, whose atmospheric concentration represents a balance with the CO_2 content of the oceanic and terrestrial biosphere. When CO_2 is added to the atmosphere through anthropogenic emissions, this balance is perturbed, and gets restored through carbon uptake by the biosphere. Approximately 50% of the anthropogenic emissions remain in the atmosphere, and they remain there essentially forever until CO_2 is removed through geological processes. The scientific uncertainties concern exactly what fraction of the emitted CO_2 remains in the atmosphere, and how this fraction may depend on climate itself. Overall, it is expected that if anything the retained fraction will increase as climate warms, representing a positive feedback.

Other greenhouse gases have finite atmospheric lifetimes (typically several decades); in those cases, a constant atmospheric abundance could be maintained with non-zero emissions. But for CO_2, the emissions accumulate and the only way to maintain a constant atmospheric abundance is to reach zero emissions. That is why CO_2 is at the heart of the climate-change problem.

Because CO_2 is a greenhouse gas, increasing CO_2 increases the longwave (infrared) opacity of the atmosphere. This forces the longwave emission to space to occur at higher altitudes and, therefore, at lower temperatures (Fig. 1.18). Since the longwave emission is lower for lower temperatures (from the Stefan–Boltzmann or blackbody radiation law), this decreases the longwave emission to space. This decrease in longwave emission is what is called "radiative forcing", and has been very well constrained for a very long time.

As a result of decreased longwave emission, the balance between ASR and OLR is perturbed, and more energy is absorbed than emitted.

---

[1] Available for free download from https://www.nap.edu/catalog/12181/carbon-dioxide-and-climate-a-scientific-assessment.

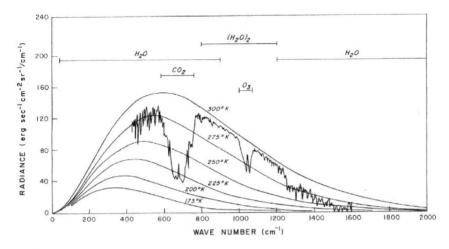

Figure 1.18.   The spectrum of OLR at the top of the atmosphere shows minima, corresponding to emission at lower temperatures (see corresponding blackbody spectra for various temperatures), from opaque regions of the spectrum associated with particular greenhouse gases, as indicated. As the greenhouse gas concentrations increase, the emission to space in those spectral regions gets pushed to higher altitudes and lower temperatures, which reduces the net emission. Radiative balance can only be restored by increasing the surface temperature, which shifts the whole emission spectrum up to compensate for the lower emission in the opaque spectral regions.

Conservation of energy then implies that the climate system warms up until the OLR (which increases as the atmosphere warms) is restored to its original value, and radiative balance is restored. This feedback process is known as the Planck response. As part of this process, sea level rises, both because water expands as it is warmed, and because of the melting of glaciers and ice sheets. All this is very basic physics. A systematic warming of the climate system that is attributable to the anthropogenic increase in greenhouse gases is seen across the entire globe (Fig. 1.19).

There is scientific uncertainty in the time needed to restore radiative balance, because the ocean needs to warm up and this requires at least several decades and perhaps much longer. This lagged response is captured in a metric known as transient climate sensitivity. The time delay has important implications, because it means that future warming of the atmosphere is committed before we see it manifested in a rise in surface temperatures. It also means that a lack of increase in surface temperatures would not disprove the greenhouse effect, so long as the climate system continues to warm as a whole.

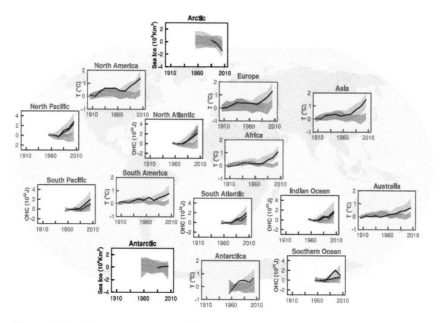

Figure 1.19.   Long-term warming of the climate system on the scale of entire continents (surface air temperature) or ocean basins (upper ocean heat content) is of anthropogenic origin. The purple band shows the range of climate model simulations without anthropogenic forcing, and pink the model simulations with anthropogenic forcing. In practically every case the observations (black curve) are consistent with the models with anthropogenic forcing, and inconsistent with the models without anthropogenic forcing. Reproduced from Intergovernmental Panel on Climate Change (IPCC), Fifth Assessment Report 2013.

There is also scientific uncertainty in the magnitude of the surface temperature changes that are needed to restore radiative balance. This is captured in a metric known as equilibrium climate sensitivity. But there is no uncertainty in the fact that CO_2 emissions will accumulate and continue to warm the planet, accompanied by rising sea levels. If warming is to be limited at some point, it is necessary to move to a state of zero net CO_2 emissions. The only question is how fast this must be done.

There are a number of feedbacks in the climate system that affect the amount of surface warming needed to restore radiative balance for a given radiative forcing. A positive feedback comes from water vapour; water vapour is itself a very important greenhouse gas, and assuming no change in relative humidity, water vapour increases by about 7% for every 1 K of warming via the Clausius–Clapeyron equation, which relates the saturation vapour pressure of water to the ambient air temperature and pressure based on an assumption of thermodynamic equilibrium. A negative feedback comes from the weakening of the lapse rate (decrease of temperature with altitude) because of the moister atmosphere. These two processes are obviously correlated, so the sum of the two is more constrained than each component separately. A positive feedback comes from the decrease in surface reflectivity (albedo) associated with decreased snow and ice in a warmer climate, so the climate system absorbs a larger fraction of the solar radiation. Finally, changes in clouds provide very important feedbacks through both shortwave (albedo) and longwave effects. Low clouds reflect sunlight but contribute little to the greenhouse effect, because their emission temperatures are close to surface temperatures, so a decrease in low clouds would be a positive feedback. High clouds have both a shortwave and a longwave effect, but if high clouds move to higher altitudes and lower temperatures, then this would be a positive feedback. There is very large uncertainty in the cloud feedback, but overall it is expected to be positive.

The Charney report of 1979 provided the following estimates of the global-mean equilibrium surface warming expected from a doubling of CO_2. The basic Planck response implied by the Stefan–Boltzmann law would lead to 1 K warming. The water vapour and lapse rate feedbacks together would increase this to 2 K. All this results from very basic physics. Surface albedo changes would increase this further to 2.4 K, and cloud changes to 3 K. Because of the uncertainties associated with the cloud changes, the overall uncertainty range (allowing for the possibility of a negative cloud feedback) was estimated as 1.5 K to 4.5 K. Quite remarkably, the current estimates of these terms are virtually identical to those in the Charney report.

1.6.2 Regional aspects of climate change

It is expected that the Arctic will warm much more than the global-mean temperature under climate change. This is observed (Fig. 1.20), and again is understandable from some basic physics. (Whilst the same physics applies to the Antarctic, the Antarctic is cooled by the Southern Ocean and this delays the emergence of polar amplification.) The first factor is that the Planck response via the Stefan–Boltzmann law is weaker at lower temperatures, thus the warming needs to be greater at higher latitudes in order to compensate for the same radiative imbalance. The second factor is that the lapse-rate feedback is positive rather than negative at high latitudes, because the stable atmospheric boundary layer acts to isolate the surface

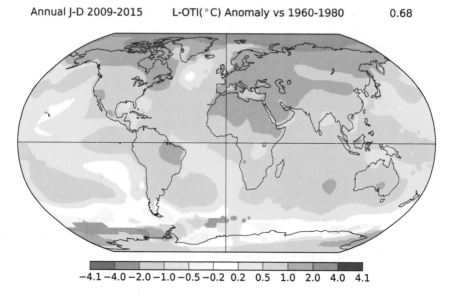

Annual J-D 2009-2015   L-OTI(°C) Anomaly vs 1960-1980   0.68

−4.1 −4.0 −2.0 −1.0 −0.5 −0.2 0.2 0.5 1.0 2.0 4.0 4.1

Figure 1.20.   Observed annual-mean surface temperature changes (2009–2015 compared with 1960–1980). The Arctic is warming faster than any region on Earth. Reproduced from NASA GISS (https://data.giss.nasa.gov/gistemp/).

layer from the upper troposphere. (Over the rest of the atmosphere, moist convective adjustment keeps the two layers tightly coupled.) The third factor is the albedo feedback from melting ice and snow. Together, these three factors account for virtually all of the Arctic amplification in climate models, although other processes can also contribute.

Since a warmer atmosphere holds more moisture, it might seem logical that one would expect an increase in precipitation. However this is not so obvious; moisture is a concentration, whereas precipitation is a rate of change of moisture, so the dimensions are different. Precipitation is associated with latent heat release, which warms the atmosphere, and in the global mean this warming needs to be balanced by longwave cooling. For example, in the Hadley circulation, moist ascent is convectively driven with $S\overline{w} \approx Q_{lh}$, where S is the static stability, whilst the dry descent is radiatively driven with $S\overline{w} \approx Q_{\mathrm{rad}}$. Apart from spatial variations in S, mass conservation then implies that Q_{lh} balances Q_{rad} across the tropics.

Thus, on a global scale, precipitation is constrained by the energy balance. We can make a very simple estimate of how this changes with warming. The longwave cooling follows the Stefan–Boltzmann law σT^4. Hence, its relative change with warming is given by

$$\frac{1}{\sigma T^4} \frac{d\left(\sigma T^4\right)}{dT} = \frac{4}{T}. \tag{1.45}$$

Assuming an upper tropospheric temperature of $230\,\mathrm{K}$, this implies a precipitation increase of 1.7% per degree of warming, which is well below the 7% increase in moisture per degree of warming predicted by Clausius–Clapeyron. Climate models suggest the global precipitation increase is roughly 2% per degree of warming, which is close to this rough estimate. The implication is that the atmospheric overturning circulation actually weakens under climate change, which is rather counter-intuitive.

Along similar lines, although a warmer climate has more energy, this does not necessarily translate into more instability. Atmospheric motions are driven by temperature contrasts (according to the second law of thermodynamics), and in the lower atmosphere, these tend to weaken under climate change (except in the upper troposphere, where they tend to strengthen). For extratropical storms, polar amplification thus acts to reduce the baroclinicity, but a moister atmosphere increases the latent heat release; the overall outcome remains unclear.

If the atmospheric circulation did not change, then a moister atmosphere would imply an enhancement of the moisture transport carried by the atmosphere, and thus an enhancement of the surface exchange of moisture, i.e., the difference between precipitation and evaporation. This is often summarised as "wet regions get wetter, dry regions get drier". However, it has already been noted that the overall atmospheric overturning circulation should weaken, and in general this argument seems inapplicable over land, which is where we most care about precipitation.

Thus, whilst the global aspects of climate change are anchored in basic physics and reflected in a hierarchy of models, for many regional aspects there remain considerable uncertainties and an overall lack of physical understanding. This is reflected in the striking contrast between the robustness of model projections of changes in surface temperature and in precipitation (Fig. 1.21). Further discussion can be found in Shepherd (2014).

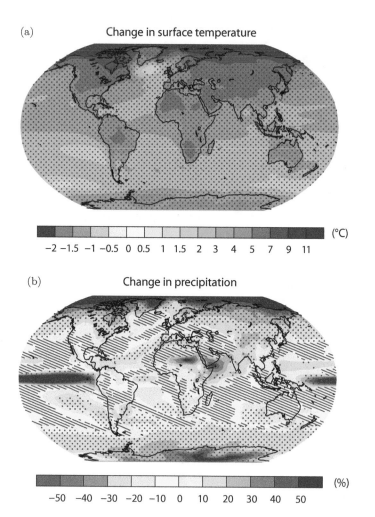

Figure 1.21.   Contrast between the robustness of projected changes in surface temperature (a) and in precipitation (b), over the 21st century by the CMIP5 model ensemble according to the RCP 8.5 scenario. Hatching indicates where the multi-model mean change is small compared to natural internal variability. Stippling indicates where the multi-model mean change is large compared to natural internal variability and where at least 90% of models agree on the sign of change. Although temperature changes are robust over all land areas, the mean precipitation changes over many populated regions are non-robust either because of natural variability or because of model discrepancies. Reproduced from Intergovernmental Panel on Climate Change (IPCC), Fifth Assessment Report 2013.

Appendix

Some basic concepts in meteorology are briefly described here.

A.1 *Hydrostatic balance*

The vertical component of the momentum equation is

$$\frac{Dw}{Dt} + \text{Sphericity terms} + \text{Coriolis term} = -\frac{1}{\rho}\frac{\partial p}{\partial z} - g + F_z. \qquad (A.1)$$

The dominant balance in this equation is hydrostatic balance, $\frac{\partial p}{\partial z} = -\rho g$ (see Chapter 2, Sec. 2.5). Combining this with the ideal gas law $p = \rho RT$, where R is the gas constant, leads to $\frac{\partial p}{\partial z} = -\frac{pg}{RT}$. If the atmosphere were isothermal, with $T = T_0 = \text{const.}$, then this equation has the solution $p(z) = p(0)\exp(-z/H)$ where $H = \frac{RT_0}{g}$ is the scale height. In practice, T does vary with z, but since T is measured in degrees Kelvin, its relative variation is fairly small and pressure decays nearly exponentially with altitude. For typical atmospheric parameters, $H \approx 7\,\text{km}$. Because the variation of pressure with altitude is monotonic, pressure can be used as a vertical coordinate (assuming hydrostatic balance).

A.2 *Thermal-wind balance*

For large-scale motions, the dominant balance in the horizontal component of the momentum equation is geostrophic balance, namely the horizontal component of $2\overrightarrow{\Omega} \times \overrightarrow{u} = -\frac{\nabla p}{\rho}$ (see Chapter 2, Sec. 2.5). For the zonal flow u this implies

$$2\Omega \sin(\phi)u = -\frac{1}{a\rho}\frac{\partial p}{\partial \phi}. \qquad (A.2)$$

But

$$\left.\frac{\partial p}{\partial \phi}\right|_z = \frac{\partial(p,z)}{\partial(\phi,z)} = \frac{\partial(p,z)}{\partial(p,\phi)}\frac{\partial(p,\phi)}{\partial(\phi,z)} = \left.\frac{\partial z}{\partial \phi}\right|_p \left(-\left.\frac{\partial p}{\partial z}\right|_\phi\right). \qquad (A.3)$$

From hydrostatic balance, the final factor, within the parentheses, equals ρg. Substituting this expression into the geostrophic balance relation then gives

$$2\Omega \sin(\phi) u = -\frac{1}{a} \frac{\partial \Phi}{\partial \phi}\Big|_p, \tag{A.4}$$

where $\Phi = gz$, whence (again using hydrostatic balance).

$$2\Omega \sin(\phi) \frac{\partial u}{\partial p} = -\frac{1}{a} \frac{\partial}{\partial \phi}\Big|_p \left(\frac{\partial \Phi}{\partial p}\right) = -\frac{1}{a} \frac{\partial}{\partial \phi}\Big|_p \left(-\frac{RT}{p}\right) = \frac{R}{ap} \frac{\partial T}{\partial \phi}\Big|_p. \tag{A.5}$$

Thus, if T increases toward the equator (at constant pressure), then u increases as p decreases (and as z increases). This thermal-wind balance is the basic reason for the eastward zonal winds (or westerlies) in the extratropical upper troposphere.

A.3   Adiabatic relations

When the diabatic heat input $dQ = 0$, the 1st law of thermodynamics for an ideal gas implies $c_p dT = \frac{dp}{\rho}$. These define adiabats, with

$$\frac{dT}{dz}\Big|_{\text{adiabat}} = \frac{1}{\rho c_p} \frac{\partial p}{\partial z} = -\frac{g}{c_p} \approx -10\,\text{K/km}. \tag{A.6}$$

Thus, under adiabatic conditions, the temperature decreases rapidly with altitude. This accounts for the fact that the upper troposphere is colder than the surface. Along an adiabat,

$$dT = \frac{1}{\rho c_p} dp \Rightarrow \frac{dT}{T} = \frac{R}{c_p} \frac{dp}{p} \Rightarrow \log T = \frac{R}{c_p} \log p + \text{const.} \tag{A.7}$$

(again using the ideal gas law), whence

$$\theta = T\left(\frac{p}{p_{00}}\right)^{-\frac{R}{c_p}} = \text{const.} \tag{A.8}$$

for some constant p_{00}. This quantity θ is called the potential temperature, and represents the temperature that a parcel of air would have if brought adiabatically from pressure p to the reference pressure p_{00} (normally taken to be $1000\,\text{hPa}$, approximately the surface pressure). It is equivalent to the parcel's entropy, so surfaces of constant θ are isentropic surfaces.

Chapter 2

Partial Differential Equations

Beatrice Pelloni[*] and Darryl Holm[†]

Heriot-Watt University, UK
†Imperial College London, UK

2.1 Preliminaries

In this chapter, we introduce the notion of partial differential equations (PDEs), as well as the basic definitions and results concerning such equations. We then describe the governing equations for an ideal fluid, and give the fundamental properties, before analysing in greater detail the case of one-dimensional flows. To end the first part, an example of a PDE governing large-scale atmospheric flows, of particular mathematical interest, is introduced and briefly discussed. In the second part, we discuss various representations and approximations of ideal shallow water dynamics in a rotating frame. The rotating shallow water (RSW) equations possess a slow + fast decomposition in which they reduce approximately to quasi-geostrophic (QG) motion (conservation of energy and potential vorticity) plus nearly decoupled equations for gravity waves in an asymptotic expansion in small Rossby number, $\epsilon \ll 1$. The solution properties of the RSW equations are discussed, and alternative representations of them which highlight their slow + fast interactions are given. Finally, the RSW equations and thermal rotating shallow water (TRSW) equations are derived as Euler–Poincaré equations from Hamilton's principle in the Eulerian description.[1]

---

[1]BP acknowledges inspiration and selected material from Olver (2014); Marsden and Chorin (1993); Majda and Bertozzi (2001).

Notation:

The derivative of a function $\mathbf{u} = \mathbf{u}(\mathbf{x}) = \mathbf{u}(x_1, x_2, \ldots)$ with respect to one of its variables x will be denoted interchangeably as

$$\frac{\partial \mathbf{u}}{\partial x_i}, \quad \partial_{x_i} \mathbf{u}, \quad \mathbf{u}_{x_i}.$$

For the components of vector, we use the usual notation:

$$\mathbf{u} = (u_1, u_2, u_3); \quad \mathbf{x} = (x_1, x_2, x_3, \ldots).$$

2.1.1 Grad, Div, and Curl

These are basic definitions that we will use throughout. We denote by ∇ the differential operator *nabla* defined as

$$\nabla = \left(\frac{\partial}{\partial x_1}, \frac{\partial}{\partial x_2}, \frac{\partial}{\partial x_3} \right).$$

Definition 2.1. Let $\Phi : \mathbb{R}^3 \to \mathbb{R}$ be a differentiable function. The **gradient** of Φ is the **vector** in $\mathbb{R}^3$ given by

$$\operatorname{grad} \Phi = \nabla \Phi = \left(\frac{\partial \Phi}{\partial x_1}, \frac{\partial \Phi}{\partial x_2}, \frac{\partial \Phi}{\partial x_3} \right).$$

Definition 2.2. Let $\Psi : \mathbb{R}^3 \to \mathbb{R}^3$ be a differentiable function. The **divergence** of Ψ is the **scalar** quantity (to be understood as a *density*) given by

$$\operatorname{div} \Psi = \nabla \cdot \Psi = \frac{\partial \Psi_1}{\partial x_1} + \frac{\partial \Psi_2}{\partial x_2} + \frac{\partial \Psi_3}{\partial x_3}.$$

Definition 2.3. Let $\Psi : \mathbb{R}^3 \to \mathbb{R}^3$ be a differentiable function. The **curl** of the vector Ψ is the element of $\mathbb{R}^3$ given by

$$\operatorname{curl} \Psi = \nabla \times \Psi = \left(\frac{\partial \Psi_3}{\partial x_2} - \frac{\partial \Psi_2}{\partial x_3}, \frac{\partial \Psi_1}{\partial x_3} - \frac{\partial \Psi_3}{\partial x_1}, \frac{\partial \Psi_2}{\partial x_1} - \frac{\partial \Psi_1}{\partial x_2} \right).$$

This can be obtained by formally computing the determinant of the matrix

$$\begin{pmatrix} i & j & k \\ \dfrac{\partial}{\partial x_1} & \dfrac{\partial}{\partial x_2} & \dfrac{\partial}{\partial x_3} \\ \Psi_1 & \Psi_2 & \Psi_3 \end{pmatrix}.$$

2.1.2 The Levi-Civita symbol

The *Levi-Civita symbol* ϵ_{ijk} with $i,j,k = 1,2,3$, may be used to define both the determinant of a 3×3 matrix, and the vector cross product in $\mathbb{R}^3$. Here, $\epsilon_{ijk} = +1$ for even permutations of 123 and $\epsilon_{ijk} = -1$ for odd permutations of 123. Being totally antisymmetric means the quantity ϵ_{ijk} vanishes if any two of its indices are the same. A few familiar examples of its use are listed below.

- The determinant of a 3×3 matrix may be defined using the Levi-Civita symbol ϵ_{ijk} as the product,

$$\epsilon_{ijk} \det A = \epsilon_{abc} A_{ai} A_{bj} A_{ck}.$$

- The three components of the vector cross product of two vectors **a** and **b** in $\mathbb{R}^3$ are defined using the Levi-Civita symbol ϵ_{ijk} as

$$(\mathbf{a} \times \mathbf{b})_i = \epsilon_{ijk} a_j b_k.$$

- The Levi-Civita symbol ϵ_{ijk} is related to the Kronecker delta δ_i^j by the identities,

$$\epsilon_{ijk} \epsilon_{klm} = \delta_{il} \delta_{jm} - \delta_{im} \delta_{jl},$$
$$\epsilon_{ijk} \epsilon_{kjm} = \delta_{ij} \delta_{jm} - \delta_{im} \delta_{jj} = -2\delta_{im},$$
$$\epsilon_{ijk} \epsilon_{kji} = -2\delta_{ii} = -6.$$

- Thus, the double cross product of three vectors **a**, **b**, **c** in $\mathbb{R}^3$ may be expressed using the Levi-Civita symbol ϵ_{ijk} as

$$(\mathbf{a} \times (\mathbf{b} \times \mathbf{c}))_i = \epsilon_{ijk} a_j \epsilon_{klm} b_l c_m = \epsilon_{ijk} \epsilon_{klm} (a_j b_l c_m)$$
$$= (\delta_{il} \delta_{jm} - \delta_{im} \delta_{jl})(a_j b_l c_m) = b_i a_j c_j - c_i a_j b_j$$
$$= (\mathbf{b}(\mathbf{a} \cdot \mathbf{c}) - \mathbf{c}(\mathbf{a} \cdot \mathbf{b}))_i.$$

This is the "BAC minus CAB" rule for double cross products.

The Levi-Civita symbol ϵ_{ijk} is also an invaluable vector calculus tool for GFD. For example, it can be used to derive the vector calculus identities in Eqs. (2.97) and (2.100).

Exercise 2.1. *Use the Levi-Civita symbol ϵ_{ijk} to prove the vector calculus identities in Eqs. (2.97) and (2.100). These identities will be very useful in what follows.*

Answer. The curl of the cross product of two vector functions **a** and **b** in $\mathbb{R}^3$ may be written in terms of the Levi–Civita symbol ϵ_{ijk} as

$$(\nabla \times (\mathbf{b} \times \mathbf{c}))_i = \epsilon_{ijk}\partial_j\epsilon_{klm}b_lc_m = \epsilon_{ijk}\epsilon_{klm}\partial_j(b_lc_m)$$
$$= \partial_j(b_ic_j) - \partial_j(c_ib_j) = b_{i,j}c_j + b_ic_{j,j} - c_{i,j}b_j - c_ib_{j,j}.$$

This formula recovers the ith component of Eq. (2.100) when we replace $(\mathbf{b}, \mathbf{c})$ by $(\mathbf{a}, \mathbf{b})$, respectively.

As for the fundamental vector identity of fluid dynamics in Eq. (2.97), we compute

$$-(\mathbf{b} \times \text{curl } \mathbf{a})_i = -\epsilon_{ijk}b_j\epsilon_{klm}a_{m,l} = -b_j(\delta_{il}\delta_{jm} - \delta_{im}\delta_{jl})a_{m,l}$$
$$= -b_j(a_{j,i} - a_{i,j}) = -\partial_i(a_jb_j) + b_ja_{i,j} + a_jb_{j,i}$$
$$= (-\nabla(\mathbf{a}\cdot\mathbf{b}) + (\mathbf{b}\cdot\nabla)\mathbf{a} + a_j\nabla b^j)_i.$$

2.1.3 *Differential form notation for vector calculus*

Differential forms are objects from integral calculus. For example, Stokes theorem from integral calculus yields the following familiar results in $\mathbb{R}^3$, with line element $d\mathbf{x}$, oriented surface area element $d\mathbf{S}$ and oriented volume element d^3x:

- The *fundamental theorem of calculus*, upon integrating df along a curve in $\mathbb{R}^3$ starting at point a and ending at point b:

$$\int_a^b df = \int_a^b \nabla f \cdot d\mathbf{x} = f(b) - f(a).$$

- The *classical Stokes theorem*, for a compact surface S with boundary ∂S:

$$\int_S (\text{curl}\,\mathbf{v}) \cdot d\mathbf{S} = \oint_{\partial S} \mathbf{v} \cdot d\mathbf{x}.$$

  (For a planar surface $S \in \mathbb{R}^2$, this is *Green's theorem*.)
- The *Gauss divergence theorem*, for a compact spatial domain D with boundary ∂D:

$$\int_D (\text{div}\,\mathbf{A}) \, d^3x = \oint_{\partial D} \mathbf{A} \cdot d\mathbf{S}.$$

These examples illustrate the following theorem.

Theorem 2.1 (Stokes' theorem). *Suppose M is a compact oriented k-dimensional manifold with boundary ∂M and α is a smooth $(k-1)$-form on M. Then,*

$$\int_M d\alpha = \oint_{\partial M} \alpha,$$

where $d\alpha$ denotes the differential of α.

Differential form relations in vector calculus notation

The spatial differential d (often called the exterior derivative) and wedge product $\wedge$ defined by $dx^i \wedge dx^j = -dx^j \wedge dx^i$ for an oriented basis with line elements dx^i, with $i = 1, 2, 3$ in $\mathbb{R}^3$ satisfy the following relations in components and in three-dimensional vector notation:

$$df = f_{,j}\, dx^j =: \nabla f \cdot d\mathbf{x},$$
$$0 = d^2 f = f_{,jk}\, dx^k \wedge dx^j \quad (\text{tr(even times odd)} = 0),$$
$$df \wedge dg = f_{,j}\, dx^j \wedge g_{,k}\, dx^k$$
$$=: (\nabla f \times \nabla g) \cdot d\mathbf{S},$$
$$df \wedge dg \wedge dh = f_{,j}\, dx^j \wedge g_{,k}\, dx^k \wedge h_{,l}\, dx^l$$
$$=: (\nabla f \cdot \nabla g \times \nabla h)\, d^3 x = J\left(\frac{\partial f,\, \partial g,\, \partial h}{\partial x^1, \partial x^2, \partial x^3}\right) d^3 x.$$

Exercise 2.2 (Vector calculus formulas). *Show that the definition of exterior derivative d defines grad, div and curl from vector calculus as the following:*

$$df = \nabla f \cdot d\mathbf{x},$$
$$d(\mathbf{v} \cdot d\mathbf{x}) = (\operatorname{curl} \mathbf{v}) \cdot d\mathbf{S},$$
$$d(\mathbf{A} \cdot d\mathbf{S}) = (\operatorname{div} \mathbf{A})\, d^3 x.$$

Show that the compatibility condition $d^2 = 0$ for exterior derivative d implies the following familiar vector calculus relations.

$$0 = d^2 f = d(\nabla f \cdot d\mathbf{x}) = (\operatorname{curl} \operatorname{grad} f) \cdot d\mathbf{S},$$
$$0 = d^2(\mathbf{v} \cdot d\mathbf{x}) = d((\operatorname{curl} \mathbf{v}) \cdot d\mathbf{S}) = (\operatorname{div} \operatorname{curl} \mathbf{v})\, d^3 x.$$

Show that the product rule from vector calculus may be written for these forms as

$$d(f(\mathbf{A} \cdot d\mathbf{x})) = df \wedge \mathbf{A} \cdot d\mathbf{x} + f \operatorname{curl} \mathbf{A} \cdot d\mathbf{S}$$

$$= \left(\nabla f \times \mathbf{A} + f \operatorname{curl} \mathbf{A} \right) \cdot d\mathbf{S}$$
$$= \operatorname{curl}(f\mathbf{A}) \cdot d\mathbf{S},$$
$$d\left((\mathbf{A} \cdot d\mathbf{x}) \wedge (\mathbf{B} \cdot d\mathbf{x}) \right) = (\operatorname{curl} \mathbf{A}) \cdot d\mathbf{S} \wedge \mathbf{B} \cdot d\mathbf{x} - \mathbf{A} \cdot d\mathbf{x} \wedge (\operatorname{curl} \mathbf{B}) \cdot d\mathbf{S}$$
$$= \left(\mathbf{B} \cdot \operatorname{curl} \mathbf{A} - \mathbf{A} \cdot \operatorname{curl} \mathbf{B} \right) d^3 x$$
$$= d\left((\mathbf{A} \times \mathbf{B}) \cdot d\mathbf{S} \right)$$
$$= \operatorname{div}(\mathbf{A} \times \mathbf{B}) \, d^3 x.$$

These calculations with exterior derivative d using its compatibility condition $d^2 = 0$ and its product rule with the antisymmetric wedge product operation $\wedge$ illustrated in the Stokes Theorem have recovered the familiar formulas from vector calculus for quantities $\operatorname{curl}(\operatorname{grad})$, $\operatorname{div}(\operatorname{curl})$, $\operatorname{curl}(f\mathbf{A})$ *and* $\operatorname{div}(\mathbf{A} \times \mathbf{B})$.

2.1.4   What is a Partial Differential Equation (PDE)?

Let $u(x_1, x_2, \ldots, x_n)$ be a function defined e.g., on $\mathbb{R}^n$, and differentiable k times with respect to each variable.

A *differential equation* of degree k for u is an equation satisfied by u and by all its derivatives up to degree k. If $n \geqslant 2$, the derivatives involved are partial derivatives, and the equation is called a partial differential equation, or PDE.

Example:

$$u = u(x,t); \qquad u_t + u_x + u_{xxx} + u u_x = 0 \quad \text{(the KdV equation)},$$
$$u = u(x,y,z); \qquad u_{xx} + u_{yy} + u_{zz} = 0 \quad \text{(the Laplace equation)}.$$

A (classical) solution of a given PDE is any function with sufficient differentiability (i.e., such that all the derivatives appearing in the PDE are well defined) that satisfies the equation.

2.1.5   General properties of solutions to PDEs — Symmetry and similarity

For an arbitrary PDE, there are some general considerations that can be made to understand or simplify its form without explicitly solving anything.

The first important consideration to be made concerns any *symmetry property* that the PDE may possess.

Definition 2.4. A *symmetry transformation* is a smooth transformation of dependent and independent variables that leaves invariant the space of solutions of a given PDE.

Exercise 2.3. *Prove that transformations which leave a given PDE invariant in form also leave its space of solutions invariant.*

Common examples of symmetry groups which apply in fluid dynamics are space and time translations, rotations, changes of Galilean reference frame and scaling transformations of the units of mass, length and time.

As an illustration, suppose we are dealing with a PDE for a dependent variable $u = u(x,t)$ in the two independent variables (x,t) (usually space and time).

The most physically natural and common symmetry groups to act on the space of solutions are:

- **Translation**

$$x \to x+a, \qquad t \to t+b \implies u \to U(x,t) = u(x-a,t-b).$$

In this case, derivatives do not change, hence u and U satisfy the same PDE. Hence, given one solution, one can find a 2-parameter family of solutions simply by translation.

Note that all PDEs with constant coefficients are translation invariant.

- **Scaling**

$$x \to ax, \qquad t \to bt \implies u \to U(x,t) = u(x/a,t/b).$$

Scaling changes the differentiation operators by a constant:

$$\frac{\partial U}{\partial x} = \frac{1}{a}\frac{\partial u}{\partial x}, \quad \text{similarly} \quad \frac{\partial U}{\partial t} = \frac{1}{b}\frac{\partial u}{\partial t}.$$

Hence, u and U in general satisfy different PDEs, and it will only certain specific scalings that preserve the solutions.

Symmetries are useful because they allow the problem to be simplified from a PDE (in two independent variables) to an ODE, for which it is (sometimes) possible to find explicit solutions. We will give standard examples of this procedure below.

Example 2.1. Consider the heat equation, i.e., the PDE with

$$\partial_t u = \gamma \partial_{xx} u, \quad \text{where } \gamma > 0.$$

The function $u(x,t) = e^{-\gamma t}\sin x$ is a solution of the equation posed on the circle S^1, i.e., for $x \in [0,2\pi]$ and assuming 2π-periodicity.

Under translation of space and time, for any $a,b \in \mathbb{R}$ the formula

$$U_{a,b}(x,t) = e^{-\gamma(t-a)}\sin(x-b)$$

defines a 2-parameter family of solutions.

Under scaling, the PDE for U becomes

$$\partial_t U = \frac{1}{a}\partial_t u = \frac{\gamma}{a}\partial_x^2 u = \frac{\gamma b^2}{a}\partial_x^2 U.$$

If

$$\frac{b^2}{a} = 1$$

then, u and U satisfy the same PDE and we have a **scaling symmetry**, that is, the PDE keeps its form under the transformations.

Note also that since multiples of the solution are themselves a solution, we can combine the variable scalings with this to obtain the general transformation

$$U(x,t) = cu(x/a,t/b), \qquad c \neq 0.$$

A scaling symmetry is also called a **similarity transformation**.

Definition 2.5. A similarity solution of a PDE is a solution fixed by a one-parameter group of scaling symmetries.

In practice, one can exploit the scaling symmetries of a PDE to reduce the complexity of the equation from a system of PDEs in two independent variables to a system of ODEs. Then, one can find explicit *similarity solutions* by finding solutions of the ODE system.

Example 2.2 (Similarity solutions for the heat equation). Consider again the heat equation (where we set $\gamma = 1$)

$$\partial_t u = \partial_{xx} u, \qquad x \in \mathbb{R}, \quad t > 0.$$

Consider a generic scaling transformation, where the scaling constants are written for convenience (as will be clear in the course of the calculation) in terms of exponentials of a fixed quantity β:

$$t \to T = \beta^a t, \quad x \to \beta^b x,$$
$$u \to U = \beta^c u \implies u \to U(X,T) = \beta^c u(\beta^{-b}X, \beta^{-a}T).$$

This will be a scaling symmetry if $U(X,T)$ satisfies the same PDE as $u(x,t)$, hence if all terms appearing in the PDE scale to the same power of β: Since

$$\partial_T U = \beta^{c-a}\partial_t u; \quad \partial_X U = \beta^{c-b}\partial_x u; \quad \partial_X^2 U = \beta^{(c-2b)}\partial_x^2 u.$$

for a scaling symmetry we need $c - a = c - 2b$, hence $a = 2b$ (c is arbitrary).
    Therefore, a scaling symmetry is given by

$$t \to \beta^{2b}t, \quad x \to \beta^b x,$$

$$u \to \beta^c u(x,t) \implies u \to U(X,T) = \beta^c u(\beta^{-b}X, \beta^{-2b}T).$$

Choosing this particular scaling, we can now reduce the PDE to an ODE, for which we can find solutions. (With a slight abuse of notation, given that u and U satisfy the same PDE, henceforth we drop the use of capitals for the transformed variables.)
    Since β is arbitrary, choose it such that $\beta^{-2b}t = 1$, hence $\beta = t^{1/2b}$. Then,

$$u(x,t) = t^{c/2b}u(t^{-1/2}x, 1) = t^{c/2b}v(\xi), \quad \xi = t^{-1/2}x, \ v(\xi) = u(1,\xi).$$

Here, ξ is the similarity variable, and v is now a function of one variable.
    The next step is to find the ODE for v.
    Set $p = c/2b$. Then, with $u(x,t) = t^p v(\xi)$ we find

$$t^{p-1}\left(pv - \frac{\xi}{2}v' - v''\right) = 0.$$

We can make special choices of p for which this ODE is easy to solve.
    For **p** = 0, the solution is

$$v(\xi) = A\int_{-\infty}^{\xi} e^{-y^2/4}dy, \qquad A \text{ const.}$$

This yields for u the following similarity solution:

$$u(x,t) = A\int_{-\infty}^{x/t^{1/2}} e^{-y^2/4}dy = 2A\sqrt{\pi}\,\mathrm{erf}\left(\frac{x}{2\sqrt{t}}\right),$$

where erf is the so-called *error function*:

$$\mathrm{erf}(x) = \frac{1}{\sqrt{\pi}}\int_{-\infty}^{x} e^{-y^2}dy.$$

Motivated by the above similarity solution, we set $V = e^{\xi^2/4}v$. Then, V satisfies the ODE

$$V'' - \frac{\xi}{2}V' = \left(p + \frac{1}{2}\right)V,$$

which is easy to solve if $p = -\frac{1}{2}$, namely the solution is this case is $V = \kappa$, a constant. Hence, we find the similarity solution

$$v(\xi) = \kappa e^{-\xi^2/4} \Longrightarrow u(x,t) = \kappa t^{-1/2}e^{-x^2/4t}.$$

Exercise 2.4. *Fill in the details of the computation in Example 2.2 and verify that $u(x,t) = \kappa t^{-1/2}e^{-x^2/4t}$, $t > 0$, formally solves the PDE $\partial_t u = \partial_{xx} u$.*

Example 2.3 (Similarity solutions for the nonlinear transport equation). Consider the nonlinear PDE

$$\partial_t u + u\partial_x u = 0,$$

and a generic scaling transformation

$$t \to T = \beta^a t, \quad x \to X = \beta^b x,$$
$$u \to U = \beta^c u \Longrightarrow u \to U(X,T) = \beta^c u(\beta^{-b}X, \beta^{-a}T).$$

This will be a scaling symmetry if all terms scale to the same power of β:

$$\partial_T U = \beta^{c-a}\partial_t u; \quad \partial_X U = \beta^{c-b}\partial_x u;$$
$$\partial_t u + u\partial_x u = 0 \to \beta^{c-a}\partial_T U + \beta^{2c-b}U\partial_X U = 0,$$

hence if $c - a = 2c - b$ hence $c = b - a$.

Choosing this particular scaling, we can now reduce the PDE to an ODE. (As in the previous example, we revert to u, x, t for the variables.) Since β is arbitrary, we can choose it as we wish. Choose $\beta = t^{1/a}$, and $c = b - a$ (we assume $a \neq 0$). Then,

$$u(x,t) = t^{c/a}u(t^{-b/a}x, 1) = t^{c/a}v(\xi), \quad \xi = t^{-b/a}x, \; v(\xi) = u(1,\xi).$$

Here, ξ is called a similarity variable, and v is now a function of one variable. To find the ODE for v, we scale to have $a = 1$ $c = b - 1$. Then, with $u(x,t) =$

$t^{b-1}v(t^{-b}x)$ we find

$$\partial_t U + U\partial_x U = 0 \implies t^{b-2}\left\{(v - b\xi)v' + (b-1)v\right\} = 0.$$

This ODE has many solutions, but some particular ones are easy to find, e.g.,

$$v = b\xi, \quad b = 1 \implies v = \xi = \frac{x}{t}. \tag{2.1}$$

This corresponds to the similarity solution $u(x,t) = \frac{x}{t}$.

Exercise 2.5. *Fill in the details of the computation in Example 2.3 and verify that $u(x,t) = x/t$, $t > 0$, formally solves the PDE $\partial_t u + u\partial_x u = 0$.*

These ideas extend naturally to vector-valued functions, as we will see for the specific example of the Euler and Navier–Stokes equations.

Additional exercises

Exercise 2.6. *Find the scaling symmetries and the corresponding similarity solutions for the PDE*

$$u_t + u^2 u_x = 0,$$

and for the so-called Laplace equation

$$u_{xx} + u_{yy} = 0.$$

Exercise 2.7. *Consider the effect of the scaling transformation*

$$(t, x, u) \to (\alpha t, \beta x, \lambda u)$$

on Burger's equation

$$u_t + u u_x = \nu u_{xx}, \qquad \nu > 0.$$

Exercise 2.8. *Find the scaling symmetries of Burger's equation and determine the ODE satisfied by the similarity solutions.*

2.2 The Governing Equations for an Ideal Fluid

The first task is to give a description of a fluid that is mathematically well posed and consistent, and then derive a model that represents the dynamical properties of the fluid.

For this we assume we can consider the ensemble of microscopic particles that constitute a fluid as a **continuum**.

Our modelling is based on the assumption that, given a fluid, it is possibly to define a *mass density* function $\rho(\mathbf{x},t)$ at each point $\mathbf{x}$ in the fluid domain, and each time t. Hence, the mass of fluid occupying a region $D \subset \mathbb{R}^d$ ($d = 2,3$) is given by

$$M(D,t) = \int_D \rho(\mathbf{x},t)d\mathbf{x}. \tag{2.2}$$

We assume that at each point in space-time, the fluid moves with velocity $\mathbf{u}(\mathbf{x},t)$.

See Marsden and Chorin (1993) for additional fundamental descriptions of fluid dynamics.

Mathematical description of a fluid — Lagrangian

We interpret the fluid through the movement of the initial domain occupied by the fluid, also called the reference configuration of the fluid.

We model the fluid through the map $\Phi_t(\mathbf{y})$ which describes the trajectory of each space particle $\mathbf{y}$ as time varies.

Definition 2.6. Given open, smooth domains $D_s \subset \mathbb{R}^d$, $s \in [0,\infty)$, a **fluid** is a set of maps

$$\Phi_t : D_0 \to D_t \qquad (\textbf{notation: } \Phi_t(\mathbf{y}) = \Phi(\mathbf{y},t) \; \forall \mathbf{y} \in D_0)$$

such that

- $\{\Phi_t\}_{t \geqslant 0}$ are smooth functions with respect to both $\mathbf{y}$ and t.
- $\{\Phi_t\}_{t \geqslant 0}$ have continuous inverses for all $t \geqslant 0$.
- $\Phi_0 = Id$.

For each point $\mathbf{y}$ in D_0, the curve $\Phi_t(\mathbf{y})$ is the trajectory of the "particle" which is at position $\mathbf{y}$ at the initial time $t = 0$.

Mathematical description of a fluid — Eulerian

This description takes as its central object the velocity of the fluid, $\mathbf{u}(\mathbf{x},t)$, is given by

$$\begin{cases} \dfrac{d\mathbf{x}(\mathbf{y},t)}{dt} = \mathbf{u}(\mathbf{x},t), & (\mathbf{x}(\mathbf{y},t) = \Phi_t(\mathbf{y})) \\ \mathbf{x}(\mathbf{y},0) = \mathbf{y}. \end{cases}$$

In terms of the fluid map $\Phi_t(y)$, the initial value problem above is written as the equation

$$\frac{d\Phi_t(y)}{dt} = \mathbf{u}(\Phi_t(y), t).$$

Remark 2.1. If the flow map Φ_t is sufficiently smooth then the Eulerian velocity is defined in terms of the flow maps by $\mathbf{u}(\mathbf{x}, t) = d\Phi_t(\mathbf{y})/dt$ and conversely the flow maps $\Phi_t(\mathbf{y})$ is well defined as equal to the Eulerian position $\mathbf{x}(\mathbf{y}, t)$ — indeed the two descriptions above are equivalent. But this is not true if the flow map is not sufficiently smooth! The interpretation of a flow in terms of an appropriate flow map when the velocity is not smooth is an active area of current research.

2.2.1 *The governing equations*

To find these, we start from three physical principles that we assume true:

(1) conservation of mass,
(2) balance of momentum (i.e., the rate of change in momentum of a portion of fluid is equal to the force applied to it),
(3) conservation of energy.

Mass conservation

This principle says that the mass of fluid does not change in time, hence given Definition 2.2 of the mass we impose that

$$\int_{D_t} \rho(\mathbf{x}, t) d\mathbf{x} = \int_{D_0} \rho(\mathbf{y}, 0) d\mathbf{y}, \quad D_t = \Phi_t(D_0).$$

Using the change of variables $\mathbf{x} = \Phi_t(\mathbf{y})$, we can write

$$\int_{D_t} \rho(\mathbf{x}, t) d\mathbf{x} = \int_{D_0} \rho(\Phi_t(\mathbf{y}), t) J(\mathbf{y}, t) d\mathbf{y} = \int_{D_0} \rho(\mathbf{y}, 0) d\mathbf{y}, \qquad (2.3)$$

where $J(\mathbf{y}, t)$ is the Jacobian of the change of variables. Since Φ is invertible, this Jacobian is non-zero.

Exercise 2.9. *Write down the definition of $J(\mathbf{y}, t)$ in terms of the components of the map $\Phi_t(\mathbf{y})$.*

Since D_0 is arbitrary, in Lagrangian form (i.e., in terms of the map $\Phi_t(\mathbf{y})$) this reads

$$\rho(\Phi_t(\mathbf{y}),t)J(\mathbf{y},t) = \rho(\mathbf{y},0), \quad \forall t > 0.$$

For D a region in $\mathbb{R}^d$, then mass changes with time according to

$$\frac{d}{dt}M(D,t) = \frac{d}{dt}\int_D \rho(\mathbf{x},t)d\mathbf{x}.$$

Hence its mass is conserved, we must have

$$\frac{d}{dt}\int_D \rho(\mathbf{x},t)d\mathbf{x} = 0; \text{ equivalently } \frac{d}{dt}[\rho(\Phi_t(\mathbf{y}),t)J(\mathbf{y},t)] = 0.$$

If D is fixed, and ρ is a smooth function of t, then

$$\frac{d}{dt}\int_D \rho(\mathbf{x},t)d\mathbf{x} = \int_D \frac{\partial\rho(\mathbf{x},t)}{\partial t}d\mathbf{x}.$$

But this is not true if $D = D_t$ depends on time. In this case, we need the following simple result.

Reynolds transport theorem

In the Lagrangian description of a fluid, the region of integration generally varies with time.

We therefore need a formula to take into account the dependence on time of the domain when we take time derivatives of integrals evaluated in the fluid region. This formula is known as Reynolds transport theorem.

We first need to evaluate the time derivative of the Jacobian J.

Exercise 2.10. *Recalling that*

$$\frac{\partial\Phi_t(\mathbf{y})}{\partial t} = \mathbf{u}(\Phi_t(\mathbf{y}),t)$$

show by an explicit computation that

$$\frac{\partial J(\mathbf{y},t)}{\partial t} = J(\mathbf{y},t)[\operatorname{div}\mathbf{u}(\Phi_t(\mathbf{y}),t)]. \tag{2.4}$$

Theorem 2.2 (Reynolds transport theorem). *Let $D_0 \subset \mathbb{R}^3$ be an open bounded domain with smooth boundary, and let $\Phi_t(\mathbf{y})$ denote the flow map of a fluid with Eulerian velocity $\mathbf{u}$. Let $D_t = \Phi(D_0)$.*

Let $f(\cdot, t) : D_t \subset \mathbb{R}^d \to \mathbb{R}$ be an arbitrary smooth function. Then,

$$\frac{d}{dt} \int_{D_t} f d\mathbf{x} = \int_{D_t} \left(\frac{\partial f}{\partial t} + \nabla \cdot (f \mathbf{u}) \right) d\mathbf{x}.$$

Proof. Since $D_t = \Phi(D_0)$, each $\mathbf{x} \in D_t$ can be written as $\mathbf{x} = \Phi_t(\mathbf{y})$ for some $\mathbf{y} \in D_0$. Using this change of variable we find

$$\frac{d}{dt} \int_{D_t} f d\mathbf{x} = \frac{d}{dt} \int_{D_0} f(\Phi_t(\mathbf{y}), t) J(\mathbf{y}, t) d\mathbf{y} = \int_{D_0} \frac{\partial}{\partial t} [f(\Phi_t(\mathbf{y}), t) J(\mathbf{y}, t)] d\mathbf{y}$$

$$= \int_{D_0} \left\{ \left[\frac{\partial}{\partial t} f(\Phi_t(\mathbf{y}), t) + \nabla_x f(\Phi_t(\mathbf{y}), t) \frac{\partial (\Phi_t(\mathbf{y}), t)}{\partial t} \right] J(\mathbf{y}, t) \right.$$

$$\left. + f(\Phi_t(\mathbf{y}), t) \frac{d}{dt} J(\mathbf{y}, t) \right\} d\mathbf{y}$$

and using Eq. (2.4)

$$= \int_{D_0} X \left[\frac{\partial}{\partial t} f(\Phi_t(\mathbf{y}), t) + \nabla_x f(\Phi_t(\mathbf{y}), t) \frac{\partial (\Phi_t(\mathbf{y}), t)}{\partial t} \right.$$

$$\left. + f(\Phi(\mathbf{y}, t)[\operatorname{div} \mathbf{u}(\Phi_t(\mathbf{y}), t)]) \right] J(\mathbf{y}, t) d\mathbf{y}.$$

Since $\frac{\partial (\Phi_t(\mathbf{y}), t)}{\partial t} = \mathbf{u}(\Phi_t(\mathbf{y}), t)$, changing variable back to $\mathbf{x} = \Phi_t(\mathbf{y})$ this is equal to

$$\int_{D_t} \left[\frac{\partial}{\partial t} f(\mathbf{x}, t) + \nabla f(\mathbf{x}, t) \cdot \mathbf{u}(\mathbf{x}, t) \right.$$

$$\left. + f(x, t)[\nabla \cdot \mathbf{u}(\mathbf{x}, t)] \right] d\mathbf{x} = \int_{D_t} \left(\frac{\partial f}{\partial t} + \nabla \cdot f \mathbf{u} \right) d\mathbf{x}. \qquad \square$$

We can now use this theorem to deduce that conservation of mass implies that

$$0 = \frac{d}{dt} \int_{D_t} \rho(\mathbf{x}, t) d\mathbf{x} = \int_{D_t} \left(\frac{\partial \rho}{\partial t} + \nabla \cdot \rho \mathbf{u} \right) d\mathbf{x}.$$

Exercise 2.11. *Use a computation similar to the one in the proof of Reynold's transport theorem to express the conservation of mass as the* *condition*

$$\int_{D_0} \left[\frac{\partial \rho}{\partial t}(\Phi_t(\mathbf{y}), t) + \nabla \rho(\Phi_t(\mathbf{y}), t) \cdot \mathbf{u}(\Phi_t(\mathbf{y}), t) \right.$$

$$\left. + \rho(\Phi_t(\mathbf{y}))[\nabla \cdot \mathbf{u}(\Phi_t(\mathbf{y}), t)] \right] J(\mathbf{y}, t) d\mathbf{y} = 0. \tag{2.5}$$

Given that D_0 is arbitrary and $J \neq 0$ (why?) the identity in Eq. (2.5) implies the *continuity equation*, expressing **conservation of mass in Eulerian form**:

$$\frac{\partial \rho}{\partial t} + \nabla \cdot (\rho \mathbf{u}) = 0. \tag{2.6}$$

Exercise 2.12. *Use conservation of mass to prove the following form of the transport theorem:*

$$\frac{d}{dt} \int_{D_t} \rho f \, dv = \int_{D_t} \rho \frac{Df}{Dt} \, dv = \int_{D_t} \rho \left(\frac{\partial f}{\partial t} + \mathbf{u} \cdot \nabla f \right) dv.$$

where $D_t = \Phi(D_0)$ is a portion of fluid of density ρ transported with Eulerian velocity $\mathbf{u}$.

Balance of momentum

This is obtained by generalising Newton's law

$$force = mass \cdot acceleration$$

to the case of a continuum of particles.

The *acceleration* of a fluid is given by

$$a(t) = \frac{d^2}{dt^2} \mathbf{x}(t) = \frac{d}{dt} \mathbf{u}(\mathbf{x}(t), t) = \frac{\partial \mathbf{u}}{\partial x_1} \dot{x}_1(t) + \frac{\partial \mathbf{u}}{\partial x_2} \dot{x}_2(t) + \frac{\partial \mathbf{u}}{\partial x_3} \dot{x}_3(t) + \frac{\partial \mathbf{u}}{\partial t}$$

$$= \frac{\partial \mathbf{u}}{\partial x_1} u_1(t) + \frac{\partial \mathbf{u}}{\partial x_2} u_2(t) + \frac{\partial \mathbf{u}}{\partial x_3} u_3(t) + \frac{\partial \mathbf{u}}{\partial t} = \mathbf{u} \cdot \nabla \mathbf{u} + \partial_t \mathbf{u}.$$

We define the two differential operators appearing in the expression above as

$$\partial_t = \frac{\partial}{\partial t}; \quad \mathbf{u} \cdot \nabla = u_1 \frac{\partial}{\partial x_1} + u_2 \frac{\partial}{\partial x_2} + u_3 \frac{\partial}{\partial x_3}; \tag{2.7}$$

Definition 2.7. The operator D/Dt, given by

$$\frac{D}{Dt} = \partial_t + \mathbf{u} \cdot \nabla, \tag{2.8}$$

and depending on the velocity **u** of the fluid, is called the **Lagrangian (time) derivative.**

Using the Lagrangian derivative, we write the acceleration as

$$a(\mathbf{x},t) = \frac{D}{Dt}\mathbf{u}.$$

This formalises that the acceleration not only depends on the movement of the fluid but also on the fact that the position of the particles is changing.

We now need to model the *force* acting on the fluid. These will be roughly falling into one of two categories:

- External forces, such as gravity or a magnetic field.
  In the region D_t, these can modelled by assuming that $f_e(\mathbf{x},t)$ is the force per unit of mass, as

$$F_e xt = \int_{D_t} \rho f_e(\mathbf{x},t)d\mathbf{x}.$$

- Forces of stress, through which the rest of continuum acts on the fluid across its surface

We will only consider **ideal fluids:**

Definition 2.8. A fluid is called as **ideal fluid** if there exists a scalar function $p(\mathbf{x},t)$, called **pressure**, such that for any surface S in the fluid, with unit normal **n**, we have

$$\text{force across } S \text{ per unit area} = p(\mathbf{x},t)\mathbf{n}.$$

This means that forces act only normally to the surface, and that there are no tangential forces — in particular, there is no way that rotation can be initiated or stopped in the fluid.

For an ideal fluid in a region D_t, the force (stress) exerted on the fluid across the surface ∂D_t is given by

$$S = -\int_{\partial D_t} p(\mathbf{x},t)\mathbf{n}\,dA. \tag{2.9}$$

Using the divergence theorem to write this as an integral in 3-dimensional space, we obtain

$$S = -\int_{D_t} \nabla p(\mathbf{x},t)d\mathbf{x}. \tag{2.10}$$

Hence, for any particle of fluid Newton's second law implies the differential form of balance of momentum:

$$\rho \frac{D\mathbf{u}(\mathbf{x},t)}{Dt} = -\nabla p(\mathbf{x},t) + \rho f_e(\mathbf{x},t). \tag{2.11}$$

Hence,

$$\rho \frac{\partial \mathbf{u}}{\partial t} = -\rho(u \cdot \nabla \mathbf{u}) - \nabla p + \rho f_e.$$

Exercise 2.13. *Use the continuity equation to write the differential balance of momentum as*

$$\frac{\partial \rho \mathbf{u}}{\partial t} = -\nabla(\rho \mathbf{u})\mathbf{u} - \rho(\mathbf{u} \cdot \nabla)\mathbf{u} - \nabla p(\mathbf{x},t) + \rho f(\mathbf{x},t).$$

The balance of momentum therefore leads to the following fundamental **Euler equation for ideal fluids**:

$$\partial_t \mathbf{u} + (\mathbf{u} \cdot \nabla)\mathbf{u} = -\frac{1}{\rho}\nabla p + f. \tag{2.12}$$

This equation is to be paired with the continuity equation (Eq. (2.6)):

$$\frac{\partial \rho}{\partial t} + \nabla \cdot (\rho \mathbf{u}) = 0.$$

These are four equations; however the (scalar) unknowns are five, namely

$$\mathbf{u} = (u_1, u_2, u_3); \rho; p.$$

Hence, one additional equation is needed. To obtain it, we make additional assumptions.

Incompressible fluid

A fluid is incompressible if the volume it occupies does not change, only the shape of the region occupied by the fluid can change — i.e., volume is conserved. Since volume change is measured by the Jacobian J of the change of variable that follows the flow, for a fluid to be incompressible we must have

$$J(\mathbf{y},t) = 1.$$

Then,

$$0 = \frac{dJ}{dt} = (\operatorname{div}\mathbf{u})J \Longrightarrow \operatorname{div}\mathbf{u} = 0.$$

The Eulerian statement that the flow is incompressible is the requirement that $\operatorname{div}\mathbf{u} = 0$.

It follows that for incompressible fluids the continuity equation becomes the **transport equation**

$$\frac{\partial \rho}{\partial t} + \mathbf{u} \cdot \nabla \rho = \frac{D\rho}{Dt} = 0. \tag{2.13}$$

This implies that a fluid is incompressible if and only if the density is constant following the fluid.

The **Euler equation for incompressible flows** is the equation

$$\partial_t \mathbf{u} + (\mathbf{u} \cdot \nabla)\mathbf{u} = -\frac{1}{\rho}\nabla p + f, \tag{2.14}$$

that is to be coupled with mass conservation and incompressibility:

$$\begin{aligned} \partial_t \rho + (\mathbf{u} \cdot \nabla)\rho &= 0, \\ \operatorname{div}\mathbf{u} &= 0. \end{aligned} \tag{2.15}$$

This is now a system of five equations for the five unknowns $\mathbf{u} = (u_1, u_2, u_3); \rho; p$.

Exercise 2.14. *Write down the system of three equations for the velocity components. What equation is each of these scalar equations?*

Note that, if the fluid is incompressible, and $\rho(x,0) = 1$ (or any other constant), i.e., if the fluid is homogeneous initially, then since $D\rho/Dt = 0$, the fluid will stay homogeneous for all times, $\rho(x,t) = 1$. In this case, the system is called the *Euler system for an ideal gas*:

$$\begin{aligned} \partial_t \mathbf{u} + (\mathbf{u} \cdot \nabla)\mathbf{u} &= -\frac{1}{\rho}\nabla p + f, \\ \operatorname{div}\mathbf{u} &= 0. \end{aligned} \tag{2.16}$$

Remark 2.2. Another possible way to add an equation and obtain a system with the correct count of unknowns is to introduce a relation between density and pressure. In general, this is possible for a general class of fluids called *isentropic*.

Namely, a fluid is isentropic if there exists a function w such that

$$\nabla w = \frac{1}{\rho} \nabla p.$$

This is particularly relevant in the theory of gas (a particular type of fluid). The (incompressible) Euler equations in this case are

$$\partial_t \mathbf{u} + (\mathbf{u} \cdot \nabla) \mathbf{u} = -\nabla w + f,$$
$$\partial_t \rho + (\mathbf{u} \cdot \nabla) \rho = 0. \tag{2.17}$$

Symmetry transformations of the Euler equation.

We consider the symmetry groups of the Euler equation for incompressible, homogeneous fluids, in the absence of external forces, namely the equation

$$\partial_t \mathbf{u} + (\mathbf{u} \cdot \nabla) \mathbf{u} = -\nabla p. \tag{2.18}$$

Proposition 2.1. *The Euler equation for an ideal gas (Eq. (2.18)) admits the following symmetries. Let $(\mathbf{u}, p)$ be a solution pair (consisting of velocity and pressure).*

- *Galilean invariance: for any $\mathbf{c} \in \mathbb{R}^d$*

$$\mathbf{U}(\mathbf{x}, t) = \mathbf{u}(\mathbf{x} - \mathbf{c}t, t) + \mathbf{c}, \quad P(\mathbf{x}, t) = p(\mathbf{x} - \mathbf{c}t, t)$$

  is a solution pair.
- *Rotation symmetry: for any orthogonal matrix M (i.e., $d \times d$ matrices such that $M^t = M^{-1}$; such matrices represent a rotation of $\mathbb{R}^d$),*

$$\mathbf{U}(\mathbf{x}, t) = M^t \mathbf{u}(M\mathbf{x}, t), \quad P(\mathbf{x}, t) = p(M\mathbf{x}, t)$$

  is a solution pair.
- *Scaling symmetry: for any $\lambda, \tau \in \mathbb{R}$*

$$\mathbf{U}(\mathbf{x}, t) = \frac{\lambda}{\tau} \mathbf{u}\left(\frac{\mathbf{x}}{\lambda}, \frac{t}{\tau}\right), \quad P(\mathbf{x}, t) = \frac{\lambda^2}{\tau^2} p\left(\frac{\mathbf{x}}{\lambda}, \frac{t}{\tau}\right)$$

  is a solution pair.

Proof. Exercise. □

Viscosity and the Navier–Stokes equations

To model the shear effects observed in fluids, Navier (in 1822) and Stokes (in 1845) proposed a specific form for the internal force term, hence modifying the equation to

$$\rho\frac{D\mathbf{u}}{Dt} + \nabla p = (\lambda+\mu)\nabla(\nabla\cdot\mathbf{u}) + \mu\Delta\mathbf{u} + \rho f,$$

where Δ denotes the Laplacian operator.

In case the fluid is incompressible ($\nabla\cdot\mathbf{u}=0$), and $\mu=$ const. we obtain what is known as the Navier–Stokes equation

$$\rho\frac{D\mathbf{u}}{Dt} + \nabla p = \mu\Delta\mathbf{u} + \rho f. \tag{2.19}$$

For $\rho=$ const. (the homogeneous case) and no external forces ($f=0$) this becomes

$$\frac{D\mathbf{u}}{Dt} + \nabla p = \nu\Delta\mathbf{u}, \tag{2.20}$$

where $\nu=\mu/\rho$ is the *kinematic viscosity coefficient*.

Exercise 2.15. *Show that the first two symmetries of the Navier–Stokes equation (Eq. (2.20)) are the same as for the Euler equation, while the scaling symmetry becomes:*
for any $\tau\in\mathbb{R}$

$$\mathbf{U}(\mathbf{x},t) = \tau^{-1/2}\mathbf{u}\left(\frac{\mathbf{x}}{\tau^{1/2}},\frac{t}{\tau}\right), \quad P(\mathbf{x},t) = \tau^{-1}p\left(\frac{\mathbf{x}}{\tau^{1/2}},\frac{t}{\tau}\right)$$

is a solution pair.

In particular, this yields a one-parameter (rather than two-parameter as for Euler equation) family of similarity solutions.

Dimensionless variables

For a given problem, there are three fundamental physical parameters:

- $L=$ characteristic length;
- $U=$ characteristic velocity.
- $T=$ characteristic time scale;

These are not mathematically well defined — rather they are reasonable, or average, length and velocity scales typical for the particular flow under consideration.

To be consistent, we also set $T = \frac{L}{U}$ as characteristic time scale.

Parametrising by these characteristic scales, we can obtain dimensionless equations. Let us consider the example of the homogeneous incompressible Navier–Stokes. Consider the dimensionless quantities

$$\mathbf{u}' = \frac{\mathbf{u}}{U}; \quad \mathbf{x}' = \frac{\mathbf{x}}{L}; \quad t' = \frac{t}{T}.$$

Exercise 2.16. *Compute the result of this change of variable for each component of the incompressible homogeneous NS equation*

$$\frac{D\mathbf{u}}{Dt} = -\nabla p + \nu \Delta \mathbf{u}.$$

The change of variable, and dividing by UL, yields

$$\frac{\partial \mathbf{u}'}{\partial t'} + (\mathbf{u}' \cdot \nabla')\mathbf{u}' = -\nabla' p + \frac{\nu}{UL}\Delta' \mathbf{u},$$

where the primes denote the differentiation with respect to the new variables.

The parameter UL/ν appearing in this dimensionless form of the equation is an important scaling parameter, important enough to merit a name:

$$R = \frac{UL}{\nu} \quad \text{is called the Reynolds number.}$$

Initial and boundary conditions

To solve these evolution equations, initial conditions (i.e., the state of the system at $t = 0$) and conditions at the boundary of the region D must be prescribed.

We only need initial conditions (i.e., conditions at $t = 0$) on the velocity and the density, as pressure does not have an explicit time-evolution.

For the Euler equation, as we already implicitly assumed, **one boundary condition** is needed. The simplest boundary condition models the fact that fluid does not flow through the boundary (the boundary is impermeable). This condition, which we will use, is written as

$$\mathbf{u} \cdot \mathbf{n} = 0, \quad \text{on } \partial D.$$

However, for the Navier–Stokes, which has a second order (Laplacian) term, **two boundary conditions** are needed. The second most natural condition is the no-slip condition $u = 0$ on ∂D, which models the case that the fluid sticks to the boundary. This boundary condition also provides a mechanism by which the boundary introduces vorticity into the fluid.

The need for a second boundary condition is the most direct mathematical expression of the fact that the two equations, namely Euler and Navier–Stokes, model very different phenomena.

The pressure term

The Navier–Stokes equation for incompressible fluids does not have an explicit evolution equation for the pressure. Hence, it seems natural to try to eliminate pressure from the evolution equation.

To understand the role of the pressure, we need a general result in differential geometry, called in the present context *Helmholtz Theorem.*

Theorem 2.3 (Helmholtz–Hodge decomposition theorem). *A smooth* (C^2) *vector field* **w** *on a bounded open domain* $D \subset \mathbb{R}^3$ *can be uniquely decomposed in the form*

$$\mathbf{w} = \mathbf{u} + \operatorname{grad} p,$$

where p *is a scalar function,* **u** *has zero divergence and is parallel to* ∂D:

$$\nabla \cdot \mathbf{u} = 0; \qquad \mathbf{u} \cdot \mathbf{n} = 0 \quad on \ \partial D.$$

Remark 2.3. The regularity assumptions can be considerably weakened, but then one needs to use the notion of Sobolev spaces and weak solutions.

Remark 2.4. The theorem is also true in unbounded domains or in the whole space $\mathbb{R}^3$, but then one needs suitable assumption on the decay of the vector field at infinity.

This theorem yields an orthogonal decomposition for arbitrary vector fields, hence gives a natural projection $\mathcal{P}$ of any vector field **w** onto its divergence-free, parallel-to-the-boundary part. Using this projection on all terms in the incompressible Navier–Stokes equation (in dimensionless coordinates), we find (formally — some smoothness is required) since $\operatorname{div} \mathbf{u} = 0$ and $\mathbf{u} \cdot \mathbf{n} = 0$ on ∂D, then,

$$\mathcal{P}\mathbf{u} = \mathbf{u}; \quad \mathcal{P}\partial_t \mathbf{u} = \partial_t \mathbf{u}; \quad \mathcal{P}(\operatorname{grad} p) = 0$$

so that the projected equation does not involve the pressure:

$$\partial_t \mathbf{u} = \mathcal{P}\left(-(\mathbf{u} \cdot \nabla)\mathbf{u} + \frac{1}{R}\Delta \mathbf{u}\right).$$

After solving this equation for $\mathbf{u}$, the pressure can be recovered as the gradient part of $\mathbf{w} = -(\mathbf{u} \cdot \nabla)\mathbf{u} + \frac{1}{R}\Delta\mathbf{u}$.

Proof of Theorem 2.3. If $\mathbf{w} = \mathbf{u} + \operatorname{grad} p$ with $\mathbf{u}$ as in the statement of the theorem, then,

$$\operatorname{div} \mathbf{w} = \operatorname{div} \operatorname{grad} p = \Delta p, \qquad \mathbf{w} \cdot \mathbf{n} = \mathbf{n} \cdot \operatorname{grad} p.$$

Hence given $\mathbf{w}$, consider the solution of the following (Neumann) boundary problem for the Laplace equation:

$$\Delta p = \operatorname{div} \mathbf{w}, \qquad \frac{\partial p}{\partial n} = \mathbf{w} \cdot \mathbf{n} \quad \text{on } \partial D.$$

The solution of this problem can be shown to exists and is unique up to a constant. This holds provided the domain is simply connected, open, with a smooth boundary, and can be shown e.g., by using the appropriate Green's function or using Fourier transform techniques.

Given this solution p, one defines $\mathbf{u} = \mathbf{w} - \operatorname{grad} p$.

We still need to show that the decomposition is unique. Given the identity

$$\nabla \cdot (p\mathbf{u}) = (\nabla \cdot \mathbf{u})p + \mathbf{u} \cdot \nabla p (= (\operatorname{div} \mathbf{u})p + \mathbf{u} \cdot \operatorname{grad} p))$$

assuming that $\operatorname{div} \mathbf{u} = 0$ we have that

$$\int_D \mathbf{u} \cdot \nabla p \, dv = \int_D \nabla \cdot (p\mathbf{u}) dv = \int_{\partial D} p\mathbf{u} \cdot \mathbf{n} \, da = 0,$$

where the second identity follows from the divergence theorem and the last from the boundary condition $\mathbf{u} \cdot \mathbf{n} = 0$. Hence, if $\mathbf{u} + \operatorname{grad} p = \tilde{\mathbf{u}} + \operatorname{grad} \tilde{p}$, then,

$$0 = (\mathbf{u} - \tilde{\mathbf{u}}) + \operatorname{grad}(p - \tilde{p}).$$

From this, taking the inner product with $\mathbf{u} - \tilde{\mathbf{u}}$, it follows that

$$0 = \int_D \left[|\mathbf{u} - \tilde{\mathbf{u}}|^2 + (\mathbf{u} - \tilde{\mathbf{u}}) \cdot \operatorname{grad}(p - \tilde{p}) \right] dv = \int |\mathbf{u} - \tilde{\mathbf{u}}|^2 dv,$$

hence $|\mathbf{u} - \tilde{\mathbf{u}}|^2 = 0$, which implies $\mathbf{u} = \tilde{\mathbf{u}}$, and in turn $\operatorname{grad} p = \operatorname{grad} \tilde{p}$, but the pressure is not uniquely defined. $\qquad \square$

On $\mathbb{R}^3$, the explicit solution of the Neumann–Laplace in terms of the Green's kernel yields

$$\nabla p(x) = \frac{\mathbf{x}}{|\mathbf{x}|^3} \int \operatorname{div} \mathbf{w} \, d\mathbf{x}.$$

This yields Leray's formulation of Navier–Stokes equation. (In this formulation, the trace $\operatorname{tr}(A)$ of a matrix A is the sum of its diagonal elements.)

Theorem 2.4 (Leray). *Solving the incompressible Navier–Stokes equation posed on* $\mathbb{R}^3$, *with initial smooth velocity* $\mathbf{u}(\mathbf{x},0) = u_0(\mathbf{x})$ *is equivalent to solving the following evolution equation for the velocity:*

$$\frac{D\mathbf{u}}{Dt} = \text{const.} \int_{\mathbb{R}^3} \frac{\mathbf{x} - \tilde{\mathbf{x}}}{|\mathbf{x} - \tilde{\mathbf{x}}|^3} \operatorname{tr}(\nabla \mathbf{u}(\tilde{\mathbf{x}},t))^2 d\tilde{\mathbf{x}} + \nu \Delta \mathbf{u}.$$

Kinetic energy

For a fluid moving in a domain $D \subset \mathbb{R}^3$ the kinetic energy in $D_t \subset D$ is given by

$$E_k = \frac{1}{2} \int_{D_t} \rho |\mathbf{u}|^2 dv \qquad |\mathbf{u}|^2 = u_1^2 + u_2^2 + u_3^2.$$

then, a tedious calculation yields

$$\frac{dE_k}{dt} = \frac{d}{dt}\left(\frac{1}{2}\int_{D_t} \rho|\mathbf{u}|^2 dv\right) = \frac{1}{2}\int_{D_t} \rho \frac{D|\mathbf{u}|^2}{Dt} d\mathbf{x} = \frac{1}{2}\int_{D_t} \rho\left(\mathbf{u}\cdot\frac{D\mathbf{u}}{Dt}\right)d\mathbf{x}.$$

Euler equations

Proposition 2.2. *If the fluid is incompressible* $(\nabla \cdot \mathbf{u} = 0)$ *and homogeneous* $(\rho = const.)$ *the kinetic energy is conserved:*

$$\frac{dE_k}{dt} = 0.$$

Indeed, using the Euler equation (with $f = 0$, hence no external forces) using Reynold's theorem, we find for the total kinetic energy

$$\frac{dE_k}{dt} = \frac{1}{2}\int_D \rho\left(\mathbf{u}\cdot\frac{D\mathbf{u}}{Dt}\right)d\mathbf{x} = -\frac{\rho}{2}\int_D (\mathbf{u}\cdot\nabla p)\, d\mathbf{x}.$$

Since $\nabla \cdot \mathbf{u} = 0$ by the divergence theorem, we have already seen that

$$\int_D (\mathbf{u}\cdot\nabla p)\, d\mathbf{x} = \int_D \nabla(\mathbf{u}p)d\mathbf{x} = \int_{\partial D} p\mathbf{u}\cdot n\, da.$$

Assuming $\mathbf{u} \cdot \mathbf{n} = 0$ on ∂D (so that no fluid escapes the region D), we obtain (as in the uniqueness proof of Helmholtz–Hodge theorem) the **conservation of kinetic energy**.

One could also consider the equation as posed on the whole of the space $\mathbb{R}^3$. As long as the product of velocity and pressure vanishes fast enough at infinity, the same proof as above yields energy conservation, by invoking this lemma.

Lemma 2.1. *Let u be a smooth vector field such that $\operatorname{div} \mathbf{u} = 0$, and p a smooth scalar function satisfying*

$$|\mathbf{u}(\mathbf{x})| |p(\mathbf{x})| = O\left(\frac{1}{|\mathbf{x}|^2}\right) \quad as \ |\mathbf{x}| \to \infty.$$

Then,

$$\int_{\mathbb{R}^3} \mathbf{u} \cdot \nabla p \, d\mathbf{x} = 0.$$

This lemma is a consequence of the divergence's theorem on a sphere of radius r, by considering that the **surface of this sphere** is proportional to r^2, and taking the limit as $r \to \infty$. (Fill in the details!)

Navier–Stokes equation

Now we consider the equation for a viscous fluid

$$\frac{D\mathbf{u}}{Dt} = -\nabla p + \nu \Delta \mathbf{u}.$$

The viscosity affects the kinetic energy, which is not conserved anymore, but rather dissipated.

Proposition 2.3. *Let $\mathbf{u}$ be a smooth solution of the Navier–Stokes equation (vanishing sufficiently rapidly as $|\mathbf{x}| \to \infty$ if the fluid domain is infinite). Then, the kinetic energy E_k satisfies the decay condition*

$$\frac{d}{dt} E_k = \nu \int \mathbf{u} \cdot \Delta \mathbf{u} \, d\mathbf{x} = -\nu \int |\nabla \mathbf{u}|^2 d\mathbf{x}.$$

where $\nabla \mathbf{u}$ is the Jacobian matrix of $\mathbf{u}$

$$\nabla \mathbf{u} = \begin{pmatrix} \partial_{x_1} u_1 & \partial_{x_2} u_1 & \partial_{x_3} u_1 \\ \partial_{x_1} u_2 & \partial_{x_2} u_2 & \partial_{x_3} u_2 \\ \partial_{x_1} u_3 & \partial_{x_2} u_3 & \partial_{x_3} u_3 \end{pmatrix}, \tag{2.21}$$

hence,

$$|\nabla \mathbf{u}|^2 = |\nabla u_1|^2 + |\nabla u_2|^2 + |\nabla u_3|^2.$$

This formula, proved by using Green's identity, is very important in the study of mathematical properties of the solution of the Navier–Stokes equation.

2.2.2 Rotation and vorticity formulation of the Euler equation

Given a fluid moving with velocity $\mathbf{u} = (u_1, u_2, u_3)$, we define the *vorticity field* of the flow as the vector field

$$\omega = \nabla \times \mathbf{u} \; (= \operatorname{curl} \mathbf{u})$$
$$= (\partial_{x_2} u_3 - \partial_{x_3} u_2, \partial_{x_3} u_1 - \partial_{x_1} u_3, \partial_{x_1} u_2 - \partial_{x_2} u_1). \qquad (2.22)$$

It is an important general fact that locally, i.e., in a neighbourhood of each fluid particle, the velocity field $\mathbf{u}$ is the sum of a *translation*, a *deformation* and a *rotation*.

Proposition 2.4. *A vector field $\mathbf{u}$ can be written in a neighbourhood $N(\mathbf{x})$ of a point $\mathbf{x} \in \mathbb{R}^3$ as*

$$\mathbf{u}(\mathbf{y}) = \mathbf{u}(\mathbf{x}) + D(\mathbf{x}) \cdot \mathbf{h} + \tfrac{1}{2}\omega(\mathbf{x}) \times \mathbf{h} + O(|\mathbf{h}|^2),$$

where D a symmetric 3×3 matrix, $\mathbf{y} = \mathbf{x} + \mathbf{h}$ and $\mathbf{h}$ is such that $B_x(|\mathbf{h}|) \subset N(\mathbf{x})$.

Proof. Let $\nabla \mathbf{u}$ denote the Jacobian matrix of $\mathbf{u}$ given by Eq. (2.21). Then, $\nabla \mathbf{u} = D + A$, with D the symmetric and A the antisymmetric part:

$$D = \tfrac{1}{2}[\nabla \mathbf{u} + (\nabla \mathbf{u})^T]; \qquad A = \tfrac{1}{2}[\nabla \mathbf{u} - (\nabla \mathbf{u})^T].$$

D is the so-called *deformation matrix*. As for A we can write (exercise: check!)

$$A = \frac{1}{2}\begin{pmatrix} 0 & -\omega_3 & \omega_2 \\ \omega_3 & 0 & -\omega_1 \\ \omega_2 & \omega_1 & 0 \end{pmatrix}, \qquad \omega = \operatorname{curl} \mathbf{u} = (\omega_1, \omega_2, \omega_3).$$

By Taylor's formula, for $|\mathbf{h}|$ small enough

$$\mathbf{u}(\mathbf{y}) = \mathbf{u}(\mathbf{x}) + \nabla \mathbf{u}(\mathbf{x}) \cdot \mathbf{h} + O(|\mathbf{h}|^2).$$

Substituting $\nabla \mathbf{u} = D + A$ in this expression yields the statement. $\qquad \square$

Exercise 2.17. *Verify the following two vector identities*

$$(\mathbf{u} \cdot \nabla)\mathbf{u} = (\nabla \times \mathbf{u}) \times \mathbf{u} + \nabla \left(\frac{|\mathbf{u}|^2}{2} \right) = \omega \times \mathbf{u} + \nabla \left(\frac{|\mathbf{u}|^2}{2} \right),$$

$$\text{curl}(\mathbf{u} \times \omega) = (\omega \cdot \nabla)\mathbf{u} - (\mathbf{u} \cdot \nabla)\omega + \mathbf{u}(\nabla \cdot \omega) - \omega(\nabla \cdot \mathbf{u}).$$

Using these identities, one can express the Euler equation for an ideal incompressible fluid (in the absence of external forces) as a **evolution law for the vorticity**.

Proposition 2.5. *Given an ideal incompressible fluid, the vorticity evolves according to the equation*

$$\frac{D\omega}{Dt} = (\omega \cdot \nabla)\mathbf{u}.$$

Proof. Exercise. □

The above identities yield a **vorticity–velocity formulation of Euler equations**:

$$\begin{cases} \dfrac{D\omega}{Dt} = (\omega \cdot \nabla)\mathbf{u} \\ \text{curl}\,\mathbf{u} = \omega \\ \nabla \cdot \mathbf{u} = 0. \end{cases}$$

For **two-dimensional flows**, so assuming $\mathbf{u} = (u_1, u_2, 0)$ the vorticity has only one non-zero component, $\omega = (0, 0, \partial_{x_1} u_2 - \partial_{x_2} u_1)$, and the equations reduce to

$$\begin{cases} \dfrac{D\omega}{Dt} = 0, \\ \omega = (0, 0, \partial_{x_1} u_2 - \partial_{x_2} u_1), \\ \nabla \cdot \mathbf{u} = 0. \end{cases}$$

In particular, the *vorticity is transported by the flow*:

$$\omega(\Phi_t(\mathbf{y}), t) = \omega(\mathbf{y}, 0).$$

Remark 2.5 (Known and unknown mathematical results). The Navier–Stokes equations are considered the most fundamental mathematical description of fluid dynamics. Yet, it is a challenging mathematical problem even to show that they admit a solution, and to understand the nature of the solution.

In three dimensions, the incompressible equation has been proved to admit a smooth solution for short time, depending continuously on the data. No solution, smooth or otherwise, is know to exist for all times — this is a major open problem, and one of the Clay million-dollar problems. The problem with controlling the terms appearing in the equations is the formation of **turbulence**. Turbulence implies that the behaviour of the three-dimensional Navier–Stokes equations at fine scales is much more nonlinear than at large scales. This poses a formidable challenge to the analysis (and the numerics).

In two dimensions, it has been proved that solutions exist for all time for both the Euler and the Navier–Stokes equations.

The literature on these equations is vast, and we refer to Chemin *et al.* (2006) for an exhaustive bibliography.

Additional exercises

Exercise 2.18. *Suppose a velocity field is $\mathbb{R}^3$ is given by*

$$(u_1, u_2, u_3) = (-\Omega y, \Omega x, 0), \quad \Omega > 0 \, constant$$

Find the trajectory of a particle starting at the point (x_0, y_0, z_0).

Exercise 2.19. *The Euler equations for a perfect fluid in $\mathbb{R}^3$, in the absence of external forces, is*

$$\partial_t \mathbf{u} + (\mathbf{u} \cdot \nabla)\mathbf{u} = -\frac{1}{\rho}\nabla p.$$

Write down the system of three equations for the velocity components $\mathbf{u} = (u, v, w)$. What equation is obtained in the case of one space dimension? Do the same for the incompressible, homogeneous Navier–Stokes equation

$$\partial_t \mathbf{u} + (\mathbf{u} \cdot \nabla)\mathbf{u} = -\nabla p + \nu \Delta \mathbf{u}.$$

Which unidimensional PDE do you obtain in this case?

Exercise 2.20. *Show that if an incompressible fluid is initially homogeneous, e.g., is $\rho(x, 0) = 1$, then, it remains homogeneous for all times.*

Exercise 2.21. *Prove that if u is a smooth vector field such that $\operatorname{div} \mathbf{u} = 0$, and p a smooth scalar function satisfying*

$$|\mathbf{u}(\mathbf{x})||p(\mathbf{x})| = O\left(\frac{1}{|\mathbf{x}|^2}\right) \quad as \ |\mathbf{x}| \to \infty,$$

then,

$$\int_{\mathbb{R}^3} \mathbf{u} \cdot \nabla p d\mathbf{x} = 0.$$

Exercise 2.22. *In a small enough neighbourhood of a point* $\mathbf{x} \in \mathbb{R}^3$, *the velocity vector field* $\mathbf{u}(\mathbf{x}, t_0)$ *at fixed time* t_0 *can be written as*

$$\mathbf{u}(\mathbf{x}, t_0) = \mathbf{u}(\mathbf{x}_0, t_0) + D(\mathbf{x}_0) \cdot (\mathbf{h}) + \tfrac{1}{2}\omega(\mathbf{x}_0, t_0) \times \mathbf{h} + O(|\mathbf{h}|^2),$$

$$\mathbf{h} = \mathbf{x} - \mathbf{x}_0,$$

(2.23)

where D a symmetric 3×3 matrix.

- *Calculate* $\mathrm{tr}(D)$.
- *If retaining only the zero-order term in Eq. (2.23), show that the fluid flow map* $\Phi_t(x_0) = \mathbf{x}(t)$ *satisfies*

$$\frac{d\Phi_t}{dt} = \mathbf{u}(\mathbf{x}_0, t_0),$$

hence it is given by

$$\Phi_{t-t_0}(\mathbf{x}_0) = \mathbf{x}_0 + \mathbf{u}(\mathbf{x}_0, t_0)(t - t_0), \qquad t > t_0$$

describing a translation.

- *Choose a coordinate system in $\mathbb{R}^3$ in which the symmetric matrix D is diagonal, $D = \mathrm{diag}(d_1, d_2, d_3)$.*
  Retaining only the first order term in Eq. (2.23), show that

$$\frac{d\Phi}{dt} = D(\mathbf{x}_0) \cdot \mathbf{h}.$$

Hence, in the chosen basis, this decouples into the 3 scalar deformations

$$\frac{d\Phi_i}{dt} = d_i h_i, \quad i = 1, 2, 3.$$

2.3   Linear and Nonlinear Transport in One-Dimensional Flows

In one dimension, i.e., for $u \in \mathbb{R}$, the equations for a homogenous ideal incompressible fluid become the so-called **nonlinear transport equation**

$$\partial_t u + u \partial_x u = F, \qquad (F \; known).$$

We start with the mathematical analysis of this one-dimensional problem, which can be unravelled explicitly and gives some intuition of the kind of phenomena that can arise.

Linear transport

Consider the linear homogeneous problem

$$\partial_t u + c\partial_x u = 0, \quad u = u(x,t), \quad (x,t) \in \Omega.$$

A **classical (strong) solution** of this PDE is a solution for which all derivatives in the equation make sense in the domain Ω. Depending on Ω, it will be necessary to specify also initial and/or boundary conditions. For now, we just consider formally the terms in the PDE.

$c = 0$ Then, the equation is simply $\partial_t u = 0$. If Ω is simply connected, a family of classical solutions is given by

$$u(x,t) = f(x),$$

where $f(x)$ is an arbitrary C^1 function of x. (Find a counterexample for the case that Ω is not simply connected.)
The function $f(x)$ is uniquely defined if the value of u is known at one point in time — for example, if an initial condition $u(x,0)$ is prescribed.

$c \neq 0$ We reduce this case to the previous one by changing coordinates: seek a new space coordinate $x(t)$ satisfying the condition

$$u(t,x(t)) \text{ is constant in time}: \ \partial_t u(x(t),t) = 0.$$

Using the chain rule, we need,

$$0 = \partial_t u + \frac{dx(t)}{dt} \partial_x u,$$

which is true if we choose

$$\frac{dx(t)}{dt} = c \implies x(t) = ct + x(0).$$

This calculation justifies the introduction of the **characteristic variable**

$$\xi = x - ct.$$

Define v through $u(x,t) = v(\xi,t)$. Then, since v is constant with time, we have

$$u(x,t) = v(\xi,0) = v(x - ct).$$

The lines $\xi = const.$ are called **characteristics**: along these lines the solution is stationary. But to fix a unique solution, we need to know on which characteristic we are travelling.

Proposition 2.6. *The unique classical solution of the linear transport equation*

$$\partial_t u + c\partial_x u = 0, \quad x \in \mathbb{R}, \, t > 0$$

with given initial condition $u(x,0) = f(x)$ *is given by*

$$u(x,t) = f(x - ct).$$

$c = c(x)$ Again, one looks for a characteristic curves $x = x(t)$ along which the solution is constant.

Namely, we want

$$\frac{du(x(t),t)}{dt} = \partial_t u + \dot{x}(t)\partial_x u = 0.$$

Again this is true if $\dot{x}(t) = c(x)$. This equation admits a unique solution if $c(x)$ is smooth enough (Lipschitz continuous). The uniqueness of the solution implies that the characteristics do not intersect. Now the characteristic variable is $\xi = \beta(x) - t$, where

$$\beta(x) = \int \frac{dx}{c(x)}, \quad \text{if } c(x) \neq 0.$$

Note that if initial data are given in such a way that the initial data curve intersects each characteristics in only one point (the initial data curve is *transverse* to the characteristic. i.e., it is nowhere tangent to them), then the solution will be well defined by the initial values and unique.

Remark 2.6. It is usually physically relevant and customary to prescribe data on the line $t = 0$ — the so-called initial conditions. But data that specify the solution uniquely can be given on a different curve as well.

A general definition of characteristic curve can be given as follows: A curve is *characteristic* if when initial data are given on the curve then these data do not determine the solution at any point outside the characteristic.

Convince yourself that all examples of characteristics given above satisfy this definition.

Exercise 2.23. *Find the classical solution of the transport problem*

$$\partial_t u + c\partial_x u + au = 0, \quad a \in \mathbb{R}^+.$$

Exercise 2.24. *Find the characteristics of the equation*

$$\partial_t u + (x^2 - 1)\partial_x u = 0, \quad (x \neq \pm 1).$$

Exercise 2.25. *Find the characteristics of the equation*

$$\partial_t u - x\partial_x u = -u^2, \quad (x \neq \pm 1).$$

Hence, find the solution if $u(x,0) = \sin^2 x$ and $u(x,0) = \sin x$, and compare their behaviour with respect to t.

Nonlinear transport

We now seek the solution of the following nonlinear problem:

$$\partial_t u + u\partial_x u = 0, \quad x \in \mathbb{R}; t > 0. \tag{2.24}$$

We assume that a suitable initial condition $u(x,0) = f(x)$ is prescribed (for example, we may assume that $f \in C^1(\mathbb{R})$). To find a solution, we use again the idea of characteristic curves.

If $x(t)$ parametrises one such curve, then is must hold that $u(x(t),t)$ is constant, hence,

$$\frac{du(x(t),t)}{dt} = \partial_t u + \frac{dx(t)}{dt}\partial_x u = 0.$$

Since $u(x(t),t) = $ const. we can determine $x(t)$ by the condition

$$\frac{dx(t)}{dt} = u(x(t),t).$$

Characteristic curves are therefore the **lines** given by

$$x(t) = ut + \kappa, \quad \kappa = x(0).$$

The main difference with the linear case is that now characteristics line may intersect, even if initial data are compatibly prescribed. When this happens the solution is not well defined. The point of intersection of characteristics issuing for distinct initial values is called a **shock**.

Where it is well defined, the solution is given implicitly by the relation

$$u(x,t) = f(x - ut), \quad f(x) = u(x,0). \tag{2.25}$$

Exercise 2.26. *Assume that $f \in C^1(\mathbb{R})$ and prove that the function u defined by Eq. (2.25) solves the PDE at least for some time. Namely, show that the equation for the function F given by*

$$F(x,t) = u(x,t) - f(x - tu) = 0$$

can be solved uniquely for u if $f' \neq -\frac{1}{t}$.

It transpires from the exercise that, given a smooth initial condition $u(x,0)$, the equation does not necessarily admit a solution for all times.

Hence, in general the problem **does not have a classical solution defined for all times.**

Along an arbitrary characteristic, we have

$$\partial_x u = \partial_x \left[f(x - tu) \right] = f'(x - tu) \cdot (1 - t\partial_x u),$$

hence

$$\partial_x u = \frac{f'}{1 + tf'}$$

is not well defined if $f' = -1/t$.

Weak solutions

Note that the nonlinear transport equation can be written in the following *conservation form*

$$\partial_t u + \partial_x (f(u)) = 0. \tag{2.26}$$

Indeed, if $f(u) = \frac{1}{2}u^2$ then,

$$\partial_x (f(u)) = \partial_x \left(\frac{1}{2}u^2 \right) = u\partial_x u.$$

This can be written concisely as follows:

$$\nabla \cdot F = 0, \quad \text{where} \quad F : \mathbb{R}^2 \to \mathbb{R}^2, \ F(x,t) = (f(u), u), \quad (\text{and } \nabla = (\partial_x, \partial_t)).$$

Hence, Eq. (2.26) implies the integral condition

$$\int \varphi (\nabla \cdot F) dx dt = 0 \quad \forall \varphi \in C_c^1(\mathbb{R}^2),$$

where the test function φ is any smooth function with compact support in the (x,t) plane.

By the divergence theorem (or integrating by parts in each variable separately), using the fact that $\varphi \in C_c^1$ (the space of continuously differentiable functions with compact support), we find

$$0 = \int \varphi(\nabla \cdot F)dxdt = \int (\text{grad } \varphi \cdot F)dxdt = 0. \tag{2.27}$$

If u is smooth enough, Eq. (2.26) makes sense and it implies Eq. (2.27). However, the integral formulation (Eq. (2.27)) makes sense even if, e.g., u is not differentiable.

Definition 2.9. A weak solution of Eq. (2.26) is a function $u \in L^1(\mathbb{R}^2)$ that satisfies Eq. (2.27) for all $\varphi \in C_c^1(\mathbb{R}^2)$.

To include the contribution of the initial condition, we can replace integration over the whole space with integration on the space $t \geqslant 0$. When integrating by parts a boundary term appears at the boundary $t = 0$, and so the condition for a weak solution satisfying not just the PDE but also the given initial condition becomes

$$\int_{t \geqslant 0} \int_{\mathbb{R}} (\text{grad } \varphi \cdot F)dxdt + \int_{\mathbb{R}} \varphi(x,0)u(x,0)dx = 0,$$

$$\forall \varphi \in C_c^1(\mathbb{R} \times [0,\infty)). \tag{2.28}$$

Remark 2.7. The notion of weak solution is much general and powerful than the particular case given here. Giving it a rigorous foundations led to the definition of *Sobolev spaces*, i.e., spaces of weak derivatives. For example, for $I \subset \mathbb{R}$, the Sobolev space $W^{1,p}$ is defined as

$$W^{1,p}(I) = \left\{ u \in L^p(I) : \exists g \in L^P(I) : \int u\varphi' = -\int g\varphi, \right.$$

$$\left. \forall \varphi \in C_c(I) \right\}. \tag{2.29}$$

These are Banach spaces, and have a very rich theory. In particular, the space $W^{1,2}$ of L^2 weak derivatives is a Hilbert space.

Conservation laws

The differential formulation (Eq. (2.26)) allows the interpretation of the PDE as a *conservation law*.

Suppose u satisfies the PDE (say strongly). We can also consider its integral form:

$$\int [\partial_t u + \partial_x f(u)] dx = 0. \tag{2.30}$$

Then, for any fixed interval $[a,b] \subset \mathbb{R}$, time differentiation of the *mass* $\int u$ yields

$$\frac{d}{dt} \int_a^b u\, dx = \int_a^b \partial_t u\, dx = -\int_a^b \partial_x f(u)\, dx = f(u(a)) - f(u(b)),$$

which express conservation of "mass" (for which u represents a density). If a, b are allowed to become infinite, and assuming $u \to 0$ at infinity (necessary requirement for integrability) then this conservation is in the strict sense, as

$$\frac{d}{dt} \int_a^b u\, dx = 0.$$

The integral form of the PDE is the natural one from the point of view of the underlying physical laws.

Indeed, the weak formulation of a problem (i.e., the notion of weak solution) and the physical requirement of conservation of certain physical quantities (mass, energy, etc.) allow us to identify a unique solution even when

(1) the initial condition is not smooth, or discontinuous,
(2) the characteristics cross and create a *shock*.

To understand how this is achieved, we investigate the weak solution of a conservation law near a discontinuity.

Suppose that u is a solution of Eq. (2.26) which is discontinuous across a curve Σ (a *shock line*) in the (x,t) plane. We use the parametrisation $\Sigma = \{(x,t) : x = \sigma(t)\}$. Then, σ' denotes the speed of the discontinuity.

Let Ω be a domain that intersects Σ and let Ω_+ and Ω_- be the two subdomains in which the curves divides Ω.

Choose φ such that $\mathrm{supp}(\varphi) \subset \Omega$. Then, since u is a weak solution, we have

$$0 = \int_\Omega (\mathrm{grad}\ \varphi \cdot F) dx dt$$

$$= \int_{\Omega} (u\partial_x \varphi + f(u)\partial_t \varphi)\, dx dt$$

$$= \left\{ \int_{\Omega_+} + \int_{\Omega_-} \right\} (u\partial_x \varphi + f(u)\partial_t \varphi)\, dx dt.$$

Integrating by parts we find

$$0 = \int_{\Omega_+} \varphi(\partial_t u + \partial_x(f(u)))\, dx dt + \int_{\Omega_-} \varphi(\partial_t u + \partial_x(f(u)))\, dx dt$$

$$+ \int_{\partial\Omega_+} (U^+ \cdot n^+)\varphi\, ds + \int_{\partial\Omega_-} (U^- \cdot n^-)\varphi\, ds, \qquad (2.31)$$

where we denote

$$u^+ = u_{|\Omega_+}; \quad u^- = u_{|\Omega_-}; \quad U^+ = \begin{pmatrix} u^+ \\ f(u^+) \end{pmatrix}; \quad U^- = \begin{pmatrix} u^- \\ f(u^-) \end{pmatrix}.$$

Now using that $\varphi = 0$ on $\partial\Omega$, that u is a strong solution when restricted to either Ω_+ or Ω_-, we obtain

$$0 = \int_{\Sigma} (U^+ \cdot n^+ + U^- \cdot n^-)\varphi\, ds.$$

Since $\Sigma = \{(x,t) : x = \sigma(t)\}$, we can use this parametrisation to compute the integral, with

$$n^+ = \frac{1}{\sqrt{1+\sigma'(t)}} \begin{pmatrix} \sigma' \\ -1 \end{pmatrix}; \quad n^- = \frac{1}{\sqrt{1+\sigma'(t)}} \begin{pmatrix} -\sigma' \\ 1 \end{pmatrix}; \quad ds = \sqrt{1+\sigma'}\, dt.$$

This yields

$$0 = \int_{\Sigma} [(u^- - u^+)\sigma' - (f(u^-) - f(u^+))]\varphi\, dt$$

$$\implies \sigma' = \frac{f(u^-) - f(u^+)}{u^- - u^+} \quad \text{on } \Sigma. \qquad (2.32)$$

The constraint on σ' is called the *Rankine–Hugoniot condition*. This is the constraint that the weak form selected imposes on the speed of the shock separating the regions were a strong solution is defined. For our explicit example when $f(u) = u^2/2$, we get

$$\sigma' = \frac{u^- + u^+}{2} \quad \text{when } x = \sigma(t).$$

Hence, we have just proved the following.

Proposition 2.7. *Let u be a function that*

- *satisfies the differential equation where there are no discontinuities;*
- *satisfies the condition in Eq. (2.32)) across any jump discontinuity.*

Then, u satisfies the integral form and the weak form of the equation.

Remark 2.8. The conservation form (Eq. (2.30)) is consistent with the physical conservation of the mass $M = \int u \, dx$.

Sometimes the interest is on some other conserved quantity for example the energy E,

$$E = \int u^2 dx.$$

Hence, we would want to ensure that

$$\frac{d}{dt} E = \int \partial_t(u^2) dx = 2 \int u u_t dx = 0.$$

Multiplying the PDE by $2u$, we find

$$0 = 2uu_t + 2u^2 u_x = \partial_t(u^2) + \partial_x\left(\frac{2}{3}u^3\right), \tag{2.33}$$

hence a different conservation law, consistent with conservation of energy.

Exercise 2.27. *Find the shock speed condition consistent with the conservation form (Eq. (2.33)).*

We now have a criterion to select uniquely weak solutions of the nonlinear transport problem (Eq. (2.24)), depending on the particular quantity that must be conserved.

Since weak solution can be discontinuous, we can also make sense of starting from discontinuous initial condition. As an illustration of how discontinuities may propagate, we consider two fundamental examples.

Example 2.4 (Compression wave). Consider the PDE in the conservation form (Eq. (2.26)), and consider the discontinuous initial condition

$$u(x,0) = \begin{cases} 0, & x \geq 0, \\ 1, & x < 0. \end{cases}$$

We can compute the characteristics, along which u is constant, as $x = t\mathbf{u}(x_0,0) + x_0$, hence,

$$x = x_0, \qquad x_0 \geqslant 0,$$
$$x = t + x_0, \qquad x_0 < 0,$$

where different characteristics meet, the solution becomes formally multivalued. To select a well-defined unique weak solution, we introduce a shock line by the requirement that it satisfied that shock speed condition in Eq. (2.32).

Since $u = 1$ below the shock line and $u = 0$ above it, it must be

$$\sigma' = \frac{u^+ + u^-}{2} = \frac{1}{2} \implies \sigma(t) = \frac{1}{2}t,$$

therefore, the unique weak solution compatible with conservation of mass is given by

$$u(x,t) = \begin{cases} 0, & x \geqslant \frac{1}{2}t, \\ 1, & x < \frac{1}{2}t. \end{cases}$$

Example 2.5 (Rarefaction wave). We start as in the previous example, but we consider the initial condition

$$u(x,0) = \begin{cases} 1, & x \geqslant 0, \\ 0, & x < 0. \end{cases}$$

The characteristics are given by

$$x = x_0, \qquad x_0 < 0,$$
$$x = t + x_0, \quad x_0 \geqslant 0,$$

and we can verify that there is a whole region in the (x,t) plane that is not reached by any of them. Indeed the solution is

$$u(x,t) = \begin{cases} 0, & x < 0, \\ 1, & x - t \geqslant 0, \end{cases}$$

hence, it is not defined for $0 \leqslant x < t$.

To define a solution everywhere in the previous example, we could do one of several things, but there is only way to preserve the condition that the shock is determined only by the initial data, and not by data at later times. This is obtained by introducing a *rarefaction fan*, using the particular Eq. (2.1), namely

$$u(x,t) = \frac{x}{t},$$

which can be seen to solve Eq. (2.24) at least if $t > 0$. This solution interpolates between the value 0 at $x = 0$ and the value 1 at $x = t$. Then, let

$$u(x,t) = \begin{cases} 0, & x < 0, \\ \dfrac{x}{t}, & 0 \leqslant x < t, \\ 1, & x - t \geqslant 0. \end{cases}$$

This is the only solution that satisfies the entropy condition, which means that the shock is determined only by the initial data.

It is also the solution that is obtained as the limit as $\nu \to 0$ of the corresponding solution of the viscous PDE of nonlinear diffusion

$$\partial_t u + u \partial_x u = \nu u_{xx}, \quad \nu > 0.$$

Exercise 2.28. *Consider as initial condition the triangular wave*

$$u(x,0) = \begin{cases} x, & 0 \leqslant x \leqslant 1, \\ 0, & otherwise. \end{cases}$$

Find the equation of the characteristics and determine the unique weak solution consistent with mass conservation.

Systems of conservation laws

An analogous theory can be developed of systems of conservation laws. The only example, we will consider are the equations of isentropic gas, i.e., the one-dimensional version of the fluid model. These are

$$\partial_t u + u \partial_x u + \frac{\partial_x p}{\rho} = 0; \qquad \partial_t \rho + \partial_x (\rho u) = 0. \tag{2.34}$$

They can be written in conservation from by setting $m = \rho u$:

$$\begin{cases} \partial_t \rho + \partial_x m = 0, \\ \partial_t m + \partial_x \left(\dfrac{m^2}{\rho} + p \right) = 0. \end{cases}$$

Since these are only two equations for the three unknowns (u, p and ρ), one can

either assume $p = p(\rho)$

    or add the energy equation

$$\partial_t e + \partial_x \left((e + p)u \right) = 0, \quad e = \rho\varepsilon + \tfrac{1}{2}\rho u^2, \ \varepsilon \text{ given.}$$

By adding the energy equation, one has a system of three conservation laws:

$$\begin{cases} \partial_t \rho + \partial_x m = 0, \\[2mm] \partial_t m + \partial_x \left(\dfrac{m^2}{\rho} + p \right) = 0, \\[2mm] \partial_t e + \partial_x \left((e + p)\dfrac{m}{\rho} \right) = 0, \end{cases}$$

expressing conservation of mass, momentum and energy.

These two alternatives yield *different* weak solutions. We concentrate onto the second, which is arguably more fundamental.

Proceeding as for a single conservational, we find that the jump conditions across a discontinuity Σ in the (x,t) plane, with velocity σ' are the *Rankine–Hugoniot relations*

$$\sigma'[\rho] = [m]; \quad \sigma'[m] = \left[\frac{m^2}{\rho} + p \right]; \quad \sigma'[m] = [(e + p)u],$$

where $[f] = f^+ - f^-$ denotes the jump of the quantity f across Σ.

A large selection of exercises relevant to the material in this section can be found in Olver (2014), particularly Sec. 2.3.

2.4 Nonlinear Diffusion

The PDE

$$\partial_t u + u\partial_x u = \nu u_{xx}, \quad \nu > 0, \tag{2.35}$$

called also *Burger's equation*, combines transport with the damping effect of viscosity, modelled by the second-order term. By adding a forcing term modelling the pressure, we obtain the one-dimensional version of Navier–Stokes.

The presence of the the viscosity term changes entirely the tools that can be used to understand the behaviour of the solutions.

Digression on the heat equation

The heat equation is the linear version of nonlinear diffusion

$$u_t = \nu u_{xx}, \quad \nu > 0.$$

To solve the pure initial value problem ($x \in \mathbb{R}$, $t > 0$, $u(x,0) = u_0(x)$ prescribed) we use **Fourier transform** with respect to the space variable x:

$$\hat{u}(\lambda) = \int_{\mathbb{R}} e^{-i\lambda x} u(x,t)dx.$$

Then, assuming that $u \to 0$ as $|x| \to \infty$ we find

$$\widehat{\partial_t u} = \partial_t \hat{u}(\lambda); \quad \widehat{\partial_x^2 u} = -\lambda^2 \hat{u}(\lambda)$$

hence the transformed PDE is the linear first ODE

$$\partial_t \hat{u} + \lambda^2 \nu \hat{u} = 0 \Longrightarrow \hat{u}(\lambda,t) = e^{-\lambda^2 \nu t} \hat{u}(\lambda,0).$$

Inverting the transform

$$u(x,t) = \frac{1}{2\pi} \int_{\mathbb{R}} e^{i\lambda x} e^{-\lambda^2 \nu t} \hat{u}(\lambda,0)d\lambda.$$

Explicitly

$$u(x,t) = \frac{1}{2\pi} \int_{\mathbb{R}} e^{i\lambda x} e^{-\lambda^2 \nu t} \int_{\mathbb{R}} e^{-i\lambda y} u(y,0)dyd\lambda$$

$$= \frac{1}{2\pi} \int_{\mathbb{R}} \int_{\mathbb{R}} e^{i\lambda(x-y)} e^{-\lambda^2 \nu t} u(y,0)dyd\lambda.$$

Use Fubini and then compute

$$I = \int_{\mathbb{R}} e^{i\lambda(x-y)} e^{-\lambda^2 \nu t} d\lambda = \sqrt{\frac{\pi}{\nu t}} e^{-(x-y)^2/4\nu t}.$$

Hence,

$$u(x,t) = \frac{1}{2\sqrt{\pi \nu t}} \int_{\mathbb{R}} e^{-(x-y)^2/4\nu t} u(y,0)dy,$$

where the exponential in the integrand is the **fundamental solution**, or **heat kernel**.

Now consider the initial value problem for Burger's equation, with $u(x,0) = f(x)$ prescribed, smooth and decaying. We seek *travelling wave solutions*, namely solutions of the form

$$u(x,t) = v(x - ct), \quad c \text{ const.}$$

If u is a travelling wave solution, then using the definition we find

$$\partial_t u = -cv', \quad \partial_x u = v', \quad \partial_x^2 u = v''$$

so that the function v must satisfy the nonlinear ODE

$$-cv' + vv' = \nu v'' \implies \nu v' = -cv + \frac{v^2}{2} + \kappa.$$

Writing this as

$$\nu v' = (v - a)(v - b), \quad \text{with } c = \tfrac{1}{2}(a+b), \ \kappa = \tfrac{1}{2}ab \left(< \tfrac{1}{2}c^2 \text{ for } a, b \in \mathbb{R} \right)$$

we find by direct integration the expression for v, and in turns

$$u(x,t) = \frac{ae^{(b-a)\frac{x - ct + \delta}{2\nu}} + b}{e^{(b-a)\frac{x - ct + \delta}{2\nu}} + 1} \quad (a < b).$$

This is the expression for a wave travelling to the right with constant speed $c = (a+b)/2$. As $\nu \to 0$, this wave approximates the step shock.

2.4.1 *General solutions: Linearising transformation*

Burgers equation is distinguished among other nonlinear equations in that it can be linearised via an explicit transformation. Indeed, Hopf and Cole noted that the Burgers equation can be mapped to the heat equation via an explicit exponential transformation.

Again, we consider the initial value problem for Burger's:

$$u_t + uu_x = \nu u_{xx}, \quad \nu > 0; \qquad u(x,0) = f(x).$$

(i.e., we ignore boundary effects, though they are generally crucial!)

Let v satisfy the heat equation

$$v_t = \nu v_{xx}.$$

If $v > 0$ (an *a-priori* assumption, though valid if true at $t = 0$), we write

$$v(x,t) = e^{\alpha \varphi(x,t)}, \qquad \text{hence } \varphi = \frac{1}{\alpha} \ln(v).$$

Then,

$$\varphi_{tx} = \nu \varphi_{xxx} + 2\alpha \nu \varphi_x \varphi_{xx}$$

so $u = \varphi_x$ with $2\alpha\nu = -1$ satisfies the Burgers equation

$$u_t = \nu u_{xx} + u u_x.$$

Proposition 2.8. *Every smooth, positive solution of the heat equation $v_t = \nu v_{xx}$ yields a formal solution of Burger's equation u by setting*

$$u = -2\nu (\ln v)_x.$$

Conversely, if u solves the Burgers equation, then set

$$\tilde{\varphi} = \int_0^x u(y,t)dy \quad \text{so that } \tilde{\varphi}_x = u.$$

Then,

$$\tilde{\varphi}_t = \nu \tilde{\varphi}_{xx} - \tilde{\varphi}\tilde{\varphi}_x + g(t),$$

where $g(t)$ is an arbitrary function of t (constant of integration), which must be subtracted to obtain

$$\varphi = \tilde{\varphi} - \int^t g.$$

Then,

$$v = e^{-\varphi/2\nu} \quad \text{solves the heat equation.}$$

With this equivalence, one can use the representation formula for the heat equation to obtain the following explicit representation formula for the solution of the initial value problem for Burgers equation:

$$u(x,t) = \frac{\int_{\mathbb{R}} \frac{x-y}{t} e^{-G/2\nu} dy}{\int_R e^{-G/2\nu} dy}, \qquad G(y;x,t) = \int_0^y f(\xi)d\xi + \frac{(x-y)^2}{2t}.$$

$$(2.36)$$

The inviscid ($\nu \to 0$) limit of Burger's equation

In the physical application of shock waves in fluids, ν has the meaning of viscosity. Here, we are interested in understanding the inviscid limit, that is, $\nu \to 0$, of the Burgers equation. To do this, we need a method for the asymptotic evaluation of exponential integrals, the so-called Laplace's method (see e.g., Chapter 6 in Ablowitz and Fokas, 2003).

As $\nu \to 0$, we use Laplace's method to evaluate the dominant contributions to the integrals appearing in Eq. (2.36). To achieve this, we need to find the points for which $\partial G / \partial y = 0$,

$$\frac{\partial G}{\partial y} = f(y) - \frac{x-y}{t}. \tag{2.37}$$

Let $\eta = \xi(x,t)$ be such a point; that is, $\xi(x,t)$ is a solution of

$$x = \xi + f(\xi)t. \tag{2.38}$$

Then, Laplace's method implies

$$\int_{-\infty}^{\infty} \frac{x-y}{t} e^{-\frac{G}{2\nu}} dy \sim \frac{x-\xi}{t} \sqrt{\frac{4\pi\nu}{|G''(\xi)|}} e^{-\frac{G(\xi)}{2\nu}},$$

$$\int_{-\infty}^{\infty} e^{-\frac{G}{2\nu}} dy \sim \sqrt{\frac{4\pi\nu}{|G''(\xi)|}} e^{-\frac{G(\xi)}{2\nu}}.$$

Hence, if Eq. (2.38) for a given f has only one solution for ξ, then,

$$u(x,t) \sim \frac{x-\xi}{t} = f(\xi), \tag{2.39}$$

where ξ is defined by Eq. (2.38).

The above analysis shows that for appropriate f the limit of the solution of the Burgers equation is given by the solution of the limit equation, i.e., the nonlinear transport equation. However, the relationship between these two equations must be further clarified. Indeed, for some $f(x)$, Eq. (2.24) gives multivalued solutions (after the characteristics cross), while the solution we found is always single valued. This means that, out of all the possible weak solutions that Eq. (2.24) can support, there exists a unique solution that is the correct limit of the Burgers equation as $\nu \to 0$.

The asymptotic method, we just described provides us with a way of picking this correct solution: When the characteristics of Eq. (2.24) cross,

Eq. (2.38) admits two solutions, which we denote by ξ_1 and ξ_2 with $\xi_1 > \xi_2$ (both ξ_1 and ξ_2 yield the same values of x and t from Eq. (2.38). Laplace's method shows that for the sum of these contributions

$$\mathbf{u}(x,t) \sim \frac{f(\xi_1)|G''(\xi_1)|^{-\frac{1}{2}}e^{-\frac{G(\xi_1)}{2\nu}} + f(\xi_2)|G''(\xi_2)|^{-\frac{1}{2}}e^{-\frac{G(\xi_2)}{2\nu}}}{|G''(\xi_1)|^{-\frac{1}{2}}e^{-\frac{G(\xi_1)}{2\nu}} + |G''(\xi_2)|^{-\frac{1}{2}}e^{-\frac{G(\xi_2)}{2\nu}}}.$$

Hence, owing to the dominance of exponentials

$$u(x,t) \sim f(\xi_1) \quad \text{for } G(\xi_1) < G(\xi_2),$$
$$u(x,t) \sim f(\xi_2) \quad \text{for } G(\xi_1) > G(\xi_2). \tag{2.40}$$

The changeover will occur at those (x,t) for which $G(\xi_1) = G(\xi_2)$, or using the definition of G and the fact that both ξ_1, ξ_2 satisfy Eq. (2.38), these values of (x,t) satisfy (by integrating Eq. (2.38))

$$\frac{(x-\xi_1)^2}{2t} - \frac{(x-\xi_2)^2}{2t} = -\int_{\xi_2}^{\xi_1} u_0(\eta)d\eta,$$

or

$$(\xi_1 - \xi_2)\left(-\frac{x}{t} + \frac{(\xi_1+\xi_2)}{2t}\right) = -\int_{\xi_2}^{\xi_1} f(\eta)d\eta.$$

Using Eq. (2.38) for ξ_1 and ξ_2, and summing, we find that

$$-\frac{x}{t} + \frac{\xi_1+\xi_2}{2t} = -\frac{1}{2}(f(\xi_1) + f(\xi_2)),$$

hence

$$\frac{1}{2}(f(\xi_1) + f(\xi_2))(\xi_1 - \xi_2) = \int_{\xi_2}^{\xi_1} f(\eta)\eta. \tag{2.41}$$

Equation (2.40) with Eq. (2.38) shows that at $\nu \to 0$ the changeover in the behaviour of $u(x,t)$ leads to a discontinuity. In this way, the solution of the Burgers equation tends to a *shock wave* as $\gamma \to 0$. This solution is that particular solution of the limiting Eq. (2.24) that satisfies the *shock* (*Rankine–Hugoniot*) *condition* (Eq. (2.41)).

Additional exercises

Exercise 2.29. *Find an expression for the solution of the initial value problem for the heat equation*

$$u_t = u_{xx}, \quad (t,x) \in (0,\infty) \times \mathbb{R},$$

$$u(0,x) = e^{-|x|}.$$

(The point of this is to show that the heat equation has an instantaneous smoothing effect of the discontinuity in the derivative of the initial condition.)

Note: *it may be convenient to use the error function defined as*

$$\mathrm{erf}(x) = \frac{2}{\sqrt{\pi}} \int_0^x e^{-y^2} dy.$$

Exercise 2.30. *Show that the travelling wave solution of Burger's equation*

$$u(x,t) = \frac{ae^{(b-a)\frac{x-ct}{2\nu}} + b}{e^{(b-a)\frac{x-ct}{2\nu}} + 1} \quad (a < b),$$

when $\nu \to 0$ *approximates a step shock solution of the nonlinear transport equation. Write down the expression for this shock.*

Exercise 2.31. *The two-dimensional shallow water equations in a rotating frame of reference for the velocity* $\mathbf{u} = (u,v)$ *are*

$$\frac{Du}{Dt} - fv = -g\frac{\partial h}{\partial x},$$

$$\frac{Dv}{Dt} + fu = -g\frac{\partial h}{\partial y},$$

$$\frac{\partial h}{\partial t} + \frac{\partial(hu)}{\partial x} + \frac{\partial(hv)}{\partial y} = 0.$$

Here, f *is the Coriolis parameter, given by*

$$f = f_0 + \beta y, \qquad f_0, \beta \text{ positive const.}$$

If the flow is nearly geostrophic, one can neglect the terms $\frac{Du}{Dt}$, $\frac{Dv}{Dt}$. *Show that then one obtain for* h *the equation*

$$\frac{\partial h}{\partial t} + c(h)\frac{\partial h}{\partial x} = 0, \qquad c(h) = -\frac{g\beta}{f^2}h,$$

where y is a parameter. The solutions of this equation are called **Rossby waves** *(see Eq. (2.55) for much more information on them!).*
Describe which condition on the term $\frac{\partial h}{\partial x}$ *implies that the Rossby wave will break.*

2.5 Reductions of the Euler Equations for Modelling Large-Scale Flows

Consider the equations of motion for an ideal perfect incompressible in three dimensions, namely Euler equations, written in a frame of reference rotating with the Earth's angular velocity $\underline{\Omega} = (\Omega_1, \Omega_2, \Omega_3)$ and with an external potential ϕ. Specifically, this means that in Eq. (2.14) we take

$$f = -[2\underline{\Omega} \times \mathbf{u} + \nabla\phi].$$

Hence, the Euler system (Eqs. (2.14)–(2.15)) becomes

$$\frac{D\mathbf{u}}{Dt} + 2\underline{\Omega} \times \mathbf{u} + \nabla\phi + \frac{1}{\rho}\nabla p = 0, \tag{2.42}$$

$$\frac{D\rho}{Dt} = \frac{\partial\rho}{\partial t} + \mathbf{u} \cdot \nabla\rho = 0, \tag{2.43}$$

$$\nabla \cdot \mathbf{u} = 0. \tag{2.44}$$

Let,

- U be a typical scale for horizontal speed, approximately $10\,\mathrm{ms}^{-1}$ for large-scale atmospheric flow;
- L be a typical horizontal scale, approximately $10^6\,\mathrm{m}$ for large-scale atmospheric flow;
- W be a typical scale for vertical speed, approximately $10^{-2}\,\mathrm{ms}^{-1}$ for large-scale atmospheric flow;
- H be a typical vertical scale, approximately $10^4\,\mathrm{m}$ for large-scale atmospheric flow;
- $T \sim \frac{L}{U}$ be a typical time scale, approximately $10^5\,\mathrm{s}$ for large-scale atmospheric flow;
- N be the buoyancy frequency. N is approximately $10^{-2}\,\mathrm{s}^{-1}$ in the troposphere and $10^{-1}\,\mathrm{s}^{-1}$ in the stratosphere. (2.45)

By identifying parameters which are small for large-scale flow, we consider various approximations which are appropriate for large-scale flow and can be used to simplify the system (Eqs. (2.42)–(2.44)). Firstly, we note that the atmosphere is shallow; approximately 90% of the mass of the atmosphere is contained within the lowest $15\,\mathrm{km}$, which is much smaller than the radius of the Earth. Therefore, we assume for large-scale atmospheric flows that the aspect ratio H/L satisfies

$$\frac{H}{L} \ll 1. \tag{2.46}$$

Indeed, using the approximate values in (2.45), we have that $H/L \sim 10^{-2}$ so that Eq. (2.46) holds for large-scale atmospheric flow. In addition, we note that, for large-scale flow, the advective time scale is the same for horizontal and vertical motion, i.e.,

$$\frac{W}{U} \sim \frac{H}{L}. \tag{2.47}$$

Under these conditions, it can be shown that we need only consider the horizontal components of the Coriolis force $2\underline{\Omega} \times \mathbf{u}$. Therefore, we may write the Coriolis force as

$$2\underline{\Omega} \times \mathbf{u} = (-f_0 u_2, f_0 u_1, 0). \tag{2.48}$$

We will assume that f_0 is constant — note however that for large-scale approximations to be realistic, this assumption must be dropped.

We now describe mathematically some fundamental physical balances.

Hydrostatic balance

This is one of the most fundamental balances in geophysical fluid dynamics. Assume that the external potential is only due to gravity, hence that it takes the specific form

$$\phi = g_{\mathrm{grav}} x_3. \tag{2.49}$$

Combining Eqs. (2.48) and (2.49), the vertical component of the momentum Eq. (2.42) can be written

$$\frac{D u_3}{Dt} = -\frac{1}{\rho} \frac{\partial p}{\partial x_3} - g_{\mathrm{grav}}. \tag{2.50}$$

Consider the typical scales of each of the above terms;

$$\frac{W}{T} + \frac{UW}{L} + \frac{W^2}{H} \sim \left| \frac{1}{\rho} \frac{\partial p}{\partial x_3} \right| + g_{\mathrm{grav}}. \tag{2.51}$$

For most large-scale motion, the terms on the left-hand side of Eq. (2.51) are typically much smaller than those on the right-hand side. Thus, the terms on the right-hand side must be approximately equal. Indeed, considering the typical large-scale values in (2.45), and noting that g_{grav} is approximately $10\,\mathrm{ms}^{-2}$, we conclude that the pressure term is the only one that could possibly balance the gravitational term. This suggests that we may neglect the vertical acceleration term and approximate Eq. (2.50) by

$$\frac{\partial p}{\partial x_3} = -\rho g_{grav}. \tag{2.52}$$

This is known as hydrostatic balance, and states that the gravitational term is balanced by the pressure term.

Geostrophic balance

Next, we consider another fundamental balance in geophysical fluid dynamics; geostrophic balance. This physical is discussed in the context of the RSW equations later in this chapter, where you will find further information from a different point of view.

Combining Eqs. (2.48) and (2.49), the horizontal components of Eq. (2.42) take the form

$$\frac{Du_1}{Dt} - f_0 u_2 = -\frac{1}{\rho}\frac{\partial p}{\partial x_1}, \tag{2.53}$$

$$\frac{Du_2}{Dt} + f_0 u_1 = -\frac{1}{\rho}\frac{\partial p}{\partial x_2}. \tag{2.54}$$

Consider the typical scales of the terms on the left-hand side of the above equations, namely

$$\frac{U^2}{L}; \quad f_0 U.$$

Taking the ratio of the sizes of the momentum term and the Coriolis term, we obtain a fundamental parameter for large-scale flows, known as the **Rossby number**

$$\epsilon = \frac{U}{f_0 L}. \tag{2.55}$$

In view of the typical scales in (2.45), we can also express the Rossby number as $\epsilon = \frac{1}{f_0 T}$.

The Rossby number is a dimensionless number that describes the importance of the effects of rotation on the flow. A small Rossby number means that the effects of rotation are important (this is the case for the large scales that we are interested in, where the Rossby number is typically approximately 10^{-1}). When the Rossby number is sufficiently small, the rotation terms in Eqs. (2.53) and (2.54) will dominate over the acceleration terms. Thus, the rotation terms must be balanced by the terms on the right-hand side, i.e.,

$$\frac{1}{\rho}\frac{\partial p}{\partial x_1} = f_0 u_2, \qquad \frac{1}{\rho}\frac{\partial p}{\partial x_2} = -f_0 u_1. \tag{2.56}$$

This is known as geostrophic balance and represents an exact balance between the Coriolis effect and the pressure gradient force.

2.5.1 *Geostrophic reductions*

For flows with a sufficiently small Rossby number, the horizontal velocity terms u_1, u_2 are close to their geostrophic values. This leads us to define the so-called **geostrophic velocity** as $\mathbf{u}^g = (u_1^g, u_2^g, 0)$, where,

$$f_0 u_1^g = -\frac{1}{\rho}\frac{\partial p}{\partial x_2}, \qquad f_0 u_2^g = \frac{1}{\rho}\frac{\partial p}{\partial x_1}. \tag{2.57}$$

We also introduce the Froude number, defined by

$$\mathcal{F} = \frac{U}{NH}. \tag{2.58}$$

This is another dimensionless number and measures the stratification of the flow. For our approximations to be useful, we are concerned only with cases where either the Rossby number ϵ or the Froude number $\mathcal{F}$ is small. Indeed, in order to find simplified models which can be solved for sufficiently long times to be useful, we need only explore the following three possibilities:

$$\begin{aligned}
&\text{(i)} \quad \mathcal{F} < \epsilon \ll 1, \\
&\text{(ii)} \quad \mathcal{F} = \epsilon \ll 1, \\
&\text{(iii)} \quad \epsilon < \mathcal{F}, \epsilon \ll 1.
\end{aligned} \tag{2.59}$$

These cases correspond to aspect ratios $\frac{H}{L}$ greater than, equal to, or less than $\frac{f_0}{N}$, respectively. In the first case, the flow is stratification-dominated, while in the other two, the dominant effect is rotation. Other balances, valid in the shallow water regime, will be investigated in the second part of this chapter, where a more detailed analysis of geostrophic balance and velocity will be given.

We only consider the case of rotation-dominated flows; coupled with a variable rotation Coriolis coefficient, this yields a model valid for large scales.

Case (ii) in (2.59) is the traditional scaling introduced by Charney and corresponds to the so-called **quasigeostrophic equations**, which are treated in Sec. 2.7. The assumption $\epsilon \ll 1$ means that we can approximate Eqs. (2.53) and (2.54) using Eq. (2.57). We replace $\frac{D}{Dt}$ in Eqs. (2.53) and (2.54) with

$$\frac{D_\mathrm{g}}{Dt} := \frac{\partial}{\partial t} + u_1^g \frac{\partial}{\partial x_1} + u_2^g \frac{\partial}{\partial x_2} + u_3 \frac{\partial}{\partial x_3}.$$

In addition, the assumption $\mathcal{F} \ll 1$ allows us to approximate Eq. (2.44) by replacing $\frac{D}{Dt}$ with $\frac{D_\mathrm{g}}{Dt}$. The resulting model is an exact conservation law (of the corresponding potential vorticity). The mathematical structure of the quasigeostrophic equations as a conservation law enables extensive studies of their analytic behaviour and they have proved extremely useful in understanding extra-tropical weather systems. However, this approximation depends on constant f_0, hence it is not valid on large scales. Thus, the quasigeostrophic equations are unsuitable for large-scale flow, which led to their failure as a basis for numerical weather forecasting as early as the 1950s.

Case (iii) in (2.59) corresponds to solutions on large scales with small aspect ratio. The assumption $\epsilon \ll 1$ means that we can approximate the horizontal momentum equations (2.53) and (2.54) by the geostrophic relations in Eq. (2.56). However, this is only a first order approximation in ϵ (called the **planetary geostrophic equations**).

An exhaustive reference for the derivation and physical motivation of all these equation is Cullen (2006).

The semi-geostrophic system

A more useful set of limiting equations for describing large-scale flows is obtained by seeking a more accurate approximation. The so-called **semigeostrophic approximation** is accurate to the next order and gives a more useful and mathematically well-posed system for describing the atmosphere.

The semi-geostrophic equations were first introduced by Eliassen and then rediscovered by Hoskins as a model for frontogenesis — they have been studied extensively in the last 20 years after their rich mathematical structure, in particular the connection with optimal transport theory, was revealed.

Firstly, we define the geostrophic velocity as Eq. (2.57). The semi-geostrophic approximation then amounts to replacing the acceleration terms $\frac{Du_1}{Dt}$, $\frac{Du_2}{Dt}$ in momentum equations (2.53) and (2.54) with their geostrophic values $\frac{Du_1^g}{Dt}$, $\frac{Du_2^g}{Dt}$, but without modifying the definition of $\frac{D}{Dt}$. The remaining equations are not approximated. Since the thermodynamics terms in the equations have not been approximated, the equations are also asymptotically correct in the limit $\mathcal{F} \to 0$, even if there is no rotation. The semi-geostrophic equations, however, do not describe the flow accurately in this limit and case (i) in (2.59) above should instead be used. By rearranging $\epsilon < \mathcal{F}$, we conclude that the semi-geostrophic approximation is accurate when

$$\frac{H}{L} < \frac{f_0}{N}. \tag{2.60}$$

The semi-geostrophic equations are not a good model when Eq. (2.60) does not hold.

We will now derive the semi-geostrophic approximation in more detail. Let $\mathbf{u}_{\mathrm{H}} := (u_1, u_2)$ and let

$$J\mathcal{H} := \begin{pmatrix} 0 & -1 \\ 1 & 0 \end{pmatrix}.$$

We rearrange Eqs. (2.53) and (2.54) to give

$$\mathbf{u}_{\mathrm{H}} = \frac{1}{f_0} J\mathcal{H} \frac{1}{\rho} \nabla p + \frac{1}{f_0} J\mathcal{H} \frac{D\mathbf{u}_{\mathrm{H}}}{Dt}.$$

Then, using Eq. (2.57) we have

$$\mathbf{u}_{\mathrm{H}} = \mathbf{u}^g + \frac{1}{f_0} J\mathcal{H} \frac{D\mathbf{u}_{\mathrm{H}}}{Dt}. \tag{2.61}$$

Considering the typical scales of the above terms and using Eq. (2.55), we see that Eq. (2.61) is an order ϵ approximation to the velocity. Applying $\frac{D}{Dt} = \partial_t + \mathbf{u} \cdot \nabla$ to both sides gives

$$\frac{D\mathbf{u}_{\mathrm{H}}}{Dt} = \frac{D\mathbf{u}^g}{Dt} + \frac{1}{f_0} J\mathcal{H} \frac{D^2\mathbf{u}_{\mathrm{H}}}{Dt^2} \tag{2.62}$$

and we substitute Eq. (2.62) into Eq. (2.61) to obtain

$$
\begin{aligned}
\mathbf{u}_{\mathrm{H}} &= \mathbf{u}^g + \frac{1}{f_0} J\mathcal{H} \left[\frac{D\mathbf{u}^g}{Dt} + \frac{1}{f_0} J\mathcal{H} \frac{D^2 \mathbf{u}_{\mathrm{H}}}{Dt^2} \right] \\
&= \mathbf{u}^g + \frac{1}{f_0} J\mathcal{H} \frac{D\mathbf{u}^g}{Dt} - \frac{1}{f_0^2} \frac{D^2 \mathbf{u}_{\mathrm{H}}}{Dt^2}.
\end{aligned}
\tag{2.63}
$$

Recall from Eq. (2.55) that the Rossby number is defined as $\epsilon = \frac{U}{f_0 L} = \frac{1}{f_0 T}$. Now, we set $\bar{t} = \frac{t}{T}$ and rewrite Eq. (2.63) in the dimensionless form in time:

$$
\mathbf{u}_{\mathrm{H}} = \mathbf{u}^g + \epsilon J\mathcal{H} \frac{D\mathbf{u}^g}{D\bar{t}} - \epsilon^2 \frac{D^2 \mathbf{u}_{\mathrm{H}}}{D\bar{t}^2}.
$$

For flow satisfying the case (iii) in (2.59) above, we have $\epsilon \ll 1$, hence we now consider only terms of first order in ϵ, and neglect the last term. This yields the following equation for $\mathbf{u}_{\mathrm{H}}$:

$$
\mathbf{u}_{\mathrm{H}} = \mathbf{u}^g + \epsilon J\mathcal{H} \frac{D\mathbf{u}^g}{D\bar{t}} = \mathbf{u}^g + \frac{1}{f_0} J\mathcal{H} \frac{D\mathbf{u}^g}{Dt}.
$$

Rearranging and using Eq. (2.57) we obtain

$$
\frac{Du_1^g}{Dt} - f_0 u_2 + \frac{1}{\rho} \frac{\partial p}{\partial x_1} = 0,
\tag{2.64}
$$

$$
\frac{Du_2^g}{Dt} + f_0 u_1 + \frac{1}{\rho} \frac{\partial p}{\partial x_2} = 0.
\tag{2.65}
$$

Now consider also the Boussinesq approximation, which states that variations in density are sufficiently small to be neglected, unless they appear in terms multiplied by g_{grav}.

The set of all these approximations yields the final system:

$$
\frac{Du_1^g}{Dt} - f_0 u_2 + \frac{1}{\rho_0} \frac{\partial p}{\partial x_1} = 0,
\tag{2.66}
$$

$$
\frac{Du_2^g}{Dt} + f_0 u_1 + \frac{1}{\rho_0} \frac{\partial p}{\partial x_2} = 0,
\tag{2.67}
$$

$$
\frac{D\rho}{Dt} = 0,
\tag{2.68}
$$

$$
\nabla \cdot \mathbf{u} = 0,
\tag{2.69}
$$

$$\frac{\partial p}{\partial x_3} = -\rho g_{\text{grav}}, \tag{2.70}$$

$$f_0 u_1^g = -\frac{1}{\rho_0}\frac{\partial p}{\partial x_2}, \tag{2.71}$$

$$f_0 u_2^g = \frac{1}{\rho_0}\frac{\partial p}{\partial x_1}. \tag{2.72}$$

The unknowns in the above equations are $\mathbf{u}^g = (u_1^g, u_2^g, 0)$, $\mathbf{u} = (u_1, u_2, u_3)$, p, ρ.

This provides us with a complete system which is a reasonable simple model for the large-scale incompressible flows (reasonable for the ocean, but not an accurate model for the atmosphere, which is highly compressible).

2.5.1.1 *Analysis of the semigeostrophic system: Dual formulation*

We consider the system of incompressible semi-geostrophic equations derived in the previous section in the bounded domain $[0,\tau) \times \Omega$, where τ is some fixed final time. We prescribe *rigid boundary* conditions, namely

$$\mathbf{u} \cdot \mathbf{n} = 0 \qquad \text{on } [0,\tau) \times \partial\Omega, \tag{2.73}$$

Here, $\partial\Omega$ represents the boundary of Ω and $\mathbf{n}$ is the outward unit normal to $\partial\Omega$.

We assume that the rotation rate f is constant, and (as common for mathematicians) scale variables in such a way that all *constants are equal to* 1. Then, the incompressible version of the system is

$$D_t(v_1^g, v_2^g) + (\partial_1 p, \partial_2 p) = (u_2, -u_1),$$

$$D_t \rho = 0, \quad \rho = -\partial_3 p,$$

$$(v_1^g, v_2^g) = (-\partial_2 p, \partial_1 p),$$

$$\nabla \cdot \mathbf{u} = 0,$$

$$p(0, x) = p_0(x).$$

Here, $D = \partial_t + \mathbf{u} \cdot \nabla$. This choice ensures that the geostrophic energy

$$E_g(t) = \int_\Omega \left(\frac{1}{2}((v_1^g)^2(\mathbf{x},t) + (v_2^g)^2(\mathbf{x},t)) + \rho(\mathbf{x},t)x_3\right) d\mathbf{x}. \tag{2.74}$$

is conserved in time.

Exercise 2.32. *Assuming that the geostrophic velocity is sufficiently smooth, show that*

$$\frac{dE_g(t)}{dt} = 0.$$

A change of coordinates

These equations don't include any explicit evolution equation for the full fluid velocity $\mathbf{u}$. To unravel the equations, and understand the time evolution, notice that since $\partial_t \mathbf{x} = 0$ and $\nabla \mathbf{x} = \mathbf{1}$, we have

$$D_t \mathbf{x} = (\partial_t + \mathbf{u} \cdot \nabla)\mathbf{x} = \mathbf{u}.$$

Hence, proceeding by formal manipulations, we can write the first equation as

$$D_t(v_1^g, v_2^g) + (\partial_1 p, \partial_2 p) = D_t(x_2, -x_1) \Longrightarrow D_t(v_1^g - x_2, v_2^g + x_1) + (\partial_1 p, \partial_2 p) = 0.$$

So it is natural to define a transformation for the first two components as

$$x \to T(x) = (v_1^g - x_2, v_2^g + x_1, \cdot).$$

Note that by setting

$$P = p + \tfrac{1}{2}(x_1^2 + x_2^2),$$

we find

$$\nabla P = \nabla p + (x_1, x_2, 0) = (v_2^g, -v_1^g, -\rho) + (x_1, x_2, 0). \qquad (2.75)$$

This justifies a change of variable, originally proposed by Hoskins and known as *geostrophic variable change*:

$$(\mathbf{x}, t) \to (\mathbf{y}, t)$$

$$\text{with } \mathbf{y} = \nabla P_t(x) = (v_2^g + x_1, -v_1^g + x_2, -\rho), \ (P_t(\mathbf{x}) = P(\mathbf{x}, t)).$$

If P is convex (this is a *key assumption*, and needs to be justified), and at least weakly differentiable, then, ∇P is a well defined change of variables. In fact, convexity also implies that this change of variable is invertible, with inverse $(\nabla P)^{-1} = \nabla P^*$, where P^* denotes the Legendre transform of P, a fundamental concept in convex analysis, defined by

$$P^*(\mathbf{y}) = \sup_{\mathbf{x}}\{\mathbf{x}\cdot\mathbf{y} - P(\mathbf{x})\}.$$

Hence,

$$\mathbf{y} = \nabla P(\mathbf{x}) \Longleftrightarrow \mathbf{x} = \nabla P^*(\mathbf{y}).$$

Dual velocity

A formal computation shows that the velocity $\mathbf{U}$ in terms of the dual variables $\mathbf{y}$, hence, $\mathbf{U} = \frac{d}{dt}\nabla P$ satisfies

$$\mathbf{U} = (v_1^g, v_2^g, 0).$$

Indeed,

$$\mathbf{U} = \frac{d}{dt}\nabla P(\mathbf{x}) = \partial_t \nabla P(\mathbf{x}) + \dot{\mathbf{x}}(t)\nabla(\nabla P(\mathbf{x}))$$

$$= \partial_t \mathbf{y} + \dot{\mathbf{x}}\cdot\nabla(\mathbf{y}) = (\partial_t + u\cdot\nabla)\mathbf{y} = D_t\mathbf{y}.$$

Moreover,

$$D_t\mathbf{y} = (-\partial_2 p, \partial_1 p, 0) = (v_1^g, v_2^g, 0) = J(v_2^g, -v_1^g, 0) = J(\mathbf{y}-\mathbf{x}),$$

where

$$J = \begin{pmatrix} 0 & -1 & 0 \\ 1 & 0 & 0 \\ 0 & 0 & 0 \end{pmatrix}.$$

Since $J(v_2^g, -v_1^g, 0) = (v_1^g, v_2^g, 0)$, we find that the *dual flow is purely geostrophic* and, by construction, also divergence free:

$$\mathbf{U}(\mathbf{y},t) = J(\mathbf{y}-\mathbf{x}) = \mathbf{v}^g; \quad \nabla\cdot\mathbf{U} = 0.$$

Dual density

Any change of variable $\mathbf{y} = T(x)$ can be stated in terms of the density $\alpha(\mathbf{y},t) = \alpha_t(\mathbf{y})$ on $\mathbb{R}^3$ such that

$$\int_\Omega f(T(\mathbf{x}))d\mathbf{x} = \int_{\mathbb{R}^3} f(\mathbf{y})d\alpha(\mathbf{y}) \quad \forall f \in \mathbf{C}_c(\mathbb{R}^3). \tag{2.76}$$

This is concisely written using the push-forward notation as

$$\alpha = T\#\chi_\Omega.$$

Given some regularity on T, the measure α is also absolutely continuous with respect to Lebesgue measure, so can be written as $d\alpha(\mathbf{y}) = \alpha(y)d\mathbf{y}$, with $\alpha(\mathbf{y})$ an integrable functions.

The function α is called **dual (or potential) density**.

Energy, minimisation and convexity

The *Cullen's stability principle* states that stable solutions of the geostrophic system must be *energy minimisers* of the geostrophic energy $E_g(t)$ with respect to the rearrangements of particles in physical space (this can be seen by a formal variational computation) (Cullen, 2006).

For now, let us see how such a statement can be interpreted mathematically.

In the new variables $\mathbf{y} = \nabla P(\mathbf{x}) = (v_2^g + x_1, -v_1^g + x_2, -\rho)$, the geostrophic energy is given by

$$E_g(\mathbf{y}(t,\mathbf{x})) = \int_\Omega \left\{ \frac{1}{2} \left[(x_1 - y_1)^2 + (x_2 - y_2)^2 \right] - x_3 y_3 \right\} d\mathbf{x}.$$

In these variables the minimiser condition formally becomes

$$P(t,x) = \tfrac{1}{2}(x_1^2 + x_2^2) + p(t,x) \quad \text{is a convex function.}$$

To justify this statement mathematically, we will need a beautiful theorem, due to Brenier, and known as the **polar factorisation theorem** (Brenier, 1991).

Recall that a *probability measure* α on $\mathbb{R}^d$ is a measure such that

$$\int_{\mathbb{R}^d} d\alpha = 1.$$

We say that a function $s(\mathbf{x})$ defined on $\Omega \subset \mathbb{R}^d$ is *measure preserving* if

$$\int_\Omega f(s(\mathbf{x}))d\mathbf{x} = \int_\Omega f(\mathbf{x})d\mathbf{x}, \quad \forall f \in C(\bar{\Omega})).$$

We now state the polar factorisation theorem in the following form (not the most general possible):

Theorem 2.5 (Polar factorisation). *For each probability measure α on $\mathbb{R}^d$ satisfying the integrability condition $\int_{\mathbb{R}^d}(1+|\mathbf{y}|)^p d\alpha(\mathbf{y}) < \infty$, there exists* *a unique* ∇P, *with* $P \in W^{1,p}(\Omega)$ *and* **convex**, *such that*

$$\int_{\Omega} f(\nabla P(\mathbf{x}))d\mathbf{x} = \int_{\mathbb{R}^d} f(\mathbf{y})d\alpha(\mathbf{y}).$$

Moreover, if $\int f(\mathbf{y})d\alpha_n(\mathbf{y}) \to \int f(\mathbf{y})d\alpha(y)$ *for all* f *continuous, then* $P_n \to P$ *in* $W^{1,p}(\Omega)$.

(The symbol $W^{1,p}(\Omega)$, already defined in Eq. (2.29), denotes the space of $L^p(\Omega)$ functions admitting a first weak derivative in L^p — this is an important but technical point, but it means that it makes sense to consider ∇P and that this function has some integrability.)

If α is the realisation of a given function T as in Eq. (2.76), then the theorem says that

$$\int_{\Omega} f(\nabla P(\mathbf{x}))d\mathbf{x} = \int_{\mathbb{R}^d} f(T(\mathbf{x}))d(\mathbf{x}),$$

i.e., that T has a rearrangement in terms of the gradient of a convex function. In addition, it can be proved that $T = \nabla P \circ s$, where s is a volume preserving map.

Semigestrophic system in dual Lagrangian variables

In the variables $(\mathbf{y}, t)$, the semigeostrophic system becomes the following

$$\frac{D}{Dt}\mathbf{y}(\mathbf{x},t) = J(\mathbf{y} - \mathbf{x}), \qquad (2.77)$$

$$\nabla \cdot \mathbf{u} = 0, \qquad (2.78)$$

$$\mathbf{y}(\mathbf{x},t) = \nabla P(\mathbf{x},t), \qquad (2.79)$$

$$\mathbf{u} \cdot \mathbf{n} = 0 \text{ on } [0,\tau) \times \partial\Omega, \qquad (2.80)$$

$$P(0,\mathbf{x}) = P_0(\mathbf{x}) := \tfrac{1}{2}(x_1^2 + x_2^2) + p_0(\mathbf{x}) \text{ in } \Omega. \qquad (2.81)$$

So far all computations have been formal, and based on the assumption that it is in fact possible to find a convex function P giving us the appropriate change of variables and the equations in the form Eqs. (2.77)–(2.81).

We now use Lagrangian coordinates to un-tangle this system and make the procedure rigorous using the polar factorisation theorem.

Define $t \to X(t,\mathbf{x})$ as the trajectory of the fluid particle positioned initially at $\mathbf{x}$, and let $T(t,\mathbf{x}) = \mathbf{y}(X(t,\mathbf{x}),t)$ be the Lagrangian variable. Then,

T satisfies

$$\partial_t T(t,\mathbf{x}) = J(T(t,\mathbf{x}) - X), \qquad \text{constrained by } T(t,\mathbf{x}) = \nabla P(t, X(t,\mathbf{x})).$$

Note that the incompressibility condition coupled with the rigid boundary condition imply that the Jacobian det $\nabla(X(t,\mathbf{x})) = 1$. Hence, $x \to X(t,\mathbf{x})$ is a *volume preserving map*.

The polar factorisation theorem gives us, for each fixed time t, the existence of a convex function P_t and of a volume preserving map $X(x,t)$ that satisfy

$$\int_\Omega f(T(t,\mathbf{x})) d\mathbf{x} = \int_\Omega f(\nabla P_t(\mathbf{x})) d\mathbf{x}, \quad \forall f \in C_c(\mathbb{R}^3). \tag{2.82}$$

This equation is the weak form of the constraint $T(t,\mathbf{x}) = \nabla P_t(X(t,\mathbf{x}))$ and guarantees the existence of a convex function whose gradient gives precisely the Lagrangian description of the semigeostrophic flow.

Setting $\alpha_t = T \# \chi_\Omega$ by definition of push forward, we have

$$\int_\Omega f(T(t,\mathbf{x})) d\mathbf{x} = \int_{\mathbb{R}^3} f(\mathbf{y}) \alpha_t(\mathbf{y}) d\mathbf{y}, \quad \forall f \in C_c(\mathbb{R}^3).$$

Then, Eq. (2.82), under the variable change $\mathbf{y} = \nabla P_t(\mathbf{x})$, becomes

$$\int_{\mathbb{R}^3} f(\mathbf{y}) \alpha_t(\mathbf{y}) d\mathbf{y} = \int_\Omega f(\mathbf{y}) \det(\nabla^2 P_t^*)(\mathbf{y}) d\mathbf{y}, \quad \forall f \in C_c(\mathbb{R}^3). \tag{2.83}$$

(This equation is the weak form of the Monge–Ampére equation det $\nabla^2 P_t^* = \alpha$).

Evolution equation for the dual density

Since $\partial_t T(t,\mathbf{x}) = J(T(t,\mathbf{x}) - X)$, we can write, for all $\xi \in C_c^\infty(\mathbb{R}^3 \times [0,\tau))$ the identity

$$\int_0^\tau \int_{\mathbb{R}^3} \partial_t T(t,\mathbf{x}) \nabla \xi(T(t,\mathbf{x}),t) d\mathbf{x} dt$$

$$= \int_0^\tau \int_{\mathbb{R}^3} J(T(t,\mathbf{x}) - X) \nabla \xi(T(t,\mathbf{x}),t) d\mathbf{x} dt.$$

Now we can compute (using the compact support of ξ which implies that $\xi(T(\tau,\mathbf{x}),\tau) = 0$)

$$\int_0^\tau \int_{\mathbb{R}^3} \partial_t T(t,\mathbf{x}) \nabla \xi(T(t,\mathbf{x}),t) d\mathbf{x} dt$$

$$= -\int_{\mathbb{R}^3} \xi(T(0,\mathbf{x}),0) d\mathbf{x} - \int_0^\tau \int_{\mathbb{R}^3} \partial_t \xi(T(t,\mathbf{x}),t) d\mathbf{x} dt.$$

Hence,

$$\int_0^\tau \int_{\mathbb{R}^3} J(T(t,\mathbf{x}) - X)\nabla\xi(T(t,\mathbf{x}),t)d\mathbf{x}dt$$

$$= -\int_{\mathbb{R}^3} \xi(T(0,\mathbf{x}),0)d\mathbf{x} - \int_0^\tau \int_{\mathbb{R}^3} \partial_t\xi(T(t,\mathbf{x}),t)d\mathbf{x}dt.$$

Recalling that $T(t,\mathbf{x}) = \nabla P(t, X(t,\mathbf{x}))$,

$$\int_0^\tau \int_{\mathbb{R}^3} J(\nabla P(t,X(t,\mathbf{x})) - X)\nabla\xi(\nabla P(t,X(t,\mathbf{x})),t)d\mathbf{x}dt$$

$$= -\int_{\mathbb{R}^3} \xi(\nabla P(0,X(0,\mathbf{x})))d\mathbf{x} - \int_0^\tau \int_{\mathbb{R}^3} \partial_t\xi(\nabla P(t,X(t,\mathbf{x})),t)d\mathbf{x}dt.$$

Changing variable to $\mathbf{y} = \nabla P(t, X(t,\mathbf{x}))$, and denoting $\mathbf{U}$ (as before) the dual velocity, we finally obtain

$$\int_0^\tau \int_{\mathbb{R}^3} \mathbf{U}(\mathbf{y},t)\nabla\xi(\mathbf{y},t)\alpha_t(\mathbf{y})d\mathbf{y}dt$$

$$= -\int_{\mathbb{R}^3} \xi(\mathbf{y}(0))\alpha_0(\mathbf{y})d\mathbf{y} - \int_0^\tau \int_{\mathbb{R}^3} \partial_t\xi(\mathbf{y},t)\alpha_t(\mathbf{y})d\mathbf{y}dt,$$

i.e.,

$$\int_0^\tau \int_{\mathbb{R}^3} [\partial_t\xi(\mathbf{y},t) + \mathbf{U}(\mathbf{y},t)\nabla\xi(\mathbf{y},t)]\,\alpha_t(\mathbf{y})d\mathbf{y}dt + \int_{\mathbb{R}^3} \xi(\mathbf{y}(0))\alpha_0(\mathbf{y})d\mathbf{y} = 0.$$

This equation is the weak form of the evolution equation

$$\partial_t\alpha + \nabla \cdot (\mathbf{U}\alpha) = 0.$$

In summary, we have reformulated the semigeostrophic system in the following form:

$$\frac{\partial}{\partial t}\alpha(\mathbf{y},t) + \nabla \cdot (\mathbf{U}(\mathbf{y})\alpha(\mathbf{y},t)) = 0, \qquad \text{in } [0,\tau) \times \mathbb{R}^3, \qquad (2.84)$$

$$\mathbf{U}(\mathbf{y},t) = J(\mathbf{y} - \nabla P_t^*(\mathbf{y})), \qquad \text{in } [0,\tau) \times \mathbb{R}^3, \qquad (2.85)$$

$$\nabla P_t\#\chi_\Omega = \alpha(\cdot,t), \qquad \text{for any } t \in [0,\tau). \qquad (2.86)$$

In these equations, for each fixed $t \in [0,\tau)$, P_t is a convex function and P_t^* is its Legendre transform.

*Lagrangian solution(s) of the semigeostrophic system
in dual variables*

Benamou and Brenier proved in groundbreaking work (Benamou and Brenier, 1998) that solutions of the system above exist. The solution process is as follows: at each instant t, we use α to compute $\mathbf{U}$ from Eq. (2.86) and Eq. (2.85). We then use $\mathbf{U}$ to advect α in time, using the transport Eq. (2.84). Due to the way in which $\mathbf{U}$ is constructed, we have that $\mathbf{U} \in L_{\mathrm{loc}}^{\infty}$ $([0,\tau) \times \mathbb{R}^3)$ and $\mathbf{U} \in L^{\infty}([0,\tau); BV_{\mathrm{loc}}(\mathbb{R}^3))$. This means that $\mathbf{U}$ does not have enough regularity to solve the transport equation classically.

Thus, in order to advect α using Eq. (2.84), we discretise the system in time and introduce a regularisation that approximates $\mathbf{U}$ with a Lipschitz continuous velocity field. This allows us to solve the transport equation. Then, using the stability property of polar factorisation, one can show that these approximate solutions converge to solutions of the system in Eqs. (2.84)–(2.86).

The optimal transport interpretation

Consider the geostrophic energy in Eq. (2.74). Cullen's stability principle, based on physical considerations, states that stable solutions of the semigeostrophic equations must correspond to solutions that, at each fixed time t, minimise the energy associated with the flow. As before, we consider this in dual variables, hence for each fixed time t, we introduce the functional

$$E_t[\mathbf{T}] := \int_{\Omega} \left(\frac{1}{2}\{|x_1 - T_1(\mathbf{x})|^2 + |x_2 - T_2(\mathbf{x})|^2\} - x_3 T_3(\mathbf{x}) \right) d\mathbf{x}.$$

(2.87)

The dual density, also called potential density, is $\alpha = \mathbf{T} \# \chi_{\Omega}$. Cullen's stability principle then corresponds to the requirement that $\mathbf{T}$ is the **optimal map** in the transport of χ_{Ω} to α with quadratic cost function

$$c(\mathbf{x},\mathbf{y}) = \frac{1}{2}\{|x_1 - y_1|^2 + |x_2 - y_2|^2\} - x_3 y_3.$$ (2.88)

The polar factorisation theorem can be reinterpreted to give existence and uniqueness of such an optimal map and implies that $\mathbf{T} = \nabla P$ with P *convex*. Note that α is to be found as part of the solution and, hence, this push forward expression for α will need to be coupled with the dual space transport equation for α, namely Eq. (2.84).

2.6   Rotating Shallow Water (RSW) Equations

The remainder of this chapter discussses various representations and approximations of ideal shallow water dynamics in a rotating frame. These rotating shallow water (RSW) equations possess a slow + fast decomposition in which they reduce approximately to quasigeostrophic (QG) motion (conservation of energy and potential vorticity) plus nearly decoupled equations for gravity waves in an asymptotic expansion in small Rossby number, $\epsilon \ll 1$. The solution properties of the RSW equations are discussed, and some alternative representations of the RSW equations which highlight the slow + fast interactions are given. At the conclusion, the RSW equations and the thermal rotating shallow water (TRSW) equations are derived as Euler–Poincaré equations from Hamilton's principle in the Eulerian fluid representation.

2.6.1   *Rotating shallow water motion equations*

We consider dynamics of RSW on a two-dimensional domain with horizontal planar coordinates $\mathbf{x} = (x, y)$. This RSW motion is governed by the following non-dimensional equations for horizontal fluid velocity vector $\mathbf{u} = (u, v)$ and the total depth η,

$$\epsilon \frac{D}{Dt}\mathbf{u} + f\hat{z} \times \mathbf{u} + \nabla h = 0, \qquad \frac{\partial \eta}{\partial t} + \nabla \cdot (\eta \mathbf{u}) = 0, \qquad (2.89)$$

with notation, cf. Eq. (2.8)

$$\frac{D}{Dt} := \left(\frac{\partial}{\partial t} + \mathbf{u} \cdot \nabla \right) \quad \text{and} \quad h := \left(\frac{\eta - b}{\epsilon \mathcal{F}} \right). \qquad (2.90)$$

These equations include a variable Coriolis parameter $f = f(\mathbf{x})$ and bottom topography $b = b(\mathbf{x})$. We will have more to say about the structure of these equations later, but for now we just think of them as nonlinear evolutionary PDEs in time and two-dimensional space, with homogeneous or periodic boundary conditions.

Dimensionless scale factors for RSW: The dimensionless scale factors appearing in the RSW equations (2.89) and (2.90) are the Rossby number ϵ and the squared external Rossby ratio $\mathcal{F}$, given in terms of dimensional scales by

$$\epsilon = \frac{\mathcal{U}_0}{f_0 L} \ll 1 \quad \text{and} \quad \mathcal{F} = \frac{L^2}{L_R^2} = O(1) \quad \text{with} \quad L_R^2 = \frac{g b_0}{f_0^2}, \qquad (2.91)$$

where L_R is the Rossby radius, which, as we will see in Sec. 2.7.4, occurs at the peak of the dispersion curve for Rossby waves arising from perturbations of quasigeostrophic fluid at rest.

The dimensional scales $(b_0, L, \mathcal{U}_0, f_0, g)$ in RSW dynamics denote equilibrium fluid depth (b_0) horizontal length (L), horizontal fluid velocity $(\mathcal{U}_0)$, reference Coriolis parameter (f_0) and gravitational acceleration (g). Dimensionless quantities in Eq. (2.89) are unadorned and are related to their dimensional counterparts (primed), according to

$$\mathbf{u}' = \mathcal{U}_0 \mathbf{u}, \quad \mathbf{x}' = L\mathbf{x}, \quad t' = \left(\frac{L}{\mathcal{U}_0}\right) t, \quad f' = f_0 f,$$

$$b' = b_0 b, \quad \eta' = b_0 \eta, \quad \text{and} \quad \eta' - b' = b_0(\eta - b). \qquad (2.92)$$

Here, dimensional quantities are: $\mathbf{u}'$, the horizontal fluid velocity; η', the fluid depth; b', the equilibrium depth; and $\eta' - b'$, the free surface elevation.

For barotropic horizontal motions at length scales L in the ocean for which $\mathcal{F}$ is order $O(1)$ — as we shall assume — the Rossby number ϵ is typically quite small ($\epsilon \ll 1$) as indicated in Eq. (2.91) and, thus, the Rossby number is a natural parameter for making asymptotic expansions. For example, we shall assume $|\nabla f| = O(\epsilon)$ and $|\nabla b| = O(\epsilon)$, so we may write $f = 1 + \epsilon f_1(\mathbf{x})$ and $b = 1 + \epsilon b_1(\mathbf{x})$.

2.6.2   *Geostrophic balance*

At leading order in $\epsilon \ll 1$ the pressure gradient force in Eq. (2.89) may balance the Coriolis force, as

$$\hat{z} \times \mathbf{u} = -\nabla h. \qquad (2.93)$$

Taking the cross product of the vertical unit vector $\hat{z}$ with this *geostrophic balance* yields the *geostrophic velocity*,

$$\mathbf{u}_G = \hat{z} \times \nabla h. \qquad (2.94)$$

Thus, it makes sense to assume that the velocity has an *ε-weighted Helmholtz decomposition*,

$$\mathbf{u} = \hat{z} \times \nabla \psi + \epsilon \nabla \chi \quad \text{with} \quad \psi = h + O(\epsilon). \qquad (2.95)$$

In the oceans and atmosphere, the geostrophic balance tends to be stable and small disturbances of it lead to waves, called Rossby waves. In seeking to establish the properties of these Rossby waves (such as their propagation velocity) we will analyse the linearised RSW equations. For this purpose, we first rewrite the nonlinear equations in their *RSW curl form*

$$\partial_t(\epsilon \mathbf{u} + \mathbf{R}(\mathbf{x})) - \mathbf{u} \times \text{curl}(\epsilon \mathbf{u} + \mathbf{R}(\mathbf{x})) + \nabla \left(h + \frac{\epsilon}{2}|\mathbf{u}|^2 \right) = 0, \qquad (2.96)$$

where $\text{curl}\,\mathbf{R}(\mathbf{x}) = f(\mathbf{x})\hat{z} = (1 + \epsilon f_1(\mathbf{x}))\hat{z}$, so that $\mathbf{R}(\mathbf{x})$ is the vector potential for the divergence free rotation rate about the vertical direction.

2.6.3 *Crucial vector calculus identities*

To check that the RSW motion equations (Eqs. (2.89) and (2.96)) are equivalent, one may use the fundamental vector identity of fluid dynamics,

$$(\nabla \times \mathbf{a}) \times \mathbf{b} + \nabla(\mathbf{a} \cdot \mathbf{b}) = (\mathbf{b} \cdot \nabla)\mathbf{a} + a_j \nabla b^j. \qquad (2.97)$$

Taking the curl of Eq. (2.96) yields the equation for vorticity dynamics,

$$\partial_t \varpi - \hat{z} \cdot \text{curl}\left(\mathbf{u} \times \varpi \hat{z} \right) = 0, \quad \text{with } \varpi := \epsilon \hat{z} \cdot \text{curl}\,\mathbf{u} + f(\mathbf{x}), \qquad (2.98)$$

where we have defined the *ϵ-weighted total vorticity* as

$$\varpi := \epsilon \hat{z} \cdot \text{curl}\,\mathbf{u} + f(\mathbf{x}) = \epsilon \hat{z} \cdot \text{curl}(\hat{z} \times \nabla \psi + \epsilon \nabla \chi) + f(\mathbf{x}) = \epsilon \Delta \psi + f(\mathbf{x}),$$

whereas the fluid vorticity is denoted as $\omega := \hat{z} \cdot \text{curl}\,\mathbf{u}$.

Expanding out the curl in Eq. (2.98) yields

$$\partial_t \varpi + \mathbf{u} \cdot \nabla \varpi + \varpi \nabla \cdot \mathbf{u} = 0. \qquad (2.99)$$

by virtue of the vector identity,

$$\text{curl}(\mathbf{a} \times \mathbf{b}) = -(\mathbf{a} \cdot \nabla)\mathbf{b} - (\nabla \cdot \mathbf{a})\mathbf{b} + (\mathbf{b} \cdot \nabla)\mathbf{a} + (\nabla \cdot \mathbf{b})\mathbf{a}. \qquad (2.100)$$

One may also write the RSW total vorticity equation (Eq. (2.99)) as a continuity equation,

$$\frac{\partial \varpi}{\partial t} + \nabla \cdot (\varpi \mathbf{u}) = 0, \qquad (2.101)$$

which implies conservation of integrated total vorticity $\int \varpi \, d^2 x$, provided $\mathbf{u} \cdot \hat{\mathbf{n}} = 0$ on the boundary with unit normal vector $\hat{\mathbf{n}}$.

Order by order in an expansion in powers of $\epsilon \ll 1$, the total vorticity equation (Eq. (2.99)) and the ϵ-weighted Helmholtz decomposition Eq. (2.95) combine to yield,

$$O(1): \quad \nabla \cdot \mathbf{u} = \nabla \cdot (\hat{z} \times \nabla \psi) = 0,$$

$$O(\epsilon): \quad \partial_t \Delta \psi + (\hat{z} \times \nabla \psi) \cdot \nabla (\Delta \psi + f_1(\mathbf{x})) + \Delta \chi = 0,$$

$$O(\epsilon^2): \quad \nabla \cdot \big((\Delta \psi + f_1(\mathbf{x})) \nabla \chi\big) = 0.$$

Later, we will deal with the terms at each order, which will be important in deriving a sequence of approximations first at the level of the $O(\epsilon)$ quasi-geostrophic approximation (QG) for Rossby waves and later at order $O(\epsilon^2)$ in discussing Poincaré gravity waves in two different decompositions of the fluid velocity.

2.6.4   *Conservation laws for RSW*

Exercise 2.33. *Verify the following four properties of the RSW equations (Eq. (2.89))*:

(a) *Energy conservation*

$$E = \int \frac{\epsilon}{2} \eta |\mathbf{u}|^2 + \frac{(\eta - b)^2}{2\epsilon \mathcal{F}} \, d^2 x. \tag{2.102}$$

(b) *Kelvin circulation theorem*

$$\frac{d}{dt} \oint_{c(u)} (\epsilon \mathbf{u} + \mathbf{R}(\mathbf{x})) \cdot d\mathbf{x} = 0, \quad \text{where } \nabla \times \mathbf{R}(\mathbf{x}) = f(\mathbf{x})\hat{z}, \tag{2.103}$$

and $c(u)$ *is a closed planar loop moving with the fluid velocity* $\mathbf{u}(\mathbf{x}, t)$.

(c) *Conservation of potential vorticity (PV) on fluid parcels*

$$\frac{Dq}{Dt} = \partial_t q + \mathbf{u} \cdot \nabla q = 0, \tag{2.104}$$

where PV (q) *is defined by*

$$q := \frac{\varpi}{\eta}, \quad \text{and} \quad \varpi := \hat{z} \cdot \nabla \times (\epsilon \mathbf{u} + \mathbf{R}(\mathbf{x})). \tag{2.105}$$

(d) *Infinite number of conserved integral quantities*

$$\frac{d}{dt} \int \eta \Phi(q) \, d^2 x = 0, \tag{2.106}$$

for any differentiable function Φ.

Hints: The following alternative form of the RSW motion equation (Eq. (2.89)) may be helpful in verifying these four properties:

$$\partial_t(\epsilon\mathbf{u}+\mathbf{R}(\mathbf{x}))-\mathbf{u}\times\nabla\times(\epsilon\mathbf{u}+\mathbf{R}(\mathbf{x}))+\nabla\left(h+\frac{\epsilon}{2}|\mathbf{u}|^2\right)=0, \qquad (2.107)$$

where $\nabla\times\mathbf{R}(\mathbf{x})=f(\mathbf{x})\hat{z}$ *and* $h:=(\eta-b)/(\epsilon F)$.

You might also keep in mind the fundamental vector identity of fluid dynamics,

$$(\nabla\times\mathbf{a})\times\mathbf{b}+\nabla(\mathbf{a}\cdot\mathbf{b})=(\mathbf{b}\cdot\nabla)\mathbf{a}+a_j\nabla b^j. \qquad (2.108)$$

Answer.

(a)

$$\begin{aligned}
\frac{dE}{dt} &= \int_{\mathcal{D}}\eta_t(\epsilon u^2/2+h)+\eta\mathbf{u}\cdot\epsilon\mathbf{u}_t\,dxdy\\
&= -\int_{\mathcal{D}}(\nabla\cdot(\eta\mathbf{u}))(\epsilon u^2/2+h)+\eta\mathbf{u}\cdot\nabla(\epsilon u^2/2+h)\,dxdy\\
&= -\int_{\mathcal{D}}\nabla\cdot\left(\eta(\epsilon u^2/2+h)\mathbf{u}\right)dxdy\\
&= -\oint_{\partial\mathcal{D}}\eta(\epsilon u^2/2+h)\mathbf{u}\cdot\hat{n}\,ds=0,
\end{aligned}$$

which vanishes for $\mathbf{u}$ tangent to the boundary $\partial\mathcal{D}$ of the domain of flow $\mathcal{D}$.

(b)

$$\frac{d}{dt}\oint_{c(u)}(\epsilon\mathbf{u}+\mathbf{R}(\mathbf{x}))\cdot d\mathbf{x}$$

(By Eq. (2.97))

$$=\oint_{c(u)}(\epsilon\partial_t\mathbf{u}+\epsilon\mathbf{u}\cdot\nabla\mathbf{u}+\epsilon u_j\nabla u^j-\mathbf{u}\times\nabla\times\mathbf{R}+\nabla(\mathbf{u}\cdot\mathbf{R}))\cdot d\mathbf{x}$$

(By Eq. (2.89))

$$=\oint_{c(u)}\nabla(\epsilon|\mathbf{u}|^2/2+h+\mathbf{u}\cdot\mathbf{R})\cdot d\mathbf{x}=0,$$

which vanishes because the integral of a gradient over a closed loop is zero, by the fundamental theorem of calculus.

(c) The curl of the alternative form of the RSW motion in Eq. (2.96) yields

$$0 = \partial_t \varpi - \nabla \times (\mathbf{u} \times \varpi) = \partial_t \varpi + \mathbf{u} \cdot \nabla \varpi + \varpi \nabla \cdot \mathbf{u}$$

$$= \frac{d\varpi}{dt} - \varpi \eta^{-1} \frac{d\eta}{dt} = \eta \frac{d}{dt}\left(\frac{\varpi}{\eta}\right) = \eta \frac{dq}{dt},$$

which verifies Eq. (2.104).

(d) PV conservation on fluid parcels in Eq. (2.104) implies that the time derivative

$$\frac{d}{dt}\int \eta \Phi(q)\,dxdy = \int \frac{\partial \eta}{\partial t}\Phi(q) + \eta \Phi'(q)\frac{\partial q}{\partial t}\,dxdy$$

$$= -\int \nabla \cdot (\eta \mathbf{u})\,\Phi(q) + \eta \mathbf{u} \cdot \nabla \Phi(q)\,dxdy$$

$$= -\int \nabla \cdot (\eta \Phi(q)\mathbf{u})\,dxdy$$

$$= -\oint \eta \Phi(q)\hat{n} \cdot \mathbf{u}\,ds$$

$$= 0,$$

vanishes, because $\hat{\mathbf{n}} \cdot \mathbf{u}$ vanishes on the boundary.

2.7   The Quasigeostrophic (QG) Approximation

2.7.1   *Derivation of QG*

In this section, we derive the well-known quasigeostrophic (QG) approximation (Pedlosky, 1987) of the equations for RSW motion in a rotating frame into a form. Consistent with the QG approximation, we assume $f(\mathbf{x}) = 1 + \epsilon f_1(\mathbf{x})$ and $b(\mathbf{x}) = 1 + \epsilon b_1(\mathbf{x})$, with $\mathbf{x} = (x, y)$. We return to the RSW motion in Eq. (2.89), rewritten as

$$\epsilon \frac{D\mathbf{u}}{Dt} = -f\hat{z} \times \mathbf{u} - \nabla h, \tag{2.109}$$

where

$$\frac{D}{Dt} = \frac{\partial}{\partial t} + \mathbf{u} \cdot \nabla, \quad h = \frac{\eta - b}{\epsilon \mathcal{F}}. \tag{2.110}$$

Operating with $\hat{z}\times$ on Eq. (2.109) and expanding in powers of ϵ yields

$$\mathbf{u} = \hat{z}\times\nabla h - \epsilon\, f_1\, \hat{z}\times\nabla h - \epsilon\left(\frac{\partial}{\partial t} + \mathbf{u}_G\cdot\nabla\right)\nabla h + O(\epsilon^2)$$

$$= \mathbf{u}_G + \epsilon\mathbf{u}_A + O(\epsilon^2), \tag{2.111}$$

where the geostrophic and ageostrophic components of the velocity are defined, respectively, by

$$\mathbf{u}_G = \hat{z}\times\nabla h \quad \text{and} \quad \mathbf{u}_A = \left(\frac{\partial}{\partial t} + \mathbf{u}_G\cdot\nabla\right)\hat{z}\times\mathbf{u}_G - f_1\,\mathbf{u}_G. \tag{2.112}$$

The remainder of this section is devoted to studying the class of RSW flows that satisfy condition Eq. (2.111). In Eq. (2.112), $\mathbf{u}_G$ is divergence free and $\mathbf{u}_A$ has divergence given by

$$\nabla\cdot\mathbf{u}_A = -\left(\frac{\partial}{\partial t} + \mathbf{u}_G\cdot\nabla\right)\Delta h - \mathbf{u}_G\cdot\nabla f_1, \tag{2.113}$$

in which Δh is the horizontal Laplacian of h. Substituting Eq. (2.113) for $\nabla\cdot\mathbf{u}_A$ into the continuity equation

$$\frac{\partial\eta}{\partial t} + \nabla\cdot(\eta\mathbf{u}) = 0, \quad \text{rewritten as } \epsilon\mathcal{F}h_{,t} = -\nabla\cdot(\eta\mathbf{u}), \tag{2.114}$$

and using the relations

$$\eta = b + \epsilon\mathcal{F}h, \quad \mathbf{u} = \mathbf{u}_G + \epsilon\mathbf{u}_A \quad \text{and} \quad b(\mathbf{x}) = 1 + \epsilon b_1(\mathbf{x}), \tag{2.115}$$

yields at order $O(\epsilon)$ the QG equation for the dimensionless free surface height (Pedlosky, 1987),

$$\left(\frac{\partial}{\partial t} + \mathbf{u}_G\cdot\nabla\right)(\mathcal{F}h - \Delta h + b_1 - f_1) = 0. \tag{2.116}$$

Thus, in the QG approximation, the potential vorticity, defined by

$$q = \mathcal{F}h - \Delta h + b_1 - f_1, \tag{2.117}$$

is advected by the divergenceless geostrophic velocity $\mathbf{u}_G = \hat{z}\times\nabla h$. That is,

$$\partial_t q + \mathbf{u}_G\cdot\nabla q = 0. \tag{2.118}$$

The positive-definite symmetric operator $\mathcal{F} - \Delta$ is non-degenerate, so its operator inverse $1/(\mathcal{F} - \Delta)$ exists and is well defined on Fourier transformable functions, say. Therefore, the surface height h and its derivatives are determined uniquely from the potential vorticity q in QG theory.

Equations (2.116) and (2.117) combine into

$$\frac{\partial q}{\partial t} = -\hat{z} \times \nabla h \cdot \nabla q = J(q,h), \tag{2.119}$$

where

$$J(q,h) := \hat{z} \cdot \nabla q \times \nabla h \tag{2.120}$$

is the Jacobian of the transformation $(x,y) \to (q,h)$

$$dq \wedge dh = J(q,h)\, dx \wedge dy = (q_{,x}h_{,y} - h_{,x}q_{,y})\, dx \wedge dy =: \{q,h\}_{\text{can}} dx \wedge dy, \tag{2.121}$$

where

$$\{q,h\}_{\text{can}} := q_{,x}h_{,y} - h_{,x}q_{,y} \tag{2.122}$$

denotes the canonical Poisson bracket.

2.7.2   The ageostrophic velocity in the QG approximation: Part I

Exercise 2.34. *Show that the QG motion in Eq. (2.116) implies*

$$\frac{\partial}{\partial t}\left(\frac{\mathcal{F}h^2}{2} + \frac{|\nabla h|^2}{2}\right) = \nabla \cdot (h\nabla h_{,t} - h\mathbf{u}_G(\mathcal{F}h - \Delta h + b_1 - f_1)). \tag{2.123}$$

As a consequence of Eq. (2.123), *QG motion conserves the positive-definite energy,*

$$
\begin{aligned}
E_{\text{QG}} &= \int \left(\frac{\mathcal{F}h^2}{2} + \frac{1}{2}|\mathbf{u}_G|^2\right) d^2x \\
&= \frac{1}{2}\int (\mathcal{F}h^2 + |\nabla h|^2) d^2x \\
&= \frac{1}{2}\int \mu(\mathcal{F} - \Delta)^{-1}\mu\, d^2x =: H(\mu)
\end{aligned} \tag{2.124}
$$

with $\mu := q + f_1 - b_1 = (\mathcal{F} - \Delta)h$, provided the vector ∇h in Eq. (2.123) is normal to the domain boundary (so $\mathbf{u}_G$ is tangential there) and also provided the boundary integral of the normal derivative of $\partial_t h$ vanishes (Pedlosky, 1987).

Exercise 2.35. *Show that the QG motion in Eq. (2.116) yields the formal expression,*

$$\frac{\partial h}{\partial t} = -(\mathcal{F} - \Delta)^{-1}(\mathbf{u}_G \cdot \nabla q) = -(\mathcal{F} - \Delta)^{-1}J(h, q), \qquad (2.125)$$

where $(\mathcal{F} - \Delta)^{-1}$ denotes integration against the Green's function kernel $K(x, y)$ of the Helmholtz operator $(\mathcal{F} - \Delta)$. That is,

$$(\mathcal{F} - \Delta)^{-1}J = \int K(x, y)J(y)\, dy \quad and \quad (\mathcal{F} - \Delta)K(x, y) = \delta(x - y),$$

where $\delta(x - y)$ is the Dirac delta distribution, defined by $\int \delta(x - y)f(y)dy = f(x)$.

2.7.3 The ageostrophic velocity in the QG approximation: Part II

The gradient of Eq. (2.125) provides an estimate for the quantity

$$\partial_t (\hat{z} \times \mathbf{u}_G) = -\nabla h_{,t},$$

appearing in expression (2.112) for $\mathbf{u}_A$,

$$\mathbf{u}_A = (\mathbf{u}_G \cdot \nabla)\hat{z} \times \mathbf{u}_G - \nabla h_{,t} - f_1\mathbf{u}_G, \qquad (2.126)$$

which may therefore be written as,

$$\mathbf{u}_A = (\mathbf{u}_G \cdot \nabla)\hat{z} \times \mathbf{u}_G + \nabla(\mathcal{F} - \Delta)^{-1}J(h, q) - f_1\mathbf{u}_G, \qquad (2.127)$$

where $q = \mathcal{F}h - \Delta h + b_1 - f_1$, according to Eq. (2.117). Thus, in the QG approximation, the ageostrophic velocity $\mathbf{u}_A$ may be expressed via (2.125) entirely in terms of the geostrophic velocity $\mathbf{u}_G$ and other spatial derivatives of surface height elevation, h.

2.7.4 Rossby waves for QG and their dispersion relation

Exercise 2.36. *Show that steady solutions (q_e, h_e) of Eq. (2.116) satisfy $J(q_e, h_e) = 0$, so that potential vorticity q_e and elevation h_e are functionally related.*

Linearise the QG potential vorticity equation (Eq. (2.116)) around a steady solution h_e with $\hat{z} \times \nabla h_e = \mathbf{U} = U\hat{\mathbf{e}}_x = const$ and find the dispersion relation for the resulting wave equation.

Answer.

(a) For an equilibrium solution q_e satisfying $\partial q_e/\partial t = 0$, Eq. (2.116) implies $J(q_e, h_e) = 0$, which means that $q_e(x, y) = \mathcal{F}h_e - \Delta h_e + b_1 - f_1$ and $h_e(x, y)$ are functionally related, so their gradients are collinear. We shall assume that

$$\nabla q_e = -U\hat{\mathbf{y}} \quad \text{and} \quad \nabla h_e = -\beta\hat{\mathbf{y}},$$

where U and β are positive constants.

(b) Linearise Eq. (2.116) using $J(q, h) = \hat{z} \cdot \nabla q \times \nabla h$ as

$$\frac{\partial q'}{\partial t} = -\hat{z} \times \nabla h_e \cdot \nabla q' + \hat{z} \times \nabla q_e \cdot \nabla h'. \tag{2.128}$$

Then, insert $\hat{z} \times \nabla h_e = U\hat{\mathbf{e}}_x$ and $\hat{z} \times \nabla q_e = \beta\hat{\mathbf{e}}_x$, and select a solution proportional to $\exp(i(\mathbf{k} \cdot \mathbf{x} - \nu t))$ with $\mathbf{k} = (k, l)$, to find, upon using $q' = (\mathcal{F} - \Delta)h'$, that

$$\nu = Uk - \frac{\beta k}{k^2 + l^2 + \mathcal{F}}. \tag{2.129}$$

This is the dispersion relation for the linearised QG equation.

(c) The corresponding phase and group velocities are

$$c_p = \frac{\nu}{k} = U - \frac{\beta}{k^2 + l^2 + \mathcal{F}},$$
$$\text{and} \quad c_g = \frac{d\nu}{dk} = U + \frac{\beta(k^2 - l^2)}{(k^2 + l^2 + \mathcal{F})^2}. \tag{2.130}$$

Exercise 2.37.

(a) *Suppose we linearise Eq. (2.116) about a state of rest. How does the dispersion relation change?*

(b) *Plot the dispersion relation, discuss its zonal phase and group speeds.*

Answer. A state of rest for Eq. (2.116) give $h_e = const$. Linearising then gives

$$\frac{\partial q'}{\partial t} + \hat{z} \times \nabla h' \cdot \nabla \beta y = 0,$$

$$\text{or} \quad \frac{\partial q'}{\partial t} + \beta\partial_x h' = \partial_t(\mathcal{F}h' - \Delta h') + \beta\partial_x h' = 0. \tag{2.131}$$

(a) For a solution proportional to $\exp(i(\mathbf{k}\cdot\mathbf{x}-\nu t))$ with $\mathbf{k}=(k,l)$ this yields

$$\nu = -\frac{\beta k}{k^2 + l^2 + \mathcal{F}} \tag{2.132}$$

Thus, the dispersion relation for the linearised QG equation about a state of rest amounts to setting $U=0$ in the dispersion relation for a state moving with constant velocity. This means that moving into a frame of motion with constant velocity produces a Doppler shift of the wave frequency, corresponding to the Galilean transformation of adding the moving frame velocity to the phase or group velocity.

(b) The corresponding zonal ($l^2 = 0$) phase and group velocities are

$$c_p = \frac{\nu}{k} = -\frac{\beta}{k^2 + \mathcal{F}} \quad \text{and} \quad c_g = \frac{d\nu}{dk} = \frac{\beta(k^2 - \mathcal{F})}{(k^2 + \mathcal{F})^2}. \tag{2.133}$$

Thus, the dispersion relation has a peak for $k^2 = \mathcal{F}$, beyond which the slope changes sign, so the group velocity changes direction.

The generation of Rossby waves is important in producing the deflections observed in the jet stream in the stratosphere, for example. See, e.g., `http://en.wikipedia.org/wiki/File:Aerial_Superhighway.ogv` for simulations of Rossby waves on the Jet Stream.

There are many good discussions of the meanings of the dispersion relation $\nu(\mathbf{k})$, phase velocity c_p, and group velocity c_g in the literature Vallis (2006); Whitham (1974).

2.8 Fundamental Conservation Laws of the QG Equations

2.8.1 *Energy and circulation*

The conservation laws for energy and circulation are of great value in the analysis and understanding of their solution behaviour of the QG equations.

First, we consider the QG energy,

$$E_{\mathrm{QG}} = \frac{1}{2}\int h\left(\mathcal{F}h - \Delta h\right)d^2x = \frac{1}{2}\int \mathbf{u}_G \cdot \left(\mathbf{u}_G - \mathcal{F}\Delta^{-1}\mathbf{u}_G\right)d^2x, \tag{2.134}$$

which is conserved, provided the vector ∇h is normal to the domain boundary, so that the QG velocity $\mathbf{u}_G = \hat{z} \times \nabla h$ is tangent to the boundary. This may be seen by direct computation, as

$$\frac{dE_{QG}}{dt} = \int_S h \frac{\partial}{\partial t} (\mathcal{F}h - \Delta h + b_1(\mathbf{x}) - f_1(\mathbf{x})) \, d^2 x$$

$$= \int_S h \frac{\partial q}{\partial t} \, d^2 x = - \int_S h(-\mathbf{u}_G \cdot \nabla q) \, d^2 x = - \int_S h \nabla \cdot (\mathbf{u}_G q) \, d^2 x$$

$$= - \int_S \nabla \cdot (h \mathbf{u}_G q) \, d^2 x + \int_S \nabla h \cdot (\hat{z} \times \nabla h) \, d^2 x$$

(By Gauss) $= - \oint_{\partial S} \hat{n} \cdot (h \mathbf{u}_G q) \, ds = 0, \quad \text{provided} \quad \hat{n} \cdot \mathbf{u}_G \big|_{\partial S} = 0.$

Second, the Kelvin circulation integral for QG is defined by

$$K = \oint_{c(u_G)} (\epsilon \mathbf{u}_G - \epsilon \mathcal{F} \Delta^{-1} \mathbf{u}_G + \mathbf{R}(\mathbf{x})) \cdot d\mathbf{x}$$

$$= \int \int_{\partial S = c(u_G)} \left(\Delta h - \mathcal{F}h - b_1 + f_1 \right) d^2 x$$

$$= \int \int_{\partial S = c(u_G)} q \, d^2 x, \tag{2.135}$$

where

$$\nabla \times \mathbf{R}(\mathbf{x}) = \left(1 + \epsilon \big(f_1(\mathbf{x}) - b_1(\mathbf{x}) \big) \right) \hat{z} \tag{2.136}$$

and $c(u_G)$ is a closed planar loop moving with the fluid velocity $\mathbf{u}_G(\mathbf{x}, t)$, which also coincides with the boundary ∂S of the surface integral. The Kelvin circulation theorem for QG may also be computed directly, as

$$\frac{dK}{dt} = \frac{d}{dt} \int \int_{\partial S = c(u_G)} q \, d^2 x$$

$$= \int \int_{\partial S = c(u_G)} (\partial_t q + \mathbf{u}_G \cdot \nabla q + \nabla \cdot \mathbf{u}_G) \, d^2 x$$

$$= 0, \tag{2.137}$$

since $\partial_t q + \mathbf{u}_G \cdot \nabla q = 0$ and $\nabla \cdot \mathbf{u}_G = 0$.

2.8.2 *Potential vorticity*

The conservation of potential vorticity (PV) on QG fluid parcels was proved as Eq. (2.118) in Sec. 2.7.1. Namely,

$$\partial_t q + \mathbf{u}_G \cdot \nabla q = 0, \tag{2.138}$$

where PV is defined by using $\nabla \times \mathbf{u}_G = \Delta h$ as

$$q := -\hat{z} \cdot \nabla \times (\epsilon \mathbf{u}_G - \epsilon \mathcal{F} \Delta^{-1} \mathbf{u}_G + \mathbf{R}(\mathbf{x})) = \mathcal{F} h - \Delta h + b_1 - f_1. \tag{2.139}$$

An alternative derivation of the QG motion in Eq. (2.138) may be obtained by taking the curl of the following motion equation for the QG velocity $\mathbf{u}_G = \hat{z} \times \nabla h$,

$$\partial_t(\epsilon \mathbf{u}_G - \epsilon \mathcal{F} \Delta^{-1} \mathbf{u}_G + \mathbf{R}(\mathbf{x}))$$
$$- \mathbf{u}_G \times \nabla \times (\epsilon \mathbf{u}_G - \epsilon \mathcal{F} \Delta^{-1} \mathbf{u}_G + \mathbf{R}(\mathbf{x})) + \nabla \varpi = 0, \tag{2.140}$$

where $\nabla \times \mathbf{R}(\mathbf{x})$ is given in Eq. (2.137) and $\nabla \varpi$ is a pressure force.

This calculation is facilitated by the fundamental vector identity of fluid dynamics,

$$(\nabla \times \mathbf{a}) \times \mathbf{b} + \nabla(\mathbf{a} \cdot \mathbf{b}) = (\mathbf{b} \cdot \nabla)\mathbf{a} + a_j \nabla b^j. \tag{2.141}$$

Remarkably, QG possesses an infinite number of conserved integral quantities (called enstrophies), in the form

$$\frac{dC_\Phi}{dt} = 0 \quad \text{for} \quad C_\Phi = \int \Phi(q)\, d^2 x, \tag{2.142}$$

for any differentiable function Φ. To prove this statement, we compute directly,

$$\frac{d}{dt} \iint_S \Phi(q)\, d^2 x = \iint_S \partial_t \Phi(q)\, d^2 x = -\iint_S \mathbf{u}_G \cdot \nabla \Phi(q)\, d^2 x$$
$$= -\iint_S \nabla \cdot (\mathbf{u}_G \Phi(q))\, d^2 x$$
$$\text{(By Gauss)} \quad = -\oint_{\partial S} \hat{n} \cdot (\mathbf{u}_G \Phi(q))\, ds = 0, \quad \text{provided} \quad \hat{n} \cdot \mathbf{u}_G \big|_{\partial S} = 0.$$

2.8.3   *Casting QG into Hamiltonian form*

In this section, we show that the QG equation (Eq. (2.118)) may be written in Hamiltonian form,

$$\partial_t q = \{q, H\}, \tag{2.143}$$

for the Hamiltonian $H(\mu) = E_{\text{QG}}$ in Eq. (2.124) and a Poisson bracket $\{F, H\}$ among functionals of $\mu := q + f_1 - b_1 = (\mathcal{F} - \Delta)h$, given by

$$\{F, H\} = \int q \left\{ \frac{\delta F}{\delta \mu}, \frac{\delta H}{\delta \mu} \right\}_{\text{can}} dx dy. \tag{2.144}$$

Here, the variational derivative $\frac{\delta H}{\delta \mu}$ of a functional of μ is defined by the limit

$$\lim_{\epsilon \to 0} (\epsilon^{-1}(H(\mu + \epsilon \delta \mu) - H(\mu))) = \frac{d}{d\epsilon} \bigg|_{\epsilon=0} H(\mu + \epsilon \delta \mu)$$

$$=: \left\langle \frac{\delta H}{\delta \mu}, \delta \mu \right\rangle_{L^2}, \tag{2.145}$$

where the angle brackets $\langle \cdot, \cdot \rangle_{L^2}$ represent L^2 pairing of real functions.

To verify the Hamiltonian form of the QG equation (Eq. (2.143)), one may begin by recalling the cyclic permutation formula

$$\int a\{b, c\}_{\text{can}} dx dy = \int b\{c, a\}_{\text{can}} dx dy$$

for real functions a, b, c, on the (x, y) plane with homogeneous or periodic boundary conditions.

Then, one may verify that the variational derivative of the QG energy $H(\mu) = E_{\text{QG}}$ in Eq. (2.124) with respect to the quantity μ yields the surface height elevation. That is, verify the relation $\delta H/\delta \mu = h$ by using the definition of the variational derivative in Eq. (2.145) with $H(\mu)$ in Eq. (2.124).

Notice that the Poisson bracket in Eq. (2.144) is bilinear, skew-symmetric, and satisfies the Leibniz and Jacobi identities. In particular, it satisfies the Jacobi identity, because it is a linear functional of the canonical Poisson bracket $\{\cdot, \cdot\}_{\text{can}}$ in Eq. (2.121) which satisfies the Jacobi identity.

Using Eq. (2.144), one may compute the Poisson brackets $\{F, C_\Phi\}$ for the conserved quantities C_Φ in Eq. (2.142) with an arbitrary smooth functional F. In the next section, we will see that critical points of the conserved functional

$$H_\Phi(q) := E_{\mathrm{QG}}(q) + C_\Phi(q)$$

are equilibrium solutions of the QG equations. Perturbations of these critical points will produce Rossby waves and Poincaré gravity waves whose stability naturally depends on the choice of the function Φ in the conserved functional $C_\Phi(q)$. Later, we will discuss transformations of variables and asymptotic expansions that will allow us to see how these waves in the RSW solutions couple to the circulations in the QG approximation.

2.9 Equilibrium Solutions of QG

2.9.1 *Critical point solutions of QG*

As discussed earlier, two important properties of the QG equation (Eq. (2.116)) are conservation of energy E_{QG} in Eq. (2.134) and enstrophy $C_\Phi(q)$ in Eq. (2.106)

$$E_{\mathrm{QG}}(h) = \frac{1}{2}\int h(\mathcal{F}h - \Delta h)\,d^2x \quad\text{and}\quad C_\Phi(q) = \int \Phi(q)\,d^2x = 0, \quad (2.146)$$

provided the vector ∇h is normal to the domain boundary, so that the QG velocity $\mathbf{u}_G = \hat{z} \times \nabla h$ is tangent to the boundary. The potential vorticity q is related to the elevation h by

$$q = \mathcal{F}h - \Delta h + b_1 - f_1, \quad (2.147)$$

and it satisfies Eq. (2.138), recalled here as

$$\partial_t q + \mathbf{u}_G \cdot \nabla q = 0. \quad (2.148)$$

Therefore, the QG energy E_{QG} may be written in terms of the potential vorticity q as a quadratic form,

$$E_{\mathrm{QG}}(q) = \frac{1}{2}\int (q - b_1 + f_1)(\mathcal{F} - \Delta)^{-1}(q - b_1 + f_1)\,d^2x, \quad (2.149)$$

where we have used $h = (\mathcal{F} - \Delta)^{-1}(q - b_1 + f_1)$. We define the conserved functional $H_\Phi(q) := E_{\mathrm{QG}}(q) + C_\Phi(q)$ and find the condition for its first variation to vanish for a function q_e to be

$$\delta H_\Phi(q_e) = \frac{1}{2} \int (h_e + \Phi'(q_e)) \delta q \, d^2 x = 0. \tag{2.150}$$

The critical point condition for $\delta H_\Phi(q_e) = 0$ is thus

$$h_e + \Phi'(q_e) = 0, \tag{2.151}$$

for a given choice of the function Φ.

Theorem 2.6. *Critical points of the conserved functional* $H_\Phi(q) := E_{QG}(q) + C_\Phi(q)$ *are equilibrium solutions of the QG equations.*

Proof. By the critical point condition in Eq. (2.151), we have $\nabla h_e \times \nabla q_e = 0$ for $\Phi''(q_e) \neq 0$, because the functional relation between h_e and q_e implies that their gradients ∇h_e and ∇q_e are collinear. Thus, the PV evolution equation (Eq. (2.148)) implies that q_e is an equilibrium solution. Namely,

$$\partial_t q_e = -\hat{z} \times \nabla h_e \cdot \nabla q_e = -\hat{z} \cdot \nabla h_e \times \nabla q_e = 0. \tag{2.152}$$

$\square$

Exercise 2.38. *Explain how the stability of a QG equilibrium solution depends on the choice of the function Φ in the conserved functional $C_\Phi(q)$.*
    Hint: the second variation of the conserved functional $H_\Phi(q)$ is the Hamiltonian for the linearised evolution of perturbations in the neighbourhood of an equilibrium solution q_e.

2.10   Alternative Representations of RSW: Part I

2.10.1   *Vorticity, divergence and depth representation*

We transform variables in Eq. (2.89) from fluid velocity and depth $(\mathbf{u}, \eta)$, to vorticity, divergence and depth, $(\omega = \hat{z} \cdot \nabla \times \mathbf{u}, \mathbb{D} = \nabla \cdot \mathbf{u}, \eta)$. After introducing the operator

$$\left(\frac{\partial}{\partial t} + \partial_j u^j \right) = \left(\frac{D}{Dt} + \mathbb{D} \right),$$

so that $\quad \dfrac{\partial \mathbb{D}}{\partial t} + \partial_j \left(u^j \mathbb{D} \right) = \dfrac{d \mathbb{D}}{dt} + \mathbb{D}^2, \tag{2.153}$

the RSW equations (Eq. (2.89)) take the following forms in the variables $\omega = \hat{z} \cdot \nabla \times \mathbf{u}$, $\mathbb{D} = \nabla \cdot \mathbf{u}$ and η,

$$\left(\frac{D}{Dt} + \mathbb{D}\right)(\epsilon\omega + f) = 0,$$

$$\left(\frac{D}{Dt} + \mathbb{D}\right)\eta = 0,$$

(2.154)

$$\left(\frac{D}{Dt} + \mathbb{D}\right)\epsilon\mathbb{D} = -\nabla \cdot [f\hat{z} \times \mathbf{u} + \nabla h] + 2\epsilon J(u,v),$$

$$= :\Omega + 2\epsilon J(u,v),$$

where

$$\Omega := -\nabla \cdot (f\hat{z} \times \mathbf{u} + \nabla h), \quad \text{and} \quad J(u,v) := \hat{z} \cdot \nabla u \times \nabla v. \quad (2.155)$$

Here, $J(u,v)$ is the Jacobian of the velocity components $u(x,y)$ and $v(x,y)$.

(a) We shall assume that the velocity has a weighted Helmholtz decomposition

$$\mathbf{u} = \hat{z} \times \nabla\psi + \epsilon\nabla\chi. \quad (2.156)$$

Inserting Eq. (2.156) into Eq. (2.154) shows that with this assumption the quantity Ω may be expressed as

$$\Omega = -\nabla \cdot [f\hat{z} \times \mathbf{u} + \nabla h] = \nabla \cdot (f\nabla\psi) + \epsilon J(f,\chi) - \Delta h, \quad (2.157)$$

which is called the *imbalance* (Lynch, 1989; Warn *et al.*, 1995; Vallis, 1996a,b). (Why is this a good name for Ω?)

(b) In addition, since $\mathbb{D} = \epsilon\Delta\chi$ is order $O(\epsilon)$ once we assume Eq. (2.156), the quantity $\Omega + 2\epsilon J(u,v)$ in Eq. (2.154) must then be of order $O(\epsilon^2)$.

Exercise 2.39. *Explicitly transform variables from Eqs. (2.89)–(2.90) to Eqs. (2.154)–(2.157). Hint: For this calculation, you may want to recall that*

$$\partial_j(u^j u^i_{,i}) - \partial_i(u^j u^i_{,j}) = 2J(u^1, u^2) \quad \text{for} \quad \mathbf{u} = (u^1, u^2) = (u,v), \quad (2.158)$$

when you are taking the divergence of Eq. (2.89) in the form,

$$\partial_t \mathbf{u} = -\mathbf{u} \cdot \nabla\mathbf{u} - (f\hat{z} \times \mathbf{u} + \nabla h)).$$

For more insight, see the standard literature Lynch (1989), Warn et al. (1995), Vallis (1996a,b).

Exercise 2.40. *Prove Eq. (2.158) explicitly.*

The operator $D/Dt + \mathbb{D}$ has an integrating factor $\exp \int \mathbb{D} dt$, where the integral is taken at constant Lagrangian fluid-parcel label $l^A(\mathbf{x}, t)$, $A = 1, 2$, which satisfies,

$$\frac{Dl^A}{Dt} = \partial_t l^A + \mathbf{u} \cdot \nabla l^A = 0, \qquad (2.159)$$

since the fluid-parcel label is a Lagrangian tracer. Using this integrating factor in Eq. (2.154) gives

$$e^{-\int \mathbb{D} dt} \frac{D}{Dt} \left(e^{\int \mathbb{D} dt} (\epsilon \omega + f) \right) = 0,$$

$$e^{-\int \mathbb{D} dt} \frac{D}{Dt} \left(e^{\int \mathbb{D} dt} \eta \right) = 0, \qquad (2.160)$$

$$e^{-\int \mathbb{D} dt} \frac{D}{Dt} \left(e^{\int \mathbb{D} dt} \epsilon \mathbb{D} \right) = \Omega + 2\epsilon J(u, v).$$

Consequently, we find

$$e^{\int \mathbb{D} dt} (\epsilon \omega + f) = c_1(l^A),$$

$$e^{\int \mathbb{D} dt} \eta = c_2(l^A), \qquad (2.161)$$

$$\eta \frac{D}{Dt} \left(\frac{\epsilon \mathbb{D}}{\eta} \right) = \Omega + 2\epsilon J(u, v),$$

where c_1 and c_2 are functions of the Lagrangian labels l^A so they satisfy $Dc_1/Dt = 0 = Dc_2/Dt$. The ratio of the first pair of equations in Eq. (2.161) yields potential vorticity conservation,

$$q := \frac{\epsilon \omega + f}{\eta} = \frac{c_1(l^A)}{c_2(l^A)} \Rightarrow \frac{Dq}{Dt} = 0. \qquad (2.162)$$

We also find from Eqs. (2.154) and (2.161) that

$$\frac{Dq}{Dt} = 0, \quad \frac{D}{Dt} \left(\frac{1}{\eta} \right) = \frac{\mathbb{D}}{\eta}, \quad \frac{D}{Dt} \left(\frac{\epsilon \mathbb{D}}{\eta} \right) = \left(\Omega + 2\epsilon J(u, v) \right) \left(\frac{1}{\eta} \right).$$

$$(2.163)$$

Thus, the RSW equations transform without approximation (but at the cost of introducing higher spatial derivatives) to

$$\frac{Dq}{Dt} = 0 \quad \text{and} \quad \epsilon \frac{D^2}{Dt^2}\left(\frac{1}{\eta}\right) = \left(\Omega + 2\epsilon J(u,v)\right)\left(\frac{1}{\eta}\right). \tag{2.164}$$

According to these equations, the potential vorticity q is constant along a fluid parcel trajectory and the inverse depth $1/\eta$ either oscillates stably or evolves exponentially, depending on the sign of the quantity $\Omega + 2\epsilon J(u,v)$. Of course, advection of q is well known. The oscillator equation for $1/\eta$ in Eq. (2.164) can be interpreted in a mixed Eulerian–Lagrangian fashion as saying that sufficient convergence of Eulerian force causes Lagrangian (advective) instability of the water depth.

Thus, the RSW equations may be separated into vortical motions and Rossby waves in q and Lagrangian oscillations in $1/\eta$, with no approximation. Recall that $\mathbb{D}$ is order $O(\epsilon)$ and the quantity $\Omega + 2\epsilon J(u,v)$ is of order $O(\epsilon^2)$. So the second equation in Eq. (2.164) is consistent with order $O(1)$ dynamics of η in Eq. (2.90) and order $O(\epsilon)$ variations in bottom topography

$$\eta = 1 + \epsilon(\mathcal{F}h + b_1). \tag{2.165}$$

Unfortunately, there is no corresponding separation of time scales in Eq. (2.164).

Reduction of the q-equation in Eq. (2.164) to quasigeostrophy (QG).

For $\epsilon \ll 1$ in the potential vorticity equation in Eq. (2.164), further assumptions can be imposed which decouple the slow vortical motion from the fast gravity waves. In this limit, one may write the Taylor expansion

$$q = \frac{1 + \epsilon(\omega + f_1)}{1 + \epsilon(\mathcal{F}h + b_1)} = 1 + \epsilon(\omega - \mathcal{F}h + f_1 - b_1) + O(\epsilon^2). \tag{2.166}$$

Thus, with $\omega = \Delta\psi$ and $\mathbf{u} = \hat{z} \times \nabla\psi + \epsilon\nabla\chi$, the equation in Eq. (2.164) for the potential vorticity becomes

$$\frac{\partial}{\partial t}\left(\Delta\psi - \mathcal{F}h\right) + J\left(\Delta\psi - \mathcal{F}h + f_1 - b_1, \psi\right) = O(\epsilon). \tag{2.167}$$

The further assumption that

$$\mathbf{u} = \mathbf{u}_G + \epsilon\mathbf{u}_A + O(\epsilon^2),$$

with $\mathbf{u}_G = \hat{z} \times \nabla h$, so that $\omega = \Delta\psi = \Delta h + O(\epsilon)$ reduces Eq. (2.167) to the QG motion equation (Allen and Holm, 1996; Pedlosky, 1987),

$$\frac{\partial}{\partial t}(\Delta h - \mathcal{F}h) + J(\Delta h - \mathcal{F}h + f_1 - b_1, h) = O(\epsilon),$$

$$\text{or} \quad \frac{\partial q}{\partial t} + \mathbf{u}_G \cdot \nabla q = O(\epsilon), \tag{2.168}$$

when terms of order $O(\epsilon)$ are neglected.

Exercise 2.41. *Follow the reasoning above to derive Eq. (2.168) explicitly and find its conserved energy.*

Estimating the imbalance $\Omega + 2\epsilon J(u,v) = O(\epsilon^2)$ using QG theory.

As we have seen, QG theory (Allen and Holm, 1996; Pedlosky, 1987) sets

$$\mathbf{u} = \mathbf{u}_G + \epsilon\mathbf{u}_A + O(\epsilon^2), \tag{2.169}$$

with geostrophic and ageostrophic velocities given, respectively by

$$\mathbf{u}_G = \hat{z} \times \nabla h \quad \text{and}$$

$$\mathbf{u}_A = -f_1\mathbf{u}_G + (\partial_t + \mathbf{u}_G \cdot \nabla)\hat{z} \times \mathbf{u}_G. \tag{2.170}$$

Consequently, one finds $\mathbb{D} = \epsilon\nabla\cdot\mathbf{u}_A$ and

$$\Omega + 2\epsilon J(u,v) = \epsilon\nabla\cdot[(\partial_t + \mathbf{u}\cdot\nabla)\mathbf{u}] + 2\epsilon J(u,v), \tag{2.171}$$

$$\text{Using Eq. (2.169)} \quad = \epsilon\nabla\cdot[(\partial_t + \mathbf{u}_G \cdot \nabla)\mathbf{u}_G]$$

$$+ \epsilon^2\nabla\cdot[(\partial_t + \mathbf{u}_G \cdot \nabla)\mathbf{u}_A + (\mathbf{u}_A \cdot \nabla)\mathbf{u}_G]$$

$$+ 2\epsilon J(u_G, v_G) + 2\epsilon^2(J(u_A, v_G) + J(u_G, v_A)) + O(\epsilon^3)$$

$$= \epsilon^2(\partial_t + \mathbf{u}_G \cdot \nabla)\nabla\cdot\mathbf{u}_A + O(\epsilon^3), \tag{2.172}$$

after cancellations at both orders $O(\epsilon)$ and $O(\epsilon^2)$.

The QG theory also gives

$$\frac{1}{\epsilon}\frac{D\eta}{Dt} = (\partial_t + \mathbf{u}_G \cdot \nabla)(\mathcal{F}h + b_1) + O(\epsilon) = -\nabla\cdot\mathbf{u}_A$$

$$= (\partial_t + \mathbf{u}_G \cdot \nabla)(\Delta h + f_1(x,y)), \tag{2.173}$$

which recovers the previous asymptotic Eq. (2.168) when terms of order $O(\epsilon)$ are neglected.

Exercise 2.42. *Verify the computations in Eqs. (2.172)) and (2.173).*

2.11 Alternative Representations of RSW: Part II

2.11.1 *Slow + fast decomposition*

Transformed variables for Rotating Shallow Water: Still with no approximation, we can transform the RSW equations into the following set of variables,

$$\omega := \hat{z} \cdot \nabla \times \mathbf{u}, \quad \mathbb{D} := \nabla \cdot \mathbf{u},$$

$$\Omega := -\nabla \cdot [f\hat{z} \times \mathbf{u} + \nabla h] = \nabla \cdot (f\nabla\psi) + \epsilon J(f,\chi) - \Delta h, \tag{2.174}$$

where $\mathbf{u} = \hat{z} \times \nabla\psi + \epsilon\nabla\chi$ as in Eq. (2.95). In these variables, the RSW equations from Eq. (2.163) may be written, without approximation, as follows:

$$\partial_t \eta = -\nabla \cdot (\eta \mathbf{u}), \quad \text{for total depth} \quad \eta = 1 + \epsilon b_1 + \epsilon \mathcal{F} h$$

$$\partial_t q = -\mathbf{u} \cdot \nabla q, \quad \text{for PV} \quad q = \frac{\epsilon\omega + f}{\eta},$$

$$\partial_t \mathbb{D} - \frac{1}{\epsilon}\Omega = -\nabla \cdot (\mathbb{D}\mathbf{u}) + 2J(u,v)$$

$$\partial_t \Omega - \frac{1}{\epsilon\mathcal{F}}(\Delta - \mathcal{F})\mathbb{D} = \Delta\nabla \cdot \left(\left(h + \frac{b_1}{\mathcal{F}}\right)\mathbf{u}\right) - \frac{1}{\epsilon}\nabla \cdot ((f^2 + \epsilon\omega f - 1)\mathbf{u})$$

$$- J(\epsilon f_1, h + |\mathbf{u}|^2). \tag{2.175}$$

Exercise 2.43. *Verify the calculation required to obtain Eq. (2.175) from Eq. (2.163).*

Hint: It may be helpful to notice that for constant rotation and flat bottom topography one has $f_1 = 0$ and $b_1 = 0$. In this case, the imbalance Ω simplifies as $\Omega \to \Delta\psi - \Delta h$, and the last equation in Eq. (2.175) simplifies to

$$\frac{\partial}{\partial t}(\Delta\psi - \Delta h) - \frac{1}{ep\mathcal{F}}(\Delta - \mathcal{F})\mathbb{D} = \Delta(\nabla \cdot (h\mathbf{u})) - \nabla \cdot ((\Delta\psi)\mathbf{u}). \tag{2.176}$$

Two other convenient formulas which are useful to prove for this exercise are

$$-\frac{\partial}{\partial t}\Delta h = \frac{1}{\epsilon\mathcal{F}}\Delta\mathbb{D} + \frac{1}{\mathcal{F}}\Delta(\nabla \cdot ((b_1 + \mathcal{F}h)\mathbf{u})),$$

$$-\frac{\partial}{\partial t}\Delta\psi = \nabla \cdot ((\omega + f_1)\mathbf{u}) + \frac{\mathbb{D}}{\epsilon}. \tag{2.177}$$

Klein–Gordon equations: In terms of the fast time variable t/ϵ and up to order $O(\epsilon)$, the quantities $\mathbb{D}$ and Ω in Eq. (2.175) satisfy identical linear Klein–Gordon equations

$$[\partial^2_{t/\epsilon} - (\mathcal{F}^{-1}\Delta - 1)]\mathbb{D} = O(\epsilon) + O(\epsilon^2),$$
$$[\partial^2_{t/\epsilon} - (\mathcal{F}^{-1}\Delta - 1)]\Omega = O(\epsilon) + O(\epsilon^2),$$
(2.178)

corresponding to rapidly fluctuating Poincaré-gravity waves of the RSW system (Eq. (2.175)), with fast frequency $\nu \approx \partial_{t/\epsilon}$ of order $O(1/\epsilon)$, driven by order $O(\epsilon)$ and $O(\epsilon^2)$ nonlinear slow + fast forcing terms on the right hand sides. Upon ignoring the nonlinear slow + fast forcing terms, the RSW linear Poincaré-gravity waves satisfy the dispersion relation,

$$\nu^2 = 1 + \mathcal{F}^{-1}k^2, \quad \text{with } c_p = \frac{\nu}{k} = \pm\sqrt{\mathcal{F}^{-1} + \frac{1}{k^2}}, \tag{2.179}$$

which admits both leftward and rightward travelling waves.

Exercise 2.44. *The longer the wavelength, the greater the group velocity of shallow water waves. Compute the group velocity for the dispersion relation in Eq. (2.179).*

The limiting linear PDE for RSW waves for $\epsilon \ll 1$ governing both $\mathbb{D}$ and Ω in Eq. (2.178)

$$\partial^2_{t/\epsilon}\phi - \mathcal{F}^{-1}\Delta\phi = \phi, \quad \text{for either} \quad \phi = \mathbb{D} \quad \text{or} \quad \phi = \Omega \tag{2.180}$$

is the celebrated *Klein–Gordon* (KG) equation. The KG equation is a relativistic version of the Schrödinger equation. Although KG was discovered a long time ago, it has received renewed interest in physics lately, because describes a spin-zero elementary particle, the famous Higgs boson, whose existence was verified at CERN in 2012. For a good account of its history and a few background references, see http://en.wikipedia.org/wiki/Klein-Gordon_equation.

Remark 2.9. The leading order fast-time equations for the $\mathbb{D}$ and Ω system in Eq. (2.175) are expressed as by

$$\frac{\partial}{\partial(t/\epsilon)}\begin{bmatrix} \mathbb{D} \\ \Omega \end{bmatrix} = \begin{bmatrix} 0 & 1 \\ -1 & 0 \end{bmatrix}\begin{bmatrix} (-\mathcal{F}^{-1}\Delta + 1)\mathbb{D} \\ \Omega \end{bmatrix} = \begin{bmatrix} 0 & 1 \\ -1 & 0 \end{bmatrix}\begin{bmatrix} \delta\mathcal{H}/\delta\mathbb{D} \\ \delta\mathcal{H}/\delta\Omega \end{bmatrix}.$$

Thus, $\mathbb{D}$ and Ω appear as canonically conjugate variables in a Hamiltonian system for the fast-time dynamics in t/ϵ with Hamiltonian $\mathcal{H}$ given by

$$\mathcal{H} = \int \left(\frac{1}{2\mathcal{F}} |\nabla \mathbb{D}|^2 + \frac{1}{2}\mathbb{D}^2 + \frac{1}{2}\Omega^2 \right) dx dy \qquad (2.181)$$

and canonical Poisson bracket given for functionals $\mathcal{G}$ and $\mathcal{H}$ of $\mathbb{D}$ and Ω by

$$\{\mathcal{G}, \mathcal{H}\} = \int \begin{bmatrix} \delta\mathcal{G}/\delta\mathbb{D} \\ \delta\mathcal{G}/\delta\Omega \end{bmatrix} \begin{bmatrix} 0 & 1 \\ -1 & 0 \end{bmatrix} \begin{bmatrix} \delta\mathcal{H}/\delta\mathbb{D} \\ \delta\mathcal{H}/\delta\Omega \end{bmatrix} dx dy. \qquad (2.182)$$

At orders $O(\epsilon)$ and $O(\epsilon^2)$, the right-hand sides for $\epsilon \neq 0$ of Eq. (2.175) are coupled to the QG equation (Eq. (2.168)). As it turns out, the coupled system of equations in Eqs. (2.168) and (2.175) is also Hamiltonian (Allen and Holm, 1996). This is not surprising, because the original RSW system (Eq. (2.89)) is Hamiltonian, as well.

Remark 2.10. This observation suggests a two-timing approach $(t, t/\epsilon)$ in which, upon averaging over the fast time, $\mathcal{H}$ in Eq. (2.181) would emerge as an adiabatic invariant. For an extensive review of multiple time scale expansions from the current viewpoint, see, for example, Klein (2010).

Remark 2.11. The slow + fast decomposition of the RSW solution into potential vorticity $q(t)$ governed by the QG equation (Eq. (2.168)) for the potential vorticity, interacting with wave variables $\mathbb{D}(t, t/\epsilon)$ and $\Omega(t, t/\epsilon)$ governed by coupled KG equations in Eq. (2.178), is the basis for many possible approximations in RSW dynamics. For example, the initialisation in a "balanced" state to suppress the time derivatives on the left sides of the last two equations in this set gives the "slow equations" due to Peter Lynch. See, for example, Lynch (1989), as well as Warn *et al.* (1995) and Browning and Kreiss (1987); Browning *et al.* (1980). For more information about such approximate "reduced" equations, read Chapter 4 and 5 of Vallis (2006). For more information about linear and nonlinear waves, see Whitham (1974).

2.12 Hamilton's Principle for Simple Ideal Fluids

2.12.1 *Preparation for fluid dynamical variational principles*

Definition 2.10. The *variational derivative* $\frac{\delta F}{\delta \psi} \in V^*(M)$ of a real functional $F[\psi]$ of smooth functions $\psi \in \mathcal{F}(M)$ taking values in a vector space

$V(M)$ over a manifold M is defined by

$$\delta F[\psi] = \lim_{\varepsilon \to 0} \left(F[\psi + \varepsilon \delta \psi] - F[\psi] \right) = \left\langle \frac{\delta F}{\delta \psi}, \delta \psi \right\rangle$$

$$\text{for} \quad \delta \psi \in \mathcal{F}(M). \tag{2.183}$$

The angle brackets $\langle \cdot, \cdot \rangle$ here denote L^2 pairing, as in

$$\langle f, g \rangle = \int \langle f(x), h(x) \rangle_{V^* \times V} \, dx, \tag{2.184}$$

for integrable real functions $f \in V$ and $h \in V^*$, and pairing $\langle \cdot, \cdot \rangle_{V^* \times V} : V^* \times V \to \mathbb{R}$, for a vector space V and its dual vector space V^*.

This definition of variational derivative applies for fluid dynamics, for example, with velocity $\mathbf{u}$ and depth (or density) $(\mathbf{u}, D) \in \mathfrak{X}(\mathbb{R}^2) \times \mathrm{Dens}(\mathbb{R}^2)$, since both $\mathfrak{X}(\mathbb{R}^2)$ and $\mathrm{Dens}(\mathbb{R}^2)$ are vector spaces.

Our strategy in applying Hamilton's principle $\delta S = 0$ with $S = \int l(\mathbf{u}, D) \, dt$ to derive ideal GFD approximations will be to perform variations at fixed $\mathbf{x}$ and t of the following action integral (Holm and Kupershmidt, 1983; Holm *et al.*, 1983; Holm, 1996),

$$S = \int_0^T l(\mathbf{u}, D) \, dt, \tag{2.185}$$

whose Lagrangian $l(\mathbf{u}, D)$ depends on the horizontal fluid velocity $\mathbf{u}$ and the total depth D. We will first do the general case for an arbitrary choice of Lagrangian $l : \mathfrak{X}(\mathbb{R}^2) \times \mathrm{Dens}(\mathbb{R}^2) \to \mathbb{R}$, for fluid velocity defined as a vector field over the plane $\mathbb{R}^2$, so that $\mathbf{u} \in \mathfrak{X}(\mathbb{R}^2)$, and depth defined as a density $D \in \mathrm{Dens}(\mathbb{R}^2)$, so that variations in depth conserve the volume of water. That is, D satisfies the continuity equation,

$$\partial_t D + \nabla \cdot (D\mathbf{u}) = 0. \tag{2.186}$$

2.12.2   *Explicitly varying the action integral*

Hamilton's principle $\delta S = 0$ for the action S in Eq. (2.185) is derived by taking the variations,

$$0 = \delta S = \int_0^T \left(\left\langle \frac{\delta l}{\delta \mathbf{u}}, \delta \mathbf{u} \right\rangle + \left\langle \frac{\delta l}{\delta D}, \delta D \right\rangle \right) dt. \tag{2.187}$$

For fluid dynamics, the variations of the fluid velocity vector field $\mathbf{u}$ and the mass density D are given by

$$\delta\mathbf{u} = \frac{d}{d\varepsilon}\bigg|_{\varepsilon=0} \mathbf{u} = \partial_t\mathbf{v} + \mathbf{u}\cdot\nabla\mathbf{v} - \mathbf{v}\cdot\nabla\mathbf{u}$$

$$\text{and}\quad \delta D = \frac{d}{d\varepsilon}\bigg|_{\varepsilon=0} D = -\nabla\cdot(D\mathbf{v}), \tag{2.188}$$

where the variational vector field $\mathbf{v} \in \mathfrak{X}(\mathbb{R}^2)$ is assumed to vanish at the endpoints in time $[0,T]$.

2.12.3 *Deriving the Euler–Poincaré motion equation*

Substituting the expressions in Eq. (2.188) for the variations $\delta\mathbf{u}$ and δD into Hamilton's principle in Eq. (2.187) yields

$$\begin{aligned}
0 = \delta S &= \int_0^T \left\langle \frac{\delta l}{\delta u^i}, \partial_t v^i + u^j\partial_j v^i - v^j\partial_j u^i \right\rangle + \left\langle \frac{\delta l}{\delta D}, -\partial_i(Dv^i) \right\rangle dt \\
&= \int_0^T \left\langle -\partial_t\frac{\delta l}{\delta u^i} - \partial_j\left(\frac{\delta l}{\delta u^i}u^j\right) - \frac{\delta l}{\delta u^k}\partial_i u^k + D\partial_i\frac{\delta l}{\delta D}, v^i \right\rangle dt \\
&\quad + \left\langle \frac{\delta l}{\delta u^i}, v^i \right\rangle\bigg|_0^T,
\end{aligned} \tag{2.189}$$

where we have invoked natural boundary conditions ($\hat{\mathbf{n}}\cdot\mathbf{u} = 0$ on the boundary) when integrating by parts in space, so the corresponding boundary terms vanish. The last term vanishes, as well, because we have assumed that the variational vector field $\mathbf{v}$ vanishes at the endpoints in time. Consequently, for independent variational vector fields $\mathbf{v}$ we find from Eq. (2.189) that Hamilton's principle for fluids implies the following *Euler–Poincaré equation* for ideal fluid dynamics

$$\partial_t\frac{\delta l}{\delta u^i} + \partial_j\left(\frac{\delta l}{\delta u^i}u^j\right) + \frac{\delta l}{\delta u^k}\partial_i u^k = D\partial_i\frac{\delta l}{\delta D}. \tag{2.190}$$

To complete the dynamical system, we also have the auxiliary equation (Eq. (2.186)), rewritten in components now in three dimensions as

$$\partial_t D + \partial_j(Du^j) = 0. \tag{2.191}$$

Exercise 2.45. *Verify Eq. (2.189) by substituting Eq. (2.188) into Eq. (2.187) and integrating by parts.*

Remark 2.12. The class of equations derived in this section are called Euler–Poincaré equations. For an extensive discussion of Euler–Poincaré equations for fluid dynamics, see Holm *et al.* (1998, 2000).

2.13 Hamilton's Principle for RSW

2.13.1 *Properties of the Euler–Poincaré motion equation*

Hamilton's principle in Eq. (2.187) has produced in Eq. (2.190) an example of the *Euler–Poincaré motion equation* for fluids (Holm *et al.*, 1998). This equation may be expressed in three-dimensional vector form as

$$\frac{D}{Dt}\frac{1}{D}\frac{\delta l}{\delta \mathbf{u}} + \frac{1}{D}\frac{\delta l}{\delta u^j}\nabla u^j - \nabla\frac{\delta l}{\delta D} = 0, \tag{2.192}$$

upon using the continuity Eq. (2.186) for D to simplify the components form of the equation in Eq. (2.190).

One may also write Eq. (2.192) equivalently in three dimensional vector notation as,

$$\frac{\partial}{\partial t}\left(\frac{1}{D}\frac{\delta l}{\delta \mathbf{u}}\right) - \mathbf{u}\times\nabla\times\left(\frac{1}{D}\frac{\delta l}{\delta \mathbf{u}}\right) + \nabla\left(\mathbf{u}\cdot\frac{1}{D}\frac{\delta l}{\delta \mathbf{u}} - \frac{\delta l}{\delta D}\right) = 0. \tag{2.193}$$

In writing the last equation, we have again used the fundamental vector identity of fluid dynamics, recalled from Eq. (2.97),

$$(\mathbf{b}\cdot\nabla)\mathbf{a} + a_j\nabla b^j = -\mathbf{b}\times(\nabla\times\mathbf{a}) + \nabla(\mathbf{b}\cdot\mathbf{a}), \tag{2.194}$$

for any three dimensional vectors $\mathbf{a}, \mathbf{b}\in\mathbb{R}^3$ with, in this case, $\mathbf{a} = (\frac{1}{D}\frac{\delta l}{\delta \mathbf{u}})$ and $\mathbf{b} = \mathbf{u}$.

Exercise 2.46 (Kelvin–Noether circulation theorem). *Prove that Eq. (2.193) implies Kelvin's conservation law for circulation*

$$\frac{d}{dt}\oint_{c(u)}\frac{1}{D}\frac{\delta l}{\delta \mathbf{u}}\cdot d\mathbf{x} = 0, \tag{2.195}$$

where $c(u)$ is a closed loop moving with the fluid velocity $\mathbf{u}(\mathbf{x},t)$ in three dimensions.

This form of the Kelvin circulation theorem (Holm et al., 1998, 2000) generalises the result for RSW in Eq. (2.103) to the case of an arbitrary ideal compressible fluid moving in three dimensions.

Exercise 2.47 (Energy conservation). *Verify that Eq. (2.192) or Eq. (2.193) with the continuity equation (Eq. (2.186)) conserve the energy*

$$E(\mathbf{u}, D) = \left\langle \frac{\delta l}{\delta u^j}, u^j \right\rangle - l(\mathbf{u}, D). \tag{2.196}$$

Explain why energy conservation is to be expected from Hamilton's principle $\delta S = 0$ for the Lagrangian $l(\mathbf{u}, D)$. Hint: a symmetry of the Lagrangian is involved.

2.13.2 Specialising to the RSW equation

We now specialise the Lagrangian to the RSW case, by choosing, cf. Eq. (2.102),

$$S = \int_0^T l(\mathbf{u}, D) \, dt$$

$$= \int_0^T \int \frac{\epsilon}{2} D|\mathbf{u}|^2 + D\mathbf{u} \cdot \mathbf{R}(\mathbf{x}) - \frac{(D - b(\mathbf{x}))^2}{2\epsilon \mathcal{F}} \, d^2x \, dt, \tag{2.197}$$

where $\nabla \times \mathbf{R}(\mathbf{x}) = f(\mathbf{x})\hat{\mathbf{z}}$. In this case, taking variations yields

$$0 = \delta S$$

$$= \int_0^T \int D(\epsilon \mathbf{u} + \mathbf{R}(\mathbf{x})) \cdot \delta \mathbf{u}$$

$$+ \left(\frac{\epsilon}{2}|\mathbf{u}|^2 + \mathbf{u} \cdot \mathbf{R}(\mathbf{x}) - \frac{D - b(\mathbf{x})}{\epsilon \mathcal{F}} \right) \delta D \, d^2x \, dt, \tag{2.198}$$

where we denote $h := (D - b(\mathbf{x}))/\epsilon \mathcal{F}$. Substituting these variational derivatives into the Euler-Poincaré motion equation (Eq. (2.193)) yields

$$\partial_t(\epsilon \mathbf{u} + \mathbf{R}(\mathbf{x})) - \mathbf{u} \times \nabla \times (\epsilon \mathbf{u} + \mathbf{R}(\mathbf{x})) + \nabla \left(\frac{\epsilon}{2}|\mathbf{u}|^2 + \frac{D - b(\mathbf{x})}{\epsilon \mathcal{F}} \right) = 0. \tag{2.199}$$

This recovers the RSW motion equation in its curl form (Eq. (2.96)). Thus, the RSW motion equation (Eq. (2.96)) is an Euler–Poincaré motion equation (Eq. (2.193)), which may be derived from Hamilton's principle $\delta S = 0$ with action integral in the form Eq. (2.185), $S = \int l(\mathbf{u}, D) \, dt$ with Lagrangian $l(\mathbf{u}, D)$ given in Eq. (2.197).

Exercise 2.48. *Verify Eq. (2.199) by substituting the variational derivatives in Eq. (2.198) into the Euler–Poincaré motion equation, Eq. (2.193).*

2.13.3   *Hamiltonian formulation of the Euler–Poincaré equations*

The conserved energy for the EP equations in Eq. (2.196) may be rewritten in the following form

$$h(\mathbf{m}, D) = \langle m_j, u^j \rangle - l(\mathbf{u}, D) \quad \text{with components} \quad m_j := \frac{\delta l}{\delta u^j}, \qquad (2.200)$$

in which the angle brackets $\langle \cdot, \cdot \rangle$ still denote the L^2 pairing in Eq. (2.184). In this form, the conserved energy provides the Legendre transformation from the Lagrangian $l(\mathbf{u}, D)$ to the Hamiltonian $h(\mathbf{m}, D)$, in which fluid momentum density is defined by $\mathbf{m} := \delta l/\delta \mathbf{u}$. Taking variations of both sides of the Legendre transformation in Eq. (2.200) yields

$$\delta h(\mathbf{m}, D) = \left\langle \frac{\delta h}{\delta m_j}, \delta m_j \right\rangle + \left\langle \frac{\delta h}{\delta D}, \delta D \right\rangle$$

$$= \langle \delta m_j, u^j \rangle + \left\langle m_j - \frac{\delta l}{\delta u^j}, \delta u^j \right\rangle + \left\langle -\frac{\delta l}{\delta D}, \delta D \right\rangle. \qquad (2.201)$$

Identifying coefficients of the independent variations δm_j, δu^j and δD in Eq. (2.201) then yields the following variational relations,

$$\frac{\delta h}{\delta m_j} = u^j, \quad \frac{\delta h}{\delta D} = -\frac{\delta l}{\delta D} \quad \text{and} \quad m_j := \frac{\delta l}{\delta u^j}. \qquad (2.202)$$

In terms of the variables m_i, $u^j = \delta h/\delta m_j$ and D, we may write the Euler–Poincaré equation (Eq. (2.190)) and its auxiliary continuity equation, respectively, as

$$\partial_t m_i + \partial_j \left(m_i \frac{\delta h}{\delta m_j} \right) + m_j \partial_i \frac{\delta h}{\delta m_j} = D \partial_i \frac{\delta l}{\delta D}, \qquad (2.203)$$

and

$$\partial_t D + \partial_j \left(D \frac{\delta h}{\delta m_j} \right) = 0. \qquad (2.204)$$

In turn, we may rewrite these component equations in skew-symmetric matrix operator form as

$$\partial_t \begin{bmatrix} m_i \\ D \end{bmatrix} = - \begin{bmatrix} (\partial_j m_i + m_j \partial_i) & D \partial_i \\ \partial_j D & 0 \end{bmatrix} \begin{bmatrix} \delta h/\delta m_j \\ \delta h/\delta D \end{bmatrix}. \qquad (2.205)$$

This skew-symmetric matrix representation implies a Hamiltonian formulation in terms of the Lie–Poisson bracket

$$\partial_t g = \{g, h\}$$

$$= -\int \begin{bmatrix} \delta g/\delta m_i \\ \delta g/\delta D \end{bmatrix}^T \begin{bmatrix} (\partial_j m_i + m_j \partial_i) & D\partial_i \\ \partial_j D & 0 \end{bmatrix} \begin{bmatrix} \delta h/\delta m_j \\ \delta h/\delta D \end{bmatrix} d^n x \quad (2.206)$$

for any differentiable functionals g and h. For the case of RSW with Lagrangian in Eq. (2.197), we have from the variational relations from Eq. (2.198) that

$$\mathbf{m} := \frac{\delta l}{\delta \mathbf{u}} = D(\epsilon \mathbf{u} + \mathbf{R}(\mathbf{x})) \quad \text{and}$$

$$\frac{\delta h}{\delta D} = -\frac{\delta l}{\delta D} = -\left(\frac{\epsilon}{2}|\mathbf{u}|^2 + \mathbf{u} \cdot \mathbf{R}(\mathbf{x}) - \frac{D - b(\mathbf{x})}{\epsilon \mathcal{F}}\right). \quad (2.207)$$

Exercise 2.49. *Derive the Hamiltonian for RSW by computing its Legendre transformation from the Lagrangian in Eq. (2.197).*

Exercise 2.50. *Verify the results of the Euler–Poincaré theory introduced here, by substituting the variational derivatives (Eq. (2.207)) into Hamiltonian matrix form in Eq. (2.205) to recover the RSW equations in Hamiltonian form.*

The Hamiltonian formulation of ideal fluid dynamics in terms of Lie–Poisson brackets has proven its utility many times in studying the qualitative and quantitative properties of GFD during the past 50 years. It is now part of the toolbox of every mathematician studying ideal fluid dynamics and it has accumulated a vast literature which is constantly being rediscovered.

The Euler–Poincaré approach discussed here for passing from Hamilton's variational principle to the Hamiltonian formulation of the fluid equations guarantees preservation of two key properties of the GFD balances which are responsible for large-scale ocean and atmosphere circulations. Namely:

(a) The Kelvin circulation theorem, leading to proper potential-vorticity (PV) dynamics; and
(b) The law of energy conservation.

These two important properties are preserved in the EP approach for Hamilton's-principle asymptotics at *every level of approximation*. However,

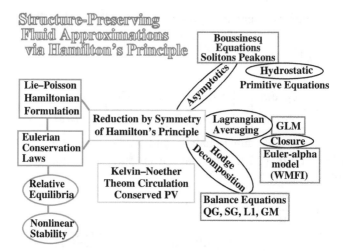

Figure 2.1. Asymptotics and averaging in Hamilton's principle in the Euler–Poincaré framework produces fluid approximations for GFD that preserve fundamental mathematical structures such as the Kelvin–Noether theorem for circulation leading to conservation of energy and potential vorticity (PV). Legendre transforming the resulting Euler–Poincaré Lagrangian yields the Lie–Poisson Hamiltonian formulation of geophysical fluid dynamics and its Eulerian conservation laws, which may be used to classify steady solutions as relative equilibria and determine sufficient conditions for their nonlinear stability(Holm *et al.*, 1985).

they have often been *lost* when using the standard asymptotic expansions of the fluid equations.

The models pictured in Fig. 2.1 form the *main sequence* of GFD model equations and they follow from the Euler–Poincaré form of Hamilton's principle for a fluid Lagrangian that depends parametrically on the advected quantities such as mass, salt and heat, all carried as material properties of the fluid's motion. Thus, the Euler–Poincaré theorem with advected quantities systematically selects and derives the useful GFD fluid models possessing the two main properties of energy balance and the circulation theorem. For fundamental references, see Holm *et al.* (1998, 2000).

2.13.4 *Thermal rotating shallow water*

While the standard RSW model applies to a single layer of homogeneous fluid moving in vertical columns, TRSW theory permits horizontal variations of the thermodynamic properties of the fluid within each layer. The TRSW model, we shall study here is a standard thermodynamic extension of the RSW model equations we have already studied in this chapter which allows for an upper active layer of fluid motion and an inert lower layer. Since the lower layer is inert, this TRSW model is often called a 1.5 layer model (Warneford and Dellar, 2013). Multilayer non-hydrostatic treatments have also been given (Cotter *et al.*, 2010).

The TRSW equations are expressed in terms of the square root $\gamma = \sqrt{\theta}$ of the (non-negative) buoyancy $\theta(\mathbf{x}, t) = (\bar{\rho} - \rho(\mathbf{x}, t))/\bar{\rho}$, where ρ is the (time and space dependent) mass density of the active upper layer, $\bar{\rho}$ is the uniform mass density of the inert lower layer, $D = D(\mathbf{x}, t)$ is the thickness of the active layer, $\mathbf{x}$ is the horizontal vector position, and t is time. The governing TRSW equations are

$$\epsilon \frac{D}{Dt} \mathbf{u} + f\hat{z} \times \mathbf{u} + \gamma \nabla(D\gamma) = 0,$$

$$\frac{\partial D}{\partial t} + \nabla \cdot (D\mathbf{u}) = 0, \quad \frac{D\gamma}{Dt} = 0 \tag{2.208}$$

with the same notation as before for Rossby number ϵ and advective time derivative $\frac{D}{Dt} = \partial_t + \mathbf{u} \cdot \nabla$, in Eq. (2.90). The boundary conditions are

$$\hat{\mathbf{n}} \cdot \mathbf{u} = 0 \quad \text{and} \quad \hat{\mathbf{n}} \times \nabla\theta = 0, \tag{2.209}$$

meaning that fluid velocity $\mathbf{u}$ is tangential and buoyancy θ is constant on the boundary of the domain of flow.

Exercise 2.51 (Conservation laws for TRSW). *Verify the following four properties of the TRSW equations (Eq. (2.208))*

(a) *Energy conservation*

$$E(\mathbf{u}, D, \gamma) = \frac{1}{2} \int \epsilon D |\mathbf{u}|^2 + \gamma^2 D^2 \, d^2 x. \tag{2.210}$$

(b) *Kelvin circulation theorem*

$$\frac{d}{dt} \oint_{c(u)} (\epsilon \mathbf{u} + \mathbf{R}(\mathbf{x})) \cdot d\mathbf{x} = \oint_{c(u)} \frac{1}{2} D\nabla\gamma^2 \cdot d\mathbf{x}, \tag{2.211}$$

where $\nabla \times \mathbf{R}(\mathbf{x}) = f(\mathbf{x})\hat{z}$ and $c(u)$ is a closed planar loop moving with the fluid velocity $\mathbf{u}(\mathbf{x}, t)$.

(c) *Evolution of potential vorticity (PV) on fluid parcels*

$$\partial_t q + \mathbf{u} \cdot \nabla q = \frac{1}{2D} J(D, \theta), \tag{2.212}$$

where PV (q) is defined by

$$q := \frac{\varpi}{\eta}, \quad and \quad \varpi := \hat{z} \cdot \nabla \times (\epsilon \mathbf{u} + \mathbf{R}(\mathbf{x})), \tag{2.213}$$

and

$$J(D, \theta) = \hat{z} \cdot \nabla D \times \nabla \theta = -\operatorname{div}(D\hat{z} \times \nabla \theta)$$

is the Jacobian of the depth $D(\mathbf{x})$ and the buoyancy is $\theta(\mathbf{x}) = \gamma^2(\mathbf{x})$.
(d) *Infinite number of conserved integral quantities*

$$C = \int Df(\gamma) + \varpi g(\gamma) \, d^2 x, \tag{2.214}$$

for boundary conditions in Eq. (2.209) and any differentiable functions f and g.

Hints: The following alternative form of the TRSW motion equation (Eq. (2.208)) may be helpful in verifying these four properties:

$$\partial_t \mathbf{v} - \mathbf{u} \times \nabla \times \mathbf{v} + \nabla \left(\frac{\epsilon}{2} |\mathbf{u}|^2 + D\gamma^2 \right) - \frac{1}{2} D\nabla\gamma^2 = 0, \tag{2.215}$$

where $\mathbf{v} := \epsilon \mathbf{u} + \mathbf{R}(\mathbf{x})$ and $\nabla \times \mathbf{R}(\mathbf{x}) = f(\mathbf{x})\hat{z}$.
You might also keep in mind the fundamental vector identity of fluid dynamics,

$$(\nabla \times \mathbf{a}) \times \mathbf{b} + \nabla(\mathbf{a} \cdot \mathbf{b}) = (\mathbf{b} \cdot \nabla)\mathbf{a} + a_j \nabla b^j. \tag{2.216}$$

Answer.

(a) The time derivative of the proposed TRSW energy is given by

$$\frac{dE}{dt} = \frac{1}{2} \frac{d}{dt} \int_{\mathcal{D}} \epsilon D |\mathbf{u}|^2 + \gamma^2 D^2 \, dx \, dy$$

$$= \int_{\mathcal{D}} D\mathbf{u} \cdot \partial_t \epsilon \mathbf{u} + \left(\frac{\epsilon}{2} |\mathbf{u}|^2 + D\gamma^2 \right) \partial_t D + (\gamma D^2) \partial_t \gamma \, dx \, dy$$

$$= -\int_{\mathcal{D}} D\mathbf{u} \cdot \nabla \left(\frac{\epsilon}{2}|\mathbf{u}|^2 + D\gamma^2\right) - \frac{1}{2}D^2\mathbf{u} \cdot \nabla\gamma^2$$

$$+ \left(\frac{\epsilon}{2}|\mathbf{u}|^2 + D\gamma^2\right)\mathrm{div}(D\mathbf{u}) - (\gamma D^2)\partial_t\gamma \, dx \, dy$$

$$= -\int_{\mathcal{D}} \mathrm{div}\left(D\mathbf{u}\left(\frac{\epsilon}{2}|\mathbf{u}|^2 + D\gamma^2\right)\right) - (\gamma D^2)(\partial_t\gamma + \mathbf{u} \cdot \nabla\gamma) \, dx \, dy,$$

which vanishes for $\mathbf{u}$ tangent to the boundary $\partial\mathcal{D}$ of the domain of flow $\mathcal{D}$.

(b) The alternative form of the TRSW motion equation in Eq. (2.215) yields

$$\frac{d}{dt}\oint_{c(u)} \mathbf{v} \cdot d\mathbf{x} = \oint_{c(u)} (\partial_t\mathbf{v} - \mathbf{u} \times \nabla \times \mathbf{v} + \nabla(\mathbf{u} \cdot \mathbf{v})) \cdot d\mathbf{x}$$

By Eq. (2.215) $= -\oint_{c(u)} \left(\nabla\left(\frac{\epsilon}{2}|\mathbf{u}|^2 + D\gamma^2 - \mathbf{u} \cdot \mathbf{v}\right) - \frac{1}{2}D\nabla\gamma^2\right) \cdot d\mathbf{x}$

$$= \oint_{c(u)} \frac{1}{2}D\nabla\gamma^2 \cdot d\mathbf{x} = -\oint_{c(u)} \frac{1}{2}\gamma^2\nabla D \cdot d\mathbf{x},$$

where we have used the fundamental theorem of calculus in setting integrals of gradients over the closed loop to zero.

(c) The curl of the alternative form of the TRSW motion equation in Eq. (2.215) yields

$$\partial_t\varpi - \nabla \times (\mathbf{u} \times \varpi) = \partial_t\varpi + \mathbf{u} \cdot \nabla\varpi + \varpi\nabla \cdot \mathbf{u} = \frac{1}{2}\hat{z} \cdot \nabla \times (D\nabla\theta).$$

Using the continuity equation to write $\nabla \cdot \mathbf{u}$ in terms of D in this equation verifies Eq. (2.212) as

$$D(\partial_t + \mathbf{u} \cdot \nabla)\frac{\varpi}{D} = \frac{1}{2}\hat{z} \cdot \nabla \times (D\nabla\theta) = \frac{1}{2}J(D, \theta).$$

(d) The expression for the Jacobian given in the remark immediately after Eq. (2.213) suggests rewriting Eq. (2.212) equivalently by using the continuity equation for D in Eq. (2.208) as

$$\partial_t\varpi + \mathrm{div}\left(\varpi\mathbf{u} + \frac{D}{2}\hat{z} \times \nabla\theta\right) = 0. \tag{2.217}$$

The infinite family of constants of motion C in Eq. (2.214) may then be verified by taking the time derivative of C and using the boundary conditions for $\mathbf{u}$ and $\hat{z} \times \nabla\theta$ in Eq. (2.209) and the defining relation, $\theta := \gamma^2$. Thus,

$$\frac{d}{dt} \int Df(\gamma) \, d^2x = -\int \mathrm{div}\left(f(\gamma)D\mathbf{u}\right) d^2x$$

$$\frac{d}{dt} \int \varpi g(\gamma) \, d^2x = -\int \mathrm{div}\left(g(\gamma)\left(\varpi\mathbf{u} + \frac{D}{2}\hat{z} \times \nabla\theta\right)\right) d^2x$$

$$+ \int \frac{D}{2}\hat{z} \times \nabla\theta \cdot \nabla g(\gamma) \, d^2x.$$

Exercise 2.52 (Euler–Poincaré equation).

(a) *Show that the Euler-Poincaré equation*

$$\partial_t \frac{\delta l}{\delta u^i} + \partial_j \left(\frac{\delta l}{\delta u^i} u^j\right) + \frac{\delta l}{\delta u^k}\partial_i u^k = D\partial_i \frac{\delta l}{\delta D} - \gamma_{,i}\frac{\delta l}{\delta \gamma}, \qquad (2.218)$$

arises from Hamilton's principle $\delta S = 0$ with action integral $S = \int l(\mathbf{u}, D, \gamma)\, dt$ given by

$$S = \int_0^T l(\mathbf{u}, D, \gamma) \, dt, \qquad (2.219)$$

for variations of the action integral which depend on the horizontal fluid velocity $\mathbf{u}$, the depth of the active layer D and square-root of buoyancy γ, as follows:

(i) *In the variational formula,*

$$0 = \delta S = \int_0^T \left(\left\langle \frac{\delta l}{\delta \mathbf{u}}, \delta\mathbf{u}\right\rangle + \left\langle \frac{\delta l}{\delta D}, \delta D\right\rangle\right.$$

$$\left. + \left\langle \frac{\delta l}{\delta \gamma}, \delta\gamma\right\rangle\right) dt, \qquad (2.220)$$

substitute variations given geometrically by the infinitesimal transformations,

$$\delta\mathbf{u} = \frac{d}{d\varepsilon}\Big|_{\varepsilon=0}\mathbf{u} = \partial_t\mathbf{v} + \mathbf{u}\cdot\nabla\mathbf{v} - \mathbf{v}\cdot\nabla\mathbf{u}, \qquad (2.221)$$

$$\delta D = \frac{d}{d\varepsilon}\Big|_{\varepsilon=0}D = -\nabla\cdot(D\mathbf{v}), \qquad (2.222)$$

$$\delta\gamma = \frac{d}{d\varepsilon}\Big|_{\varepsilon=0}\gamma = -\mathbf{v}\cdot\nabla\gamma, \qquad (2.223)$$

where the variational vector field $\mathbf{v} \in \mathfrak{X}(\mathbb{R}^2)$ which generates the flow parameterised by ϵ is assumed to vanish at the endpoints in time $[0,T]$ and the angle brackets $\langle \cdot, \cdot \rangle$ in Eq. (2.220) denote L^2 pairing, as in Eq. (2.184).

(ii) *Integrate by parts in Eq. (2.220) using these variations to obtain the Euler–Poincaré equation (Eq. (2.218)).*

Exercise 2.53 (Euler–Poincaré equation for TRSW system (Eq. (2.208))).

(a) *Verify that one may also write the Euler–Poincaré equation, Eq. (2.218), equivalently in three dimensional vector notation as,*

$$\frac{\partial}{\partial t}\left(\frac{1}{D}\frac{\delta l}{\delta \mathbf{u}}\right) - \mathbf{u} \times \nabla \times \left(\frac{1}{D}\frac{\delta l}{\delta \mathbf{u}}\right)$$

$$+ \nabla\left(\mathbf{u} \cdot \frac{1}{D}\frac{\delta l}{\delta \mathbf{u}} - \frac{\delta l}{\delta D}\right) + \frac{1}{D}\frac{\delta l}{\delta \gamma}\nabla\gamma = 0. \qquad (2.224)$$

In verifying the last equation, it may be helpful to use the fundamental vector identity of fluid dynamics, recalled from Eq. (2.97),

$$(\mathbf{b} \cdot \nabla)\mathbf{a} + a_j \nabla b^j = -\,\mathbf{b} \times (\nabla \times \mathbf{a}) + \nabla(\mathbf{b} \cdot \mathbf{a}), \qquad (2.225)$$

for any three dimensional vectors $\mathbf{a}, \mathbf{b} \in \mathbb{R}^3$ with, in this case, $\mathbf{a} = \left(\frac{1}{D}\frac{\delta l}{\delta \mathbf{u}}\right)$ and $\mathbf{b} = \mathbf{u}$.

(b) *Evaluate the variational derivatives for the Lagrangian in the following action integral*

$$S = \int_0^T l(\mathbf{u}, D, \gamma)\, dt$$

$$= \int_0^T \int \frac{\epsilon}{2}D|\mathbf{u}|^2 + D\mathbf{u} \cdot \mathbf{R}(\mathbf{x}) - \frac{1}{2}\gamma^2 D^2\, d^2x\, dt \qquad (2.226)$$

and use Eq. (2.224) with $\nabla \times \mathbf{R}(\mathbf{x}) = f(\mathbf{x})\hat{\mathbf{z}}$ to obtain the motion equation for the TRSW system in Eq. (2.208).

Exercise 2.54 (Kelvin–Noether circulation theorem).

(a) *Prove that the Euler-Poincaré equation (Eq. (2.224)) implies the following Kelvin circulation law (Holm et al., 1998, 2000)*

$$\frac{d}{dt}\oint_{c(u)} \frac{1}{D}\frac{\delta l}{\delta \mathbf{u}} \cdot d\mathbf{x} = -\oint_{c(u)} \frac{1}{D}\frac{\delta l}{\delta \gamma}\nabla\gamma \cdot d\mathbf{x}, \qquad (2.227)$$

where $c(u)$ is a closed loop moving with horizontal fluid velocity $\mathbf{u}(\mathbf{x},t)$ in two dimensions.

(b) *Evaluate this circulation law for the TRSW system (Eq. (2.208)).*

Exercise 2.55 (Energy conservation). *Verify that Eq. (2.218) or Eq. (2.224) with the auxiliary equations for D and γ in the TRSW system (Eq. (2.208)) conserve the energy*

$$E(\mathbf{u},D,\gamma) = \left\langle \frac{\delta l}{\delta u^j}, u^j \right\rangle - l(\mathbf{u},D,\gamma) = \frac{1}{2}\int \epsilon D|\mathbf{u}|^2 + \gamma^2 D^2 \, dx^2. \quad (2.228)$$

Explain why energy conservation is to be expected from Hamilton's principle $\delta S = 0$ for the Lagrangian $l(\mathbf{u},D,\gamma)$. Hint: a symmetry of the Lagrangian is involved.

Exercise 2.56. *Derive the Hamiltonian for 1.5 layer TRSW by computing its Legendre transformation from the Lagrangian in Eq. (2.226).*

Exercise 2.57 (Hamiltonian formulation of TRSW). *Following Sec. 2.13.3 in light of Eq. (2.218), we extend Eqs. (2.203), (2.204) and (2.205), as follows. In terms of the variables m_i, $u^j = \delta h/\delta m_j$, D and γ, we may write the Euler–Poincaré equation (Eq. (2.224)) and its auxiliary equations, as*

$$\partial_t m_i + \partial_j\left(m_i \frac{\delta h}{\delta m_j}\right) + m_j \partial_i \frac{\delta h}{\delta m_j} = -D\partial_i \frac{\delta h}{\delta D} + \frac{\delta h}{\delta \gamma}\partial_i \gamma, \quad (2.229)$$

$$\partial_t D + \partial_j\left(D\frac{\delta h}{\delta m_j}\right) = 0, \quad (2.230)$$

and

$$\partial_t \gamma + \frac{\delta h}{\delta m_j}\partial_j \gamma = 0. \quad (2.231)$$

In turn, we may rewrite these component equations in skew-symmetric matrix operator form as

$$\partial_t \begin{bmatrix} m_i \\ D \\ \gamma \end{bmatrix} = -\begin{bmatrix} (\partial_j m_i + m_j \partial_i) & D\partial_i & -\gamma_{,i} \\ \partial_j D & 0 & 0 \\ \gamma_{,j} & 0 & 0 \end{bmatrix}\begin{bmatrix} \delta h/\delta m_j \\ \delta h/\delta D \\ \delta h/\delta \gamma \end{bmatrix}. \quad (2.232)$$

This skew-symmetric matrix representation implies a Hamiltonian formulation in terms of the Lie–Poisson bracket

$$\partial_t g = \{g, h\}$$

$$= -\int \begin{bmatrix} \delta g/\delta m_j \\ \delta g/\delta D \\ \delta g/\delta \gamma \end{bmatrix}^T \begin{bmatrix} (\partial_j m_i + m_j \partial_i) & D\partial_i & -\gamma_{,i} \\ \partial_j D & 0 & 0 \\ \gamma_{,j} & 0 & 0 \end{bmatrix} \begin{bmatrix} \delta h/\delta m_j \\ \delta h/\delta D \\ \delta h/\delta \gamma \end{bmatrix} d^2 x,$$

$$(2.233)$$

for any differentiable functionals g and h. For the case of TRSW with Lagrangian in Eq. (2.226), we have via the variational relations from the Legendre transformation of the TRSW Lagrangian that

$$\mathbf{m} := \frac{\delta l}{\delta \mathbf{u}} = D\big(\epsilon \mathbf{u} + \mathbf{R}(\mathbf{x})\big) \quad and \quad \frac{\delta h}{\delta \mathbf{m}} = \mathbf{u},$$

$$\frac{\delta h}{\delta D} = -\frac{\delta l}{\delta D} = -\left(\frac{\epsilon}{2}|\mathbf{u}|^2 + \mathbf{u}\cdot\mathbf{R}(\mathbf{x}) - \gamma^2 D\right) \quad and \quad (2.234)$$

$$\frac{\delta h}{\delta \gamma} = -\frac{\delta l}{\delta \gamma} = \gamma D^2.$$

Exercise 2.58. *Verify the results of the Euler–Poincaré theory introduced here for TRSW, by substituting the variational derivatives (Eq. (2.234)) into the Hamiltonian matrix form in Eq. (2.232) to recover the TRSW equations in Hamiltonian form.*

Exercise 2.59 (Hamilton's principle with auxiliary Clebsch constraints). *Show that the Euler–Poincaré equation in Eq. (2.218) arises from Hamilton's principle $\delta S = 0$ with constrained action integral given by*

$$S = \int_0^T l(\mathbf{u}, D, \gamma) + \Big\langle \alpha, D_t + \nabla\cdot(D\mathbf{u})\Big\rangle + \Big\langle \beta, \gamma_t + \mathbf{u}\cdot\nabla\gamma\Big\rangle dt. \quad (2.235)$$

for free variations of the functions $\mathbf{u}$, D, γ, α and β, with L^2 pairing indicated by the angle brackets $\langle\cdot,\cdot\rangle$ as in Eq. (2.184).

 Hint: To prove this statement try taking the advective time derivative of the result for the $\delta\mathbf{u}$ variation,

$$\frac{1}{D}\frac{\delta l}{\delta \mathbf{u}}\cdot d\mathbf{x} = d\alpha - \frac{\beta}{D}d\gamma,$$

Answer.   Taking variations wrt $\mathbf{u}$, D and γ yields

$$\delta\mathbf{u}:\quad \frac{\delta l}{\delta\mathbf{u}} - D\nabla\alpha + \beta\nabla\gamma = 0,$$

$$\delta D:\quad \frac{\delta l}{\delta D} - \frac{D\alpha}{Dt} = 0,$$

$$\delta\gamma:\quad \frac{\delta l}{\delta\gamma} - \partial_t\beta - \nabla\cdot(\beta\mathbf{u}) = 0,$$

while variations with respect to α and β yield the auxiliary equations for D and γ, respectively.

Taking the advective time derivative $\frac{D}{Dt}$ along $\frac{D\mathbf{x}}{Dt} = \mathbf{u}$ of the result for the variation with respect to the fluid velocity yields

$$\frac{D}{Dt}\left(\frac{1}{D}\frac{\delta l}{\delta\mathbf{u}}\cdot d\mathbf{x}\right) = \frac{D}{Dt}\left(d\alpha - \frac{\beta}{D}d\gamma\right)$$

$$\frac{D}{Dt}\left(\frac{1}{D}\frac{\delta l}{\delta\mathbf{u}}\right)\cdot d\mathbf{x} + \left(\frac{1}{D}\frac{\delta l}{\delta\mathbf{u}}\right)\cdot d\frac{D\mathbf{x}}{Dt} = d\frac{D\alpha}{Dt} - \frac{D}{Dt}\left(\frac{\beta}{D}\right)d\gamma - \frac{\beta}{D}d\left(\frac{D\gamma}{Dt}\right)$$

$$\left(\frac{D}{Dt}\left(\frac{1}{D}\frac{\delta l}{\delta\mathbf{u}}\right) + \left(\frac{1}{D}\frac{\delta l}{\delta u^j}\right)\nabla u^j\right)\cdot d\mathbf{x} = \left(\nabla\frac{\delta l}{\delta D} - \frac{1}{D}\frac{\delta l}{\delta\gamma}\nabla\gamma\right)\cdot d\mathbf{x}$$

upon inserting the auxiliary equation for γ and the two other variational equations above. Hence, in addition to the auxiliary equations for buoyancy γ^2 and depth D, one obtains the motion equation

$$\frac{D}{Dt}\left(\frac{1}{D}\frac{\delta l}{\delta\mathbf{u}}\right) + \left(\frac{1}{D}\frac{\delta l}{\delta u^j}\right)\nabla u^j = \nabla\frac{\delta l}{\delta D} - \frac{1}{D}\frac{\delta l}{\delta\gamma}\nabla\gamma,$$

which recovers Eq. (2.218) upon using the continuity equation for depth, D. and also recovers Eq. (2.224) upon using the fundamental vector identity of fluid dynamics (Eq. (2.97)).

Chapter 3

Data and Probability

Jochen Bröcker* and Ben Calderhead[†]
*University of Reading, UK
†Imperial College London, UK

3.1 Introduction

Sections 3.1–3.9 will cover the basics of probability theory, integration, and statistics (roughly in order). The second part (Secs. 3.10–3.16) will introduce an important technique in statistics called Monte Carlo simulations. We start by motivating the use of probability theory and statistics, and in particular why we need concepts from measure theory and integration, often perceived as abstract and complicated.

3.1.1 *Probability*

It is not easy to explain what probability theory means without sounding tautological. One might say that it allows one to quantify uncertainty or chance, but then what do we mean by "uncertainty" or "chance"?

De Moivre's seminal textbook, *The Doctrine of Chances* (de Moivre, 1967), is widely considered as the first textbook on probability theory. Since then, the theory has undergone enormous developments. In particular, there is an axiomatic framework which has been universally adopted and which we will discuss in this chapter. Even though de Moivre's book was first published in 1718, there is still some debate as to the interpretation of probability theory or, in other words, to what this axiomatic framework actually pertains. Different interpretations of probability have been put forward, but somewhat fortunately, the differences matter little as far as the mathematics is concerned. Nonetheless, we will briefly mention the most prominent interpretations of probability (see the Wikipedia page on "Probability Interpretations" for more information).

- **The classical definition of probability** put forward by Laplace (Laplace, 1995), "consists in reducing all the events of the same kind to a certain number of cases equally possible, that is to say, to such as we may be equally undecided about in regard to their existence." These cases might also be termed *elementary events*, and the probability of any event A is the number of elementary events it contains, divided by the number of all elementary events. Evidently, this definition assumes that for any given problem, (i) the number of elementary events is finite, and (ii) every event can be expressed as the union of elementary events. We will see below that this theory is not powerful enough to deal with certain questions we are interested in.

- **Frequentism** considers experiments which can, at least in principle, be repeated as often as desired under constant external conditions. Internal phenomena however may lead to variable outcomes under repetition. Whether these phenomena are fundamentally deterministic (like throwing a coin) or inherently random (like radioactive decay) is irrelevant. The probability of an outcome is then defined to be the limiting observed frequency of this outcome. But the probability is only well defined if the observed frequencies actually converge, and it is very difficult to provide useful criteria as to when this is the case (other than just saying that they have to converge). We see that this theory struggles to explain exactly what experiments it applies to.

- **Subjectivism** maintains that probabilities express the degree of belief a certain individual has as to whether a certain statement about the real world is true or not. For example, I think it is likely, but not certain, that the Riemann hypothesis is correct. It is not necessary that the individual actually uses probabilities; as Savage (1971) and others have shown, these probabilities can be inferred or "elicited" from the individual's behaviour. In other words, any possible action an individual may take in the face of an uncertain event can be explained by means of a number, interpreted as that individual's subjective probability of the event. The proof of this fact however assumes that the individual acts rationally in a sense specified by Savage's axioms, and these axioms are not altogether unobjectionable. In fact, there is strong evidence that people's behaviour under uncertainty can deviate considerably from these axioms.

Just for illustrative purposes, here is a little example showing how the elicitation of (subjective) probabilities might work in practice. You can play this game with a friend of yours (if you have a few pounds to spare):

Example 3.1. The statement is: "At the time of his death, Isaac Newton had lost all but one of his teeth." This statement is known to be false; Apart from his primary teeth, Newton lost only one tooth during his lifetime, but you don't reveal this to your friend at this point. You merely say that you know the answer. You offer your friend to pay her p pounds if the statement is correct, q pounds if it is not correct, and she can choose the numbers p and q, but she will have to pay $\frac{1}{2}\left(p^2+q^2\right)$ pounds back to you in order to play the game. Once your friend has chosen p and q, you reveal the answer and exchange the money. You can convince yourself of the following facts, which support the interpretation of p and q as your friend's subjective probabilities for and against the statement:

(a) There is an incentive for your friend to play the game, as there is a way to make at least 25 p for sure.
(b) There is an incentive for your friend to state two numbers p,q which are nonnegative and so that $p+q=1$, any deviation from this will incur a certain loss.

3.1.2 Why the classical interpretation of probability is insufficient?

For the remainder of the introduction, we will discuss why the classical interpretation of probability, which has a finite number of events with equal probability as a basis, is insufficient. In fact, any theory with a finite number of events is insufficient, whatever their probability. We will see that measure theoretic probability is the right answer to this problem, so the following is really a motivation to study measure theory and integration.

Consider the following model of a *fair coin*: A *binary sequence* is a sequence $x_1, x_2, \ldots$, where $x_k = 0$ or 1 only for all $k > 0$. We identify 1 with "Head" and 0 with "Tail". Let Ω be the set of all binary sequences. A generic element is written as $\omega = (\omega_1, \omega_2, \ldots)$. An *event* is a subset of Ω.

A probability $\mathbf{P}$ is a function which assigns a number between 0 and 1 to events so that

(a) $\mathbf{P}(\Omega) = 1$,
(b) $\mathbf{P}(A \cup B) = \mathbf{P}(A) + \mathbf{P}(B)$ whenever the events A and B are disjoint.

These assumptions are reasonable for any probability. In the present example, we assume

(a) $\mathbf{P}(\{\omega \in \Omega; \omega_k = 1\}) = p$ for any k (and this implies $\mathbf{P}(\{\omega \in \Omega; \omega_k = 0\}) = 1 - p$),
(b) Different throws are independent.

The second condition means the following: Whenever $x_1, \ldots, x_n$ is a binary sequence of finite length, then,

$$\mathbf{P}(\{\omega \in \Omega; \omega_1 = x_1, \ldots, \omega_n = x_n\}) = p^{\sum_{k=1}^{n} x_k} (1 - p)^{n - \sum_{k=1}^{n} x_k}.$$

Based on these assumptions, we can now work out things like

$$\mathbf{P}(\text{No. of ``Heads'' in 10 throws}) = p^{10}$$

etc. In other words, we can deal with events that depend on finitely many throws, only. But now let $S_n = \sum_{k=1}^{n} \omega_k = $ No. of "Heads" in n throws, and consider the statement

$$\frac{S_n}{n} \to p \qquad \text{as } n \to \infty. \tag{3.1}$$

This statement, known as the *Law of Large Numbers*, "should hold", but cannot be analysed in a "finite" framework, since the very existence of the limit is a random event, depending on more than just finitely many throws (in fact, neither the value of the limit nor its existence depends on the first n throws, however large n is).

To see where the problem lies, let us delve into this a little further. We assume $p = \frac{1}{2}$ for simplicity. Fix some $\epsilon > 0$ and consider the events

$$A_k := \left\{ \text{all } \omega \text{ so that } \left| \frac{S_k}{k} - \frac{1}{2} \right| > \epsilon \right\}.$$

You might know from your previous studies that

$$\mathbf{P}(A_k) \leqslant \frac{C}{k}$$

for some constant C depending on ϵ. In fact, with a little more work you can show that

$$\mathbf{P}(A_k) \leqslant C \cdot \lambda^k \tag{3.2}$$

for some $C > 0$ and $\lambda < 1$, both depending on ϵ. Either estimate shows that $\frac{S_k}{k}$ will concentrate about $1/2$ with increasing probability. But this does not imply that the statement in Eq. (3.1) holds for any ω. In fact, if ω is so that the statement in Eq. (3.1) holds, then ω can be a member of only finitely many A_k! So we need to investigate the event $B := \{\text{all } \omega \text{ which are in infinitely many} A_k\}$. If we can show that this event has probability zero no matter how we pick ϵ, then the statement in Eq. (3.1) is true for all ω, except perhaps for some ω's in a set of probability zero.

The most challenging bit in this discussion is the construction of the event B. Consider first $B_n := A_n \cup A_{n+1} \cup \cdots$. This is the event that $\{|\frac{S_k}{k} - \frac{1}{2}| > \epsilon\}$ for *some* $k \geqslant n$. Now if there is an ω which is a member of A_k for infinitely many k's, then it must be in *all* B_n's; hence $B \subset B_n$ for any n, and therefore

$$\mathbf{P}(B) \leqslant \mathbf{P}(B_n) \quad \text{for all } n = 1, 2, \ldots.$$

We aim to show that the probability of B_n goes to zero. Now by Eq. (3.2)

$$\mathbf{P}(A_n \cup A_{n+1} \cup \cdots \cup A_{n+m}) \leqslant \sum_{k=n}^{n+m} \lambda^k = \frac{\lambda^n - \lambda^{n+m+1}}{1 - \lambda} < \frac{\lambda^n}{1 - \lambda}.$$

The left-hand side is increasing in m and bounded (by one or in fact by the right-hand side) and therefore convergent, hence

$$\lim_{m \to \infty} \mathbf{P}(A_n \cup A_{n+1} \cup \cdots \cup A_{n+m}) \leqslant \frac{\lambda^n}{1 - \lambda}. \tag{3.3}$$

And here is the point: We would like to use that the left-hand side is in fact $\mathbf{P}(B_n)$, that is

$$\lim_{m \to \infty} \mathbf{P}(A_n \cup A_{n+1} \cup \cdots \cup A_{n+m}) \stackrel{?}{=} \mathbf{P}(A_n \cup A_{n+1} \cup \ldots) = \mathbf{P}(B_n). \tag{3.4}$$

Assuming this is correct for the moment and using it in Eq. (3.3), we obtain

$$\mathbf{P}(B_n) \leqslant \frac{\lambda^n}{1 - \lambda} \to 0 \quad \text{if } n \to \infty,$$

and we can conclude that the statement in Eq. (3.1) holds for all sequences of heads and tails which are not in B, but this happens with probability zero.

So what is the problem with Eq. (3.4)? We clearly have

$$\mathbf{P}(A_n \cup A_{n+1} \cup \cdots \cup A_{n+m}) \leqslant \mathbf{P}(B_n)$$

so all we can say is that

$$\lim_{m \to \infty} \mathbf{P}(A_n \cup A_{n+1} \cup \cdots \cup A_{n+m}) \leqslant \mathbf{P}(B_n),$$

but there is no *a priori* reason why there should be equality here. If we want relations like Eq. (3.4) to be correct, we have to add this as an assumption. In other words, we will only work with probabilities where this is correct. But there is then another question: Is this consistent with our assumptions (a) and (b) made at the beginning of this section, and the other properties we would like a probability to have? This is a non-trivial question which we will address in this chapter (the answer is "yes").

3.1.3   Statistics

Statistics works with real data, that is, quantitative observations from real world experiments. The aim is to explain the data by means of models, mostly probabilistic models. So when compared to probability theory, in a sense we go in the opposite direction. Rather than investigating a given probability model (e.g., a fair coin) we ask the question: is there a probability model consistent with given data? For example, suppose we observe the two data sets

$$\{H,H,T,H,T,H,T,T,H\} \tag{3.5}$$

$$\{H,H,H,H,T,T,T,T,H\} \tag{3.6}$$

allegedly coming from tossing a fair coin. Is this data consistent with the model of a fair coin? You might observe that both data sets contain 5 heads and 4 tails. In that sense, the data sets are not entirely atypical for nine tosses of a fair coin. But you might also observe that the H's and T's are sprinkled quite evenly across the first data set while in the second, there seem to be unusually long runs of heads and tails. So the first data set might plausibly come from a fair coin but not the second. Statistics is about making these conclusions more quantitative.

3.2 Sigma Algebras and Probability Measures

In this section, we discuss probabilities and events, that is the sets we want to assign probabilities to. We start with some fundamental definitions. Let Ω, A, B be sets. Familiarity with the notations $A \subset \Omega$, $A \cup B$, $A \cap B$, $\varnothing$ is assumed. Further

$$A \setminus B := \{x \in A; x \notin B\}, \qquad \text{read "}A \text{ without } B\text{"}$$

$$A^c := \Omega \setminus A, \qquad \text{read "Complement of } A \text{ in } \Omega \text{".}$$

The notation A^c is used if Ω is clear from the context. If the elements of a set A are again sets, we call A a system or family of sets.

Definition 3.1. Let Ω be a set. A system $\mathcal{A}$ of subsets of Ω is called an *algebra* if

(a) $\varnothing \in \mathcal{A}$
(b) $A \in \mathcal{A} \Rightarrow A^c \in \mathcal{A}$.
(c) $A_1, \ldots, A_n \in \mathcal{A} \Rightarrow \bigcup_{k=1}^{n} A_k \in \mathcal{A}$.

Further, $\mathcal{A}$ is a *sigma algebra* if

(d) $A_1, A_2, \ldots \in \mathcal{A} \Rightarrow \bigcup_{k=1}^{\infty} A_k \in \mathcal{A}$.

An algebra formalises the intuition behind "events". Considering sigma algebras rather than just algebras, that is where Definition 3.1 holds for countably many A_n rather than just finitely many, is important as we have seen in the introduction. Members of $\mathcal{A}$ are called *events* or *measurable sets*.

Definition 3.2. Let $\mathcal{A}$ be an algebra. A function $\mathbf{P} : \mathcal{A} \longrightarrow [0, 1]$ is a *probability* if it satisfies

(a) *Normalisation:* $\mathbf{P}(\Omega) = 1$
(b) *Additivity:* If $A_1, \ldots, A_n \in \mathcal{A}$, with $A_i \cap A_j = \varnothing$ for $i \neq j$, then $\sum_{k=1}^{n} \mathbf{P}(A_k) = \mathbf{P}(\bigcup_{k=1}^{n} A_k)$.
(c) *Continuity at $\varnothing$:* If $A_1, A_2, \ldots \in \mathcal{A}$, with $A_1 \supset A_2 \supset \cdots$ and $\cap A_j = \varnothing$, then $\mathbf{P}(A_k) \to 0$ for $k \to \infty$.

Again, the intuition is clear. The continuity at $\varnothing$ is important for technical reasons, as we have seen in the introduction (the connection will be made clear in Exercise 3.2). It is possible to construct examples of a probability on an algebra that is not continuous at $\varnothing$. Note that a probability satisfies $\mathbf{P}(\varnothing) = 0$ (Exercise 3.2).

Definition 3.3. A pair $(\Omega, \mathcal{A})$ with Ω a set and $\mathcal{A}$ a sigma algebra is called a *measurable space*. A triple $(\Omega, \mathcal{A}, \mathbf{P})$ with Ω a set, $\mathcal{A}$ a sigma algebra, and $\mathbf{P}$ a probability is called a *probability space*.

Note that algebras are very much smaller than sigma algebras, so it should be much easier to define $\mathbf{P}$ just on an algebra (examples later).

Definition 3.4. Let $\mathcal{A}$ be an arbitrary family of subsets of Ω. Then, $\sigma(\mathcal{A})$ is defined as the smallest σ-algebra containing $\mathcal{A}$.

In Exercise 3.1 you will need to show that this concept is well defined.

Theorem 3.1 (The Measure Extension Theorem or MET for short, also known as the Hahn–Carathéodory Theorem). *Let $\mathcal{A}$ be an algebra and $\mathbf{P}$ a probability on $\mathcal{A}$. Then, there exists a unique probability $\tilde{\mathbf{P}}$ on $\sigma(\mathcal{A})$ with $\tilde{\mathbf{P}}|_{\mathcal{A}} = \mathbf{P}|_{\mathcal{A}}$. Further, if $A \in \sigma(\mathcal{A})$, then for any $\epsilon > 0$ there exist disjoint sets $A_1, \ldots, A_n \in \mathcal{A}$ with $\tilde{\mathbf{P}}(A \triangle \bigcup_{k=1}^{n} A_k) \leqslant \epsilon$.*

Sketch of a proof, see e.g., Halmos (1974). For any $Y \subset \Omega$, put

$$\mathbf{P}^*(Y) = \inf \sum_{k=1}^{\infty} \mathbf{P}(A_k),$$

inf taken over $A_1, A_2, \ldots \in \mathcal{A}$, with $Y \subset \bigcup_k A_k$. Now

(a) $\mathbf{P}^*|_{\mathcal{A}} = \mathbf{P}|_{\mathcal{A}}$ ("$\leqslant$" is trivial).
(b) Consider the family of sets $\mathcal{M}$: a set $A \subset \Omega$ is a member of $\mathcal{M}$ if $\forall E \subset \Omega$ it holds that $\mathbf{P}^*(E) \geqslant \mathbf{P}^*(E \cap A) + \mathbf{P}^*(E \setminus A)$. One then proves that $\mathcal{M}$ is a σ-algebra with $\mathcal{M} \supset \mathcal{A}$.
(c) $\mathbf{P}^*$ is a measure on $\mathcal{M}$.
(d) The approximation result is relatively straightforward from the definition of $\mathbf{P}^*$. □

We fix the uniqueness part, which is true under weaker conditions:

Theorem 3.2 (Uniqueness of probabilities). *Let $\mathcal{A}$ be a family of sets so that for any two sets $A_1 \in \mathcal{A}, A_2 \in \mathcal{A}$, also $A_1 \cap A_2 \in \mathcal{A}$. (This is true for instance if $\mathcal{A}$ is an algebra.) Further, let $\mathbf{P}, \mathbf{Q}$ be two probabilities on $\sigma(\mathcal{A})$, the sigma algebra generated by $\mathcal{A}$. Then, if $\mathbf{P}(A) = \mathbf{Q}(A)$ for any set $A \in \mathcal{A}$, they agree on $\sigma(\mathcal{A})$.*

For proof, see Proposition 2.23 in Breiman (1973). The following theorem ensures that there exists a probability on the unit interval which on any subinterval is given by the length of that subinterval. For proof, see for instance, Chapter 7 in Jacod and Protter (2000).

Theorem 3.3 (The Lebesgue measure). *A half-open interval on $(0,1]$ is a set of the form $(a,b]$, where $0 < a < b \leqslant 1$. Let $\mathcal{A}$ be the family of sets which are unions of finitely many disjoint half-open intervals. Then, $\mathcal{A}$ is an algebra. To each $A \in \mathcal{A}$, we assign $\lambda(A) :=$ the total length of A. This is a probability on $\mathcal{A}$ (the continuity at $\varnothing$ requires proof, see for instance Jacod and Protter (2000) for a somewhat more general statement). It now follows from Theorem 3.1 that λ can be extended to a probability on $\sigma(\mathcal{A})$, which is the Borel algebra (see Definition 3.5).*

Exercises for Sec. 3.2

Exercise 3.1. *Let Ω be a set.*

(a) *Show that the power set 2^{Ω} is a sigma algebra.*
(b) *Show that $\mathcal{S}_1 \cap \mathcal{S}_2$ is a sigma algebra for any two sigma algebras $\mathcal{S}_1, \mathcal{S}_2$.*
(c) *Use the previous two items to show that $\sigma(\mathcal{A})$ in Definition 3.4 makes sense, i.e., there exist sigma algebras containing $\mathcal{A}$, and among these there exists a smallest possible one.*

Exercise 3.2. *Let Ω be a set, $\mathcal{A}$ an algebra, $\mathbf{P} : \mathcal{A} \to [0,1]$ a set function satisfying properties (a) and (b) in Definition 3.2.*

(a) *Show that $\mathbf{P}(\varnothing) = 0$.*
(b) *Show that property (c) in Definition 3.2 is equivalent to sigma additivity: If $A_1, A_2, \ldots$ is a sequence of sets in $\mathcal{A}$ with $A_i \cap A_j = \varnothing$ for any $i \neq j$, and if $\bigcup_k A_k \in \mathcal{A}$ as well, then $\sum_{k=1}^{\infty} \mathbf{P}(A_k) = \mathbf{P}(\bigcup_k A_k)$.*
(c) *Show that property (c) in Definition 3.2 is equivalent to continuity from above: If $A_1, A_2, \ldots \in \mathcal{A}$, with $A_1 \supset A_2 \supset \cdots$ and $\cap A_j = A$ with $A \in \mathcal{A}$, then $\mathbf{P}(A_k) \to \mathbf{P}(A)$ for $k \to \infty$.*
(d) *Show that property (c) in Definition 3.2 is equivalent to continuity from below: If $A_1, A_2, \ldots \in \mathcal{A}$, with $A_1 \subset A_2 \subset \cdots$ and $\cup A_j = A$ with $A \in \mathcal{A}$, then $\mathbf{P}(A_k) \to \mathbf{P}(A)$ for $k \to \infty$.*
(e) *Show that for any series $A_1, A_2, \ldots$ of disjoint sets in $\mathcal{A}$, we have $\mathbf{P}(A_n) \to 0$ (in fact, $\mathbf{P}(A_n)$ must be summable).*

3.3   Measurable Functions and Integration

A probability can be seen as a generalised form of the notion of volume. As with the standard volume in Euclidean space, it is possible to integrate functions against probabilities. We want to define an integral which, to some extent, can be interchanged with pointwise limits of functions. Let $(\Omega, \mathcal{A}, \mathbf{P})$ be a probability space.

Definition 3.5.

(a) On $\mathbb{R}$ we define the Borel-algebra $\mathcal{B}$ as the smallest σ-algebra containing all open sets (see Definition 3.4).
(b) A function $f : \Omega \longrightarrow \mathbb{R}$ is *measurable* or a *random variable* if $f^{-1}(B) \in \mathcal{A}$ for all $B \in \mathcal{B}$.

The definition of a random variable guarantees that sets such as $\{\omega \in \Omega; a < f(\omega) < b\} = f^{-1}(]a,b[)$ can be assigned a probability to. To prove that a function is measurable, it is enough to check that $\{\omega; f(\omega) > a\} \in \mathcal{A}$ for any $a \in \mathbb{R}$ (see Exercise 3.4).

Theorem 3.4.

(a) *If $f_n, n \in \mathbb{N}$ are random variables, so are the pointwise* $\limsup f_n$, $\liminf f_n$, $\lim f_n$ *(if the last exists).*
(b) *If $f^{(k)}, k = 1, \ldots, d$ are random variables and $\phi : \mathbb{R}^d \to \mathbb{R}$ is a continuous function, then the function $\psi : \omega \to \phi(f^{(1)}(\omega), \ldots, f^{(d)}(\omega))$ is a random variable.*

Proof. To prove part (a), pick $a \in \mathbb{R}$. Then $\{\omega; \sup_k f_{n+k}(\omega) > a\} = \bigcup_k \{\omega; f_{n+k}(\omega) > a\} \in \mathcal{A}$, so $\sup_k f_{n+k}$ is measurable for every n by the remark after Definition 3.5. $\inf_k f_{n+k}$ is similar (take $\bigcap_k \{\ldots\}$). But

$$\liminf_n f_n = \sup_n \inf_k f_{n+k},$$

$$\limsup_n f_n = \inf_n \sup_k f_{n+k}.$$

So they are measurable. If $\lim_n f_n$ exists, it is equal to $\limsup$ and $\liminf$. To prove the second item, we note that the statement is true if $f^{(1)}, \ldots, f^{(d)}$ are simple functions. Further, we will show later on that every non-negative random variable is the pointwise limit of simple functions, and this is easily seen to extend to general (not necessarily non-negative) random variables. We can conclude that ψ is the pointwise limit of simple functions and thus a random variable by item (a). $\qquad\square$

3.3.1 *The integral*

We want to define an integral $\int f d\mathbf{P}$ for random variables, which we will also write as $\mathbb{E}(f)$, generalising the notion of expected values.

But first we make a remark about limits and increasing sequences. A sequence $\{x_n, n \in \mathbb{N}\}$ of real numbers is called *increasing* if $x_1 \leqslant x_2 \leqslant \cdots$. If $\{x_n\}$ is increasing, then $x_n \to x$ and $x_n \uparrow x$ have the same meaning, namely that $x = \sup_n x_n$. Note that x might be infinite, but if it is not, we have $x = \lim_{n \to \infty} x_n$. (We stress that per definition, the limit of a sequence is *always* finite.) If $\{x_n\}$ is not increasing though, then $x_n \uparrow x$ is meaningless, while $x_n \to x$ means that $x = \lim_n x_n$.

For a sequence $\{f_n, n \in \mathbb{N}\}$ of real valued functions on some set Ω, the limits $\lim_n f_n = f$ and $f_n \to f$ are understood pointwise (unless otherwise stated), that is $\lim_n \{f_n(\omega)\} = f(\omega)$ and also $f_n(\omega) \to f(\omega)$ for every $\omega \in \Omega$. The sequence $\{f_n\}$ is called *increasing* if $\{f_n(\omega), n \in \mathbb{N}\}$ is an increasing sequence for every $\omega \in \Omega$, and we write $f_n \uparrow f$ if $f_n(\omega) \uparrow f(\omega)$ for every $\omega \in \Omega$.

The integral of a random variable can be constructed along the following steps. See Klenke (2014); Halmos (1974); Doob (1994); Dudley (1989) for details.

(1) For $A \in \mathcal{A}$, define the *indicator function*

$$\mathbf{1}_A(\omega) = \begin{cases} 1 & \text{if } \omega \in A, \\ 0 & \text{else.} \end{cases}$$

(2) A random variable $f : \Omega \to \mathbb{R}$ is *simple* if it assumes finitely many values, say $\{f_1, \ldots, f_n\} \subset \mathbb{R}$. We can write

$$f = \sum_{l=1}^{k} f_l \cdot \mathbf{1}_{B_l}$$

with $B_l = f^{-1}(\{f_l\})$ for all $l = 1, \ldots, k$. Note that $B_l \in \mathcal{A}$ for all $l = 1, \ldots, k$ because f is assumed measurable.

(3) With f, g simple, so are $f \cdot g$, $\alpha f + \beta g$, $\alpha, \beta \in \mathbb{R}$, $\max\{f, g\}$ and $|f|$ (these operations are understood pointwise).

(4) Every non-negative random variable $f : \Omega \to \mathbb{R}_{\geqslant 0}$ is the pointwise monotone increasing limit of simple functions.

Proof. Define $g_n : \mathbb{R}_{\geqslant 0} \to \mathbb{R}_{\geqslant 0}$

$$
g_n(x) = \begin{cases} k + \dfrac{l}{2^n} & \text{if } k + \dfrac{l}{2^n} < x \leqslant k + \dfrac{l+1}{2^n}, \\[2mm] & \text{for } k = 1 \ldots n-1, l = 0 \ldots 2^n - 1, \\[2mm] n & \text{if } x > n. \end{cases}
$$

Clearly $g_n(x) \uparrow x$, $\forall\, x \in \mathbb{R}_{\geqslant 0}$. Now put $f_n := g_n \circ f$, then clearly f_n is simple and $f_n \uparrow f$. $\qquad\square$

(5) For f simple, define

$$
\int f \, d\mathbf{P} = \sum_{k=1}^{n} f_k \mathbf{P}(B_k).
$$

(6) Prove that the integral is linear, monotone (i.e., $f \leqslant g \Rightarrow \int f \, d\mathbf{P} \leqslant \int g \, d\mathbf{P}$) and $|\int f \, d\mathbf{P}| \leqslant \int |f| \, d\mathbf{P}$.

(7) If f is a non-negative random variable and $\{f_n\}$ is a sequence of simple functions and $f_n \uparrow f$ (e.g., as in setp 4), then $\{\int f_n \, d\mathbf{P}\}$ is an increasing sequence of real numbers and

$$
\sup_n \int f_n \, d\mathbf{P} = \sup_g \int g \, d\mathbf{P}, \tag{3.7}
$$

where "$\sup_g$" is over all simple g with $f \geqslant g$. This will be proved in Exercise 3.5. We define $\int f \, d\mathbf{P}$ as either side of Eq. (3.7). This might be a non-negative real number or ∞. But if $\int f \, d\mathbf{P} < \infty$, then $\int f_n \, d\mathbf{P} \to \int f \, d\mathbf{P}$ for $n \to \infty$.

(8) For a general random variable $f : \Omega \to \mathbb{R}$, put $f_+ := \max\{f, 0\}$, $f_- := f_+ - f$ (now f_+, f_- are non-negative) and set

$$
\int f \, d\mathbf{P} := \int f_+ \, d\mathbf{P} - \int f_- \, d\mathbf{P}
$$

if at least one of them is finite. If both are finite, f is called *integrable*.

We stress that the integral of a non-egative random variable is always well defined (but maybe infinite). In particular $\int |f| \, d\mathbf{P}$ is always defined for any random variable f, and f is integrable if and only if $\int |f| \, d\mathbf{P} < \infty$.

Lemma 3.1 (Properties of the integral). *The integral enjoys the properties in step 6 if both $\int |f| d\mathbf{P} < \infty$ and $\int |g| d\mathbf{P} < \infty$.*

Proof. The linearity and the monotonicity for integrals over non-negative simple functions is assumed proved in step 6. The additivity for integrals over non-negative functions f, g is shown by observing that if $f_n, g_n, n \in \mathbb{N}$ are non-negative simple functions with $f_n \uparrow f$ and $g_n \uparrow g$, then $f_n + g_n \uparrow f + g$ with $f_n + g_n$ non-negative and simple. The additivity of the integral in this case then follows from the additivity of the integral for non-negative simple functions and step 7 above. To show the monotonicity for integrals over non-negative functions $f \leqslant g$, we take non-negative simple functions $f_n, g_n, n \in \mathbb{N}$ with $f_n \uparrow f$ and $g_n \uparrow g$. Now note that $h_n = \max\{f_n, g_n\}$ is also non-negative and simple with $h_n \uparrow g$, and further $f_n \leqslant h_n$. It follows from step 7 that $\int f d\mathbf{P} \leqslant \int g d\mathbf{P}$. To prove the additivity in the general case, observe first that $|f + g| \leqslant |f| + |g|$ and hence $\int |f + g| d\mathbf{P} < \infty$ by the monotonicity for non-negative functions. From the identity

$$(f + g)_+ + f_- + g_- = (f + g)_- + f_+ + g_+$$

we obtain by the additivity for non-negative random variables that

$$\int (f + g)_+ d\mathbf{P} + \int f_- d\mathbf{P} + \int g_- d\mathbf{P} = \int (f + g)_- d\mathbf{P} + \int f_+ d\mathbf{P} + \int g_+ d\mathbf{P}.$$

Note that by integrability, all the terms in this identity are finite. Rearranging and using the definition of the integral for general f and g gives the result. To prove the monotonicity in the general case, we use the linearity (in the line marked with $(*)$) to obtain

$$\int f d\mathbf{P} = \int f_+ d\mathbf{P} - \int f_- d\mathbf{P}$$

$$\leqslant \int f_+ d\mathbf{P} + \int f_- d\mathbf{P}$$

$$= \int (f_+ + f_-) d\mathbf{P} \qquad (*)$$

$$= \int |f| d\mathbf{P}.$$

Similarly, one proves that $-\int f d\mathbf{P} \leqslant \int |f| d\mathbf{P}$ which gives the result. $\square$

3.3.2 *Interchange of integral with a.s. limits*

The most important reason for introducing this integral (as opposed to using the Riemann integral) is the nice behaviour of the integral under pointwise limits.

Theorem 3.5 (Monotone Convergence). *Suppose $\{f_n, n \in \mathbb{N}\}$ is an increasing sequence of non-negative random variable, and $f_n \uparrow f$. Then,*

$$\int f_n \, d\mathbf{P} \uparrow \int f \, d\mathbf{P}. \tag{3.8}$$

Proof. According to step 4, for every $n \in \mathbb{N}$ there exists a sequence $\{f_{n,m}, m \in \mathbb{N}\}$ of simple non-egative random variable with $\lim_{m \to \infty} f_{n,m} = f_n$. Let $g_n = \max\{f_{k,l}, k, l \leq n\}$. This is a increasing sequence of simple functions. On the one hand,

$$g_n \leq f_n \leq f \quad \text{for all } n. \tag{3.9}$$

On the other hand, if we fix $\epsilon > 0$ and $\omega \in \Omega$ we can find n and $m \geq n$ so that

$$f(\omega) \leq f_n(\omega) + \epsilon/2$$
$$f_n(\omega) \leq f_{n,m}(\omega) + \epsilon/2$$

and since $m \geq n$, we have

$$f_{n,m} \leq g_m.$$

Taking these three estimates together gives

$$f \leq g_m + \epsilon.$$

This fact together with the estimate in Eq. (3.9) proves

$$g_n \uparrow f.$$

The result now follows from the definition of the integral in step 7. $\square$

Note that the right-hand side in Eq. (3.8) might be infinity. Furthermore, the theorem remains true if the function f assumes the value ∞, but we have not quite defined the integral for such functions (the extension is not difficult). Also, it actually suffices that $\int f_n d\mathbf{P} \geq 0$ rather than $f_n \geq 0$ for the theorem to hold (Dudley, 1989).

Theorem 3.6 (Fatou Lemma). *If $\{f_n\}$ is a sequence of non-negative random variable then,*

$$\int \liminf f_n \, d\mathbf{P} \leq \liminf \int f_n \, d\mathbf{P}. \tag{3.10}$$

Before proving this, a little example for illustration.

Example 3.2. We will later see that on $\Omega = [0,1]$ equipped with the Borel algebra (i.e., the sigma algebra generated by all open sets on $[0,1]$) one can define a probability by the formula $\mathbf{P}(A) = \int_A dx$. The integral with respect to $\mathbf{P}$ is of course the standard Lebesgue integral on the unit interval (or the Riemann integral if the integrand is continuous). Define

$$f_n(x) = n \cdot \mathbf{1}_{[0, \frac{1}{n}]}(x).$$

Now $\liminf f_n = \lim f_n = 0$, and hence the left-hand side of Eq. (3.10) is zero. But $\int f_n(x)\, dx = 1$ and therefore $\liminf \int f_n(x)\, dx = 1$, hence the right-hand side is one. This helps me to remember which direction the inequality goes in Fatou's lemma. Further, the example demonstrates that the integral is in general *not* exchangeable with pointwise limits. Some additional condition (like monotonicity in Theorem 3.5) is necessary. A different but still sufficient condition will be discussed presently.

Proof of Theorem 3.6. Since

$$\inf_k f_{n+k} \leqslant f_{n+l} \quad \text{for all } l \in \mathbb{N},$$

we get by integrating that

$$\int \inf_k f_{n+k}\, d\mathbf{P} \leqslant \int f_{n+l}\, d\mathbf{P} \quad \text{for all } l \in \mathbb{N},$$

so we take the inf over l and obtain

$$\int \inf_k f_{n+k}\, d\mathbf{P} \leqslant \inf_k \int f_{n+k}\, d\mathbf{P}. \tag{3.11}$$

We now want to take the limit $n \to \infty$ on both sides of this inequality. Note that $\inf_k f_{n+k}$ is a monotone sequence in n of non-negative functions, and hence

$$\lim_n \int \inf_k f_{n+k}\, d\mathbf{P} = \int \lim_n \inf_k f_{n+k}\, d\mathbf{P} = \int \liminf_n f_{n+k}\, d\mathbf{P}$$

by monotone convergence and the definition of liminf. On the right-hand side, taking the limit simply gives $\liminf_n \int f_n\, d\mathbf{P}$. □

The next theorem shows that the integral can be interchanged with pointwise limits provided the sequence of functions is bounded. The boundedness condition replaces the monotonicity condition in the Monotone Convergence

Theorem (note that the sequence in Example 3.2 is neither bounded nor monotone).

Theorem 3.7 (Bounded Convergence). *Let* $\{f_n, n \in \mathbb{N}\}$ *be a sequence of random variable with* $|f_n| \leqslant C$, *and* $f_n \to f$ *for* $n \to \infty$. *Then,* f *is integrable and* $\int f_n \, d\mathbf{P} \longrightarrow \int f \, d\mathbf{P}$.

A more general version of this theorem goes under the name *Dominated Convergence Theorem*, in which the condition $|f_n| \leqslant C$ is replaced with $|f_n| \leqslant g$ where g is an integrable function. The conclusions are the same.

Proof. Clearly $|f| \leqslant C$ as well so we get $\int |f| \, d\mathbf{P} \leqslant C$, proving that f is integrable. Since $f_n + C$, and $f + C$ are non-negative, we can apply Fatou and get (after subtracting the constant again from both sides)

$$\int f \, d\mathbf{P} \leqslant \liminf_n \int f_n \, d\mathbf{P}.$$

The same can be done with $-f_n$ and $-f$; we get

$$\int -f \, d\mathbf{P} \leqslant \liminf_n \int -f_n \, d\mathbf{P} = -\limsup_n \int f_n d\mathbf{P},$$

or after multiplying with -1:

$$\int f \, d\mathbf{P} \geqslant \limsup_n \int f_n \, d\mathbf{P}.$$

In summary, we have shown that

$$\liminf_n \int f_n \, d\mathbf{P} \geqslant \int f \, d\mathbf{P} \geqslant \limsup_n \int f_n \, d\mathbf{P},$$

completing the proof. □

Definition 3.6 (Equivalence of random variables).

(a) Let $f_1, f_2 : \Omega \to \mathbb{R}$ functions (not necessarily measurable). We say

$$f_1 = f_2 \qquad \text{almost surely (a.s.)}$$

or f_1 and f_2 are equivalent if $f_1(\omega) = f_2(\omega)$ for all ω in a measurable set Ω_1 with $\mathbf{P}(\Omega_1) = 1$. (One can check that this is indeed an equivalence relation.)

(b) If f is an integrable random variable, we can put

$$\int \hat{f}\, d\mathbf{P} := \int f\, d\mathbf{P},$$

for any $\hat{f}$ which is equivalent to f.

(c) For integrable random variable f, we define the L_1-*norm* by $\|f\|_1 = \int f\, d\mathbf{P}$.

The L_1-norm is in fact not a norm on functions, only a pseudo-norm: $\|f\|_1 = 0$ does not quite imply $f = 0$. But by Exercise 3.6, $f = 0$ almost surely, and therefore $\|f - g\| = 0$ means that f and g are equivalent. So strictly speaking, $\|\cdot\|_1$ is a norm on equivalence classes of functions.

Definition 3.7 (The space L_1).

(a) The space of integrable functions (or strictly speaking, their equivalence classes) with the norm $\|\cdot\|_1$ is denoted as $L_1(\Omega, \mathcal{A}, \mathbf{P})$ or just L_1 if the probability space is clear from the context.

(b) If $\{f_n\}$ is a sequence of integrable random variables and f is another random variable so that $\|f_n - f\|_1 \to 0$ as $n \to \infty$, we will say that $\{f_n\}$ converges to f in L_1 or write $f_n \overset{L_1}{\to} f$.

Theorem 3.8 (Completeness of L_1). *Suppose $\{f_n\}$ is a sequence of random variable which is Cauchy with respect to $\|\cdot\|_1$. Then, there exists an integrable random variable f with $f_n \to f$ in L_1. Further, if f' is another random variable with this property, then $f = f'$ a.s.*

This result is one of the main drivers behind the development of measure and integration. With regards to Theorem 3.8 and also Definition 3.7, it has to be kept in mind that L_1 limits need not be unique; a sequence $\{f_n\}$ of random variables can converge in L_1 against two different functions f and f' at the same time, however, f and f' will be equivalent.

Exercises for Sec. 3.3

Exercise 3.3. *In this exercise, we fill in some details to Sec. 3.3. Let $(\Omega, \mathcal{A})$ be a measurable space (i.e., a set Ω with a sigma algebra $\mathcal{A}$). Consider a function $f : (\Omega, \mathcal{A}) \to (\mathbb{R}, \mathcal{B})$, where $\mathcal{B}$ is the Borel algebra.*

(a) *Consider the family $\mathcal{A}_0$ of all sets of the form $f^{-1}(B)$ where $B \in \mathcal{B}$. Show that $\mathcal{A}_0$ is a sigma algebra on Ω. ($\mathcal{A}_0$ is referred to as the sigma algebra generated by f.)*

(b) *Consider the family $\mathcal{B}_0$ of all sets $B \subset \mathbb{R}$ so that $f^{-1}(B) \in \mathcal{A}$. Show that $\mathcal{B}_0$ is a sigma algebra on $\mathbb{R}$.*

(c) *Conclude that f is a random variable if $\mathcal{B}_0$ from the previous item contains $\mathcal{B}$.*

(d) *Use the previous item and Exercise 3.4 to prove the remark after Definition 3.5 : f is a random variable if $\{\omega \in \Omega; f(\omega) > a\} \in \mathcal{A}$ for any $a \in \mathbb{R}$.*

Exercise 3.4. *In this exercise,[1] we learn more about the Borel algebra $\mathcal{B}$ on $\mathbb{R}$. (Recall that $\mathcal{B}$ is the smallest sigma algebra which contains all open sets.) Show that $\mathcal{B}$ is actually the smallest sigma algebra which contains all sets of the form $]a,\infty]$ for any $a \in \mathbb{R}$. You need to prove that if $\tilde{\mathcal{B}}$ is a sigma algebra containing all sets of the form $]a,\infty]$ for any $a \in \mathbb{R}$, then $\tilde{\mathcal{B}}$ must contain all open sets. Proceed along the following steps:*

(a) *Show that $\tilde{\mathcal{B}}$ contains all left open right closed intervals, i.e., sets of the form $]a,b]$ with $a < b$.*

(b) *Show that $\tilde{\mathcal{B}}$ contains all open intervals (Hint: $]a,b[= \cup_{n=1}^{\infty}]a, b - \frac{1}{n}]$).*

(c) *Show that $\tilde{\mathcal{B}}$ contains countable unions of open intervals.*

(d) *Show that every open set in $\mathbb{R}$ is the union of countably many open intervals (this is difficult, so skip if you want), and conclude that $\tilde{\mathcal{B}}$ contains every open set.*

Exercise 3.5. *In this exercise, we will prove step 7 in the construction of the integral.*

(a) *Because the f_n are an increasing sequence of functions, the same is true for the real numbers $\int f_n d\mathbf{P}$. Therefore, $c = \lim_n \int f_n d\mathbf{P}$ exists. Show that the following statement implies step 7: If g is simple and $g \leqslant f$, then,*

$$\int g d\mathbf{P} \leqslant c. \tag{3.12}$$

The following steps will establish this statement.

(b) *Set $\epsilon > 0$ and define the sets $M_n = \{\omega \in \Omega; f_n(\omega) > g(\omega) - \epsilon\}$. Show that these sets are measurable, that $M_1 \subset M_2 \subset \dots$, and that $\bigcup_{n=1}^{\infty} M_n = \Omega$.*

(c) *Justify all "$\geqslant$" signs in the following:*

$$\int f_n d\mathbf{P} \geqslant \int f_n \cdot \mathbf{1}_{M_n} d\mathbf{P} \geqslant \int g \cdot \mathbf{1}_{M_n} d\mathbf{P} - \epsilon \mathbf{P}(M_n). \tag{3.13}$$

---

[1] This exercise might require bookwork. Check for example Dudley (1989).

(d) *Use sigma additivity to establish that* $\mathbf{P}(M_n) \to 1$, *and that* $\int g \cdot 1_{M_n} d\mathbf{P} \to \int g d\mathbf{P}$ *(remember that g is simple). Using this in Eq. (3.13) gives*

$$c = \lim_n \int f_n d\mathbf{P} \geqslant \int g d\mathbf{P} - \epsilon$$

for any ϵ, establishing Eq. (3.12).

Exercise 3.6. *Show that if f is a non-negative random variable with $\int f d\mathbf{P} = 0$, then $f = 0$ almost surely, that is $f(\omega) = 0$ for all ω in a set Ω_1 with $\mathbf{P}(\Omega_1) = 1$. Hint: Consider the sets $A_n = \{\omega : f(\omega) > 1/n\}$ and show that $n \cdot f \geqslant 1_{A_n}$ to get an upper bound on $\mathbf{P}(A_n)$. What can you now say about $\bigcup_{n=1}^{\infty} A_n$?*

Exercise 3.7. *In this exercise, we will introduce the concept of densities. Let $(\Omega, \mathcal{A}, \mathbf{P})$ be a probability space. Let f be a non-negative random variable, and suppose that $\int f d\mathbf{P} = 1$. On $\mathcal{A}$, define the set function F by*

$$F(A) = \int 1_A \cdot f \, d\mathbf{P}.$$

(a) *Show that F is a probability on $(\Omega, \mathcal{A})$. To prove that F is sigma additive, you need to invoke the Monotone Convergence Theorem.*

(b) *Show that $\mathbf{P}(A) = 0$ implies $F(A) = 0$. (Attention: this is not immediately obvious; assume first that f is simple, then use Monotone Convergence.)*

We will say that f is a density for F. The next item will show that densities are (essentially) unique.

(c) *Using Exercise 3.6, show that if two densities f and g give rise to the same probability F, then $f = g$ almost everywhere. Hint: let $h = f - g$ and consider h_+, h_-.*

3.4   Transformations

This short chapter is devoted to transformations, the pushforward of probabilities and the transformation formula. The material is important for later parts of this chapter but also for dynamical systems.

Let $(\Omega_k, \mathcal{A}_k)$, $k = 1, 2$ be two measurable spaces. In this context, a mapping $T : \Omega_1 \to \Omega_2$ is defined as *measurable* if $T^{-1}(A) \in \mathcal{A}_1$ for all $A \in \mathcal{A}_2$.

Note that random variables as defined in Definition 3.5 are just a special case of this, namely with $(\Omega_2, \mathcal{A}_2) = (\mathbb{R}, \mathcal{B})$. Let $\mathbf{P}$ be a measure on $(\Omega_1, \mathcal{A}_1)$. Then, the formula $T_*\mathbf{P}(A) := \mathbf{P}(T^{-1}(A))$ for all $A \in \mathcal{A}_2$ defines a probability $T_*\mathbf{P}$ on $(\Omega_2, \mathcal{A}_2)$ called the *pushforward* of $\mathbf{P}$ under T. That the pushforward is a probability will be proved in Exercise 3.8.

Theorem 3.9 (Transformation formula). *If $f : (\Omega_2, \mathcal{A}_2) \to (\mathbb{R}, \mathcal{B})$ is a random variable, either positive or integrable w.r.t. $T_*\mathbf{P}$, then,*

$$\int_{\Omega_2} f d(T_*\mathbf{P}) = \int_{\Omega_1} f \circ T d\mathbf{P}.$$

Proof. We prove this for simple functions first. If $f = \sum_{k=1}^{n} f_k \cdot \mathbf{1}_{A_k}$, we have on the left-hand side

$$\int_{\Omega_2} f \, d(T_*\mathbf{P}) = \sum_{k=1}^{n} f_k \cdot \mathbf{P}(T^{-1}(A_k)).$$

On the right-hand side we obtain

$$\int_{\Omega_2} f \circ T \, d\mathbf{P} = \sum_{k=1}^{n} f_k \cdot \int \mathbf{1}_{A_k} \circ T \, d\mathbf{P},$$

$$= \sum_{k=1}^{n} f_k \cdot \int \mathbf{1}_{T^{-1}(A_k)} \, d\mathbf{P} = \sum_{k=1}^{n} f_k \cdot \mathbf{P}(T^{-1}(A_k)),$$

establishing the transformation formula for simple functions. The rest of the proof is covered in Exercise 3.9. □

Exercises for Sec. 3.4

Exercise 3.8. *This exercise fills in several details to the beginning of Sec. 3.4 in preparation of the transformation in Theorem 3.9. Let $(\Omega_k, \mathcal{A}_k), k = 1, 2$ be measurable spaces, $\mathbf{P}$ is a measure on $(\Omega_1, \mathcal{A}_1)$. Further, $T : (\Omega_1, \mathcal{A}_1) \to (\Omega_2, \mathcal{A}_2)$ is a measurable mapping and $f : (\Omega_2, \mathcal{A}_2) \to (\mathbb{R}, \mathcal{B})$ a random variable.*

(a) *Show that the pushforward $T_*\mathbf{P}$ defined by $T_*\mathbf{P}(A) := \mathbf{P}(T^{-1}(A))$ for all $A \in \mathcal{A}_2$ is a probability on the sigma algebra $\mathcal{A}_2$.*
(b) *Show that $f \circ T : (\Omega_1, \mathcal{A}_1) \to (\mathbb{R}, \mathcal{B})$ is a random variable.*
(c) *If $S : (\Omega_0, \mathcal{A}_0) \to (\Omega_1, \mathcal{A}_1)$ is another measurable mapping, show that $T \circ S : (\Omega_0, \mathcal{A}_0) \to (\Omega_2, \mathcal{A}_2)$ is measurable. (Hint: the previous item is a special case of this statement.)*

Exercise 3.9. *In this exercise, we actually prove the transformation in Theorem 3.9. The same setup is as in Theorem 3.9, and we assume it has been proved for simple functions.*

(a) *Use the Monotone Convergence Theorem and the fact that the pushforward is a probability to prove Theorem 3.9 in the case that $f \geqslant 0$.*
(b) *For integrable f prove Theorem 3.9 by considering f_+ and f_- and using the previous item.*

3.5   Products Spaces and Product Measures, Fubini-Theorem

A rectangle in $\mathbb{R}^2$ is the cartesian product of two intervals in $\mathbb{R}$, and the volume of the rectangle is the product of the volumes (i.e., lengths) of these two intervals. Rather than using the standard volume, basically the same can be done with probabilities, resulting in product probabilities. We will also look at the integral of functions against product probabilities and prove that such integrals can be computed as iterated integrals.

Consider a sequence $(\Omega_k, \mathcal{A}_k)$, $k \in \mathbb{N}$ of measurable spaces. We define the *Cartesian Product*

$$\Omega := \prod_{k \in \mathbb{N}} \Omega_k := \text{ sequences } (\omega_1, \omega_2, \dots) \text{ with } \omega_k \in \Omega_k \text{ for all } k \in \mathbb{N}.$$

(3.14)

A sigma algebra can be introduced on Ω as follows. A *rectangular cylinder* is a set of the form

$$\{\omega \in \Omega; \omega_k \in A_k, k \in \mathbb{N}\},$$

where $A_k \in \mathcal{A}_k$ for all $k \in \mathbb{N}$, and $A_k \neq \Omega_k$ for only finitely many k. Now let $\mathcal{A} :=$ smallest *sigma* algebra on Ω containing all rectangular cylinders. Notation $\mathcal{A} := \bigotimes_{k \in \mathbb{N}} \mathcal{A}_k$. The measurable space $(\Omega, \mathcal{A})$ is called the measurable product of $(\Omega_k, \mathcal{A}_k)$, $k \in \mathbb{N}$. A cartesian product over finitely many factors $(\Omega_k, \mathcal{A}_k)$, $k = 1, \dots, K$ is defined in the same way (the requirement that $\mathcal{A}_k \neq \Omega_k$ for only finitely many k in the definition of rectangular cylinders is of course not needed then).

Example 3.3. We define

$$\mathbb{R}^\infty := \prod_{k \in \mathbb{N}} \mathbb{R}, \quad \mathcal{B}_\infty := \bigotimes_{k \in \mathbb{N}} \mathcal{B}(\mathbb{R}),$$

using $(\mathbb{R}, \mathcal{B})$ for all factors. Let $(\Omega, \mathcal{A})$ be another measurable space. A mapping

$$f : (\Omega, \mathcal{A}) \longrightarrow (\mathbb{R}^\infty, \mathcal{B}_\infty)$$
$$\omega \longrightarrow f(\omega) = (f_1(\omega), f_2(\omega), \dots)$$

is measurable if and only if each component f_k is a random variable.

Proof. Exercise 3.10. □

Lemma 3.2. *The set of finite unions of all rectangular cylinders is an algebra.*

Proof. Exercise 3.11. □

Definition 3.8.

(a) For any finite $I \subset \mathbb{N}$, we define the *projections*

$$\pi_I : \prod_{k \in \mathbb{N}} \Omega_k \longrightarrow \prod_{k \in I} \Omega_k,$$

$$(\omega_1, \omega_2, \dots) \longrightarrow (\omega_{k_1}, \dots, \omega_{k_N}),$$

where $\{k_1 < \cdots < k_N\} = I$.

(b) If $\mathbf{P}$ is a probability on $(\prod_{k \in I} \Omega_k, \bigotimes_{k \in I} \mathcal{A}_k)$ we define the I-marginal as $\mathbf{P}_I := \pi_I * \mathbf{P}$, which is a probability on $(\prod_{k \in I} \Omega_k, \bigotimes_{k \in I} \mathcal{A}_k)$.

(c) $\mathbf{P}$ is called a *product probability* if for every rectangular cylinder

$$A = \{\omega \in \Omega; \ \omega_k \in A_k; \ k \in \mathbb{N}\}$$

we have

$$\mathbf{P}(A) = \prod_{k \in \mathbb{N}} \mathbf{P}_{\{k\}}(A_k), \tag{3.15}$$

where $\mathbf{P}_{\{k\}}$ is the marginal for $I = \{k\}$. Note that in Eq. (3.15), only finitely many factors are $\neq 1$. In particular for finite products

$$\Omega = \Omega_1 \times \cdots \times \Omega_N, \quad \mathcal{A} = \mathcal{A}_1 \otimes \cdots \otimes \mathcal{A}_N$$

we have that for $A_1 \in \mathcal{A}_1, \dots, A_N \in \mathcal{A}_N$

$$\mathbf{P}(A_1 \times \cdots \times A_N) = \mathbf{P}_{\{1\}}(A_1) \cdots \cdots \mathbf{P}_{\{N\}}(A_N).$$

Theorem 3.10. *Let* $\mathbf{P}, \mathbf{Q}$ *be two probabilities on*

$$(\Omega, \mathcal{A}) = \left(\prod_{k \in \mathbb{N}} \Omega_k, \bigotimes_{k \in \mathbb{N}} \mathcal{A}_k \right)$$

with all marginals being the same. Then,

$$\mathbf{P} = \mathbf{Q}.$$

Proof. The condition just means that $\mathbf{P} = \mathbf{Q}$ on rectangular cylinders. The rest of the proof is Exercise 3.12. □

Theorem 3.11 (Fubini–Tonelli theorem). *Consider* $(\Omega, \mathcal{A}) = (\Omega_1 \times \Omega_2, \mathcal{A}_1 \otimes \mathcal{A}_2)$ *with product measure* $\mathbf{P} = \mathbf{P}_1 \otimes \mathbf{P}_2$. *Then, for every random variable* $f : \Omega \to \mathbb{R}$,

(a) *For all* $\omega_1 \in \Omega_1$ *the function* $\omega_2 \to f(\omega_1, \omega_2)$ *is measurable.*
(b) *If* $f \geqslant 0$ *or if for all* $\omega_1 \in \Omega_1$ *the function* $\omega_2 \to f(\omega_1, \omega_2)$ *is* $\mathbf{P}_2$-*integrable, then the function* $\omega_1 \to \int f(\omega_1, \omega_2) \cdot d\mathbf{P}_2(\omega_2)$ *is measurable.*
(c) *If* $f \geqslant 0$ *then,*

$$\int f \, d\mathbf{P} = \int \left[\int f(\omega_1, \omega_2) \cdot d\mathbf{P}_2(\omega_2) \right] d\mathbf{P}_1(\omega_1).$$

(d) *If* f *is* $\mathbf{P}$ *integrable, then the function* $\omega_1 \to \int f(\omega_1, \omega_2) \cdot d\mathbf{P}_2(\omega_2)$ *is* $\mathbf{P}_1$-*integrable and*

$$\int f \, d\mathbf{P} = \int \left[\int f(\omega_1, \omega_2) \cdot d\mathbf{P}_2(\omega_2) \right] d\mathbf{P}_1(\omega_1).$$

We will use two lemmata.

Lemma 3.3. *Conditions* (a) *and* (b) *hold for indicators* $1_A, A \in \mathcal{A}$.

Lemma 3.4.

$$\mathbf{P}(A) = \int 1_A \, d\mathbf{P} = \int \left[\int 1_A(\omega_1, \omega_2) \cdot d\mathbf{P}_1 \right] d\mathbf{P}_2.$$

Proof of Lemma 3.3. Put $\mathcal{D} :=$ set of all $A \subset \Omega$ so that (a) and (b) hold for indicators 1_A. If $A = A_1 \times A_2$, $A_1 \in \mathcal{A}_1$, $A_2 \in \mathcal{A}_2$, then $1_A = 1_{A_1}(\omega_1) \cdot 1_{A_2}(\omega_2)$ and (a), (b) hold trivially. Thus, $\mathcal{D} \supset$ all cylinders.

Now let $A_1 \subset A_2 \subset \cdots \in \mathcal{D}$. Then,

$$\mathbf{1}_{A_k}(\omega_1,\omega_2) \uparrow \mathbf{1}_{\bigcup_{k=1}^{\infty} A_k}(\omega_1,\omega_2) \quad \forall(\omega_1,\omega_2) \in \Omega,$$

so in particular for ω_1 fixed. Hence,

$$\omega_2 \longrightarrow \mathbf{1}_{\bigcup_{k=1}^{\infty} A_k}(\omega_1,\omega_2) \quad \text{is measurable.}$$

Further

$$\omega_1 \to \int \mathbf{1}_{\bigcup_{k=1}^{\infty}}(\omega_1,\omega_2)d\mathbf{P}_2 = \int \lim_n \mathbf{1}_{A_k}(\omega_1,\omega_2)d\mathbf{P}_2$$

converges monotonically to $\lim_n \int \mathbf{1}_{A_k}(\omega_1,\omega_2)d\mathbf{P}(\omega_2)$ (measurable!) So $\bigcup_{k=1}^{\infty} A_k \in \mathcal{D}$. If $A_1 \supset A_2 \supset \cdots \in \mathcal{D}$, prove that $\bigcap_{k=1}^{\infty} A_k \in \mathcal{D}$ along similar lines, using dominated convergence. We have shown that $\mathcal{D}$ contains all rectangular cylinders and is a *monotone class*. A family $\mathcal{D}$ is a monotone class if

(a) $A_1 \subset A_2 \subset \cdots \in \mathcal{D} \Rightarrow \bigcup_k A_k \in \mathcal{D}$
(b) $A_1 \supset A_2 \supset \cdots \in \mathcal{D} \Rightarrow \bigcap_k A_k \in \mathcal{D}$.

This implies that $\mathcal{D} \supset \mathcal{A}$ (see e.g., Dudley, 1989).   □

Proof of Lemma 3.4. It is trivial to verify Lemma 3.4 for rectangular cylinders. However, the right-hand side makes sense for any $A \in \mathcal{A}$ and forms a probability (σ-additivity comes from monotone convergence). We thus get Lemma 3.4 by the MET.   □

Completion of proof of Theorem 3.11. We have shown that Theorem 3.11 holds for indicators and clearly also for simple functions. If $f \geqslant 0$ random variable, take $f_n \uparrow f$, f_n simple. Now the functions

$$\omega_2 \longrightarrow f_n(\omega_1,\omega_2) \quad \text{and} \quad \omega_1 \longrightarrow \int f_n(\omega_1,\omega_2)d\mathbf{P}_2,$$

respectively, are measurable and converge monotonically to

$$\omega_2 \longrightarrow f(\omega_1,\omega_2) \quad \forall \omega_1 \quad \text{and} \quad \omega_1 \longrightarrow \int f(\omega_1,\omega_2)d\mathbf{P}_2 \quad \forall \omega_2,$$

respectively (the second by monotone convergence). Finally, because $\int f_n d\mathbf{P} = \int[\int f_n(\omega_1,\omega_2)d\mathbf{P}_2]d\mathbf{P}_1$ and both sides converge by monotone convergence to $\int f d\mathbf{P}$ and $\int[\int f(\omega_1,\omega_2)d\mathbf{P}_2]d\mathbf{P}_1$, respectively, Theorem 3.11 is proved for random variable $f \geqslant 0$.

If f is integrable, Theorem 3.11 holds for f_+ and f_-. The only thing that needs proving is that

$$\omega_1 \longrightarrow \int f(\omega_1,\omega_2)d\mathbf{P}_2 = \int f_+(\omega_1,\omega_2)d\mathbf{P}_2 - \int f_-(\omega_1,\omega_2)d\mathbf{P}_2 \quad (3.16)$$

is well defined (no $\infty - \infty$ situation occurs). Let $N_+ = \{\omega_1 \in \Omega_1; \int f_+(\omega_1,\cdot)d\mathbf{P}_2 = \infty\}$. We must have $\mathbf{P}_1(N_+) = 0$, because

$$\infty > \int f_+ d\mathbf{P} = \int \left[\int f_+(\omega_1,\omega_2)d\mathbf{P}_2 \right] d\mathbf{P}_1.$$

Similarly, for $N_- = \{\omega_1 \in \Omega_1; \int f_-(\omega_1,\cdot)d\mathbf{P}_2 = \infty\}$. So Eq. (3.16) is well defined apart from $\omega_1 \in N_+ \cap N_-$. which has $\mathbf{P}(N_+ \cap N_-) = 0$. $\qquad \square$

Remark 3.1.

(a) The integrability condition cannot be omitted. It's not hard to find cases where $\int |f|d\mathbf{P} = \infty$ and then both sides of item (c) are well defined but fail to be equal.

(b) To verify that f is integrable, one might use item (c) of the Fubini–Tonelli theorem which says that $\int |f|d\mathbf{P} = \int [\int |f|(\omega_1,\omega_2) \cdot d\mathbf{P}_2(\omega_2)]d\mathbf{P}_1(\omega_1)$.

(c) One can use the right-hand side of Lemma 3.4 to define $\mathbf{P}$ from the marginals. We have shown that right-hand side is well defined for $A \in \mathcal{A}$ and is σ-additive. But one then need to invoke MET to show that

$$\int \left[\int 1_A(\omega_1,\omega_2)d\mathbf{P}_1 \right] d\mathbf{P}_2 = \int \left[\int 1_A(\omega_1,\omega_2)d\mathbf{P}_2 \right] d\mathbf{P}_1.$$

(d) Fubini–Tonelli extends to finite products.

Exercises for Sec. 3.5

Exercise 3.10. *Prove the statement in Example 3.3.*

Exercise 3.11. *Show Lemma 3.2: The family of sets which are finite unions of cylinders is an algebra of subsets of $\prod_{k \in \mathbb{N}} \Omega_k$. (Hint: The complement is the tricky bit. Start with assuming (and later showing) that if A and B are cylinders, then $A \setminus B$ is a finite union of cylinders.)*

Exercise 3.12. *Demonstrate Theorem 3.10. You can use without proof the Theorem 3.2.*

Exercise 3.13. *Setup is as in Sec. 3.5.*

(a) *Consider a set of the form*

$$B = \{\omega; \omega_k \in A_k \text{ for all } k \in \mathbb{N}\},$$

with $A_k \in \mathcal{A}_k$ for all $k \in \mathbb{N}$. (B is not necessarily a cylinder!) Show that B is nonetheless measurable.

(b) *Demonstrate that for a product probability $\mathbf{Q}$ (see Definition 3.8, item (c)) and with B as in the previous item, $\mathbf{Q}(B) = \lim_{n \to \infty} \prod_{k=1}^{n} \mathbf{P}_k(A_k)$.*

3.6   Distributions and Independence

In this section, we will change notation somewhat, bringing it closer to standard notation in probability theory. Further, we introduce the important concept of independence.

Let $(\Omega, \mathcal{A}, \mathbf{P})$ be a probability space. Random variables are measurable functions with values in $(\mathbb{R}^d, \mathcal{B}_d)$ (where $d = \infty$ possible) and are denoted by capital letters:

$$X : (\Omega, \mathcal{A}) \to (\mathbb{R}^d, \mathcal{B}_d).$$

If $d = 1$, we put $\mathbb{E}(X) := \int X d\mathbf{P}$ ("expectation value"). If $d < \infty$, $\mathbb{E}(X)$ is taken component wise. Throughout this section, the symbol d stands for a finite integer or for ∞, unless otherwise stated.

Definition 3.9.

(a) The *distribution* of a random variable $X : (\Omega, \mathcal{A}) \to (\mathbb{R}^d, \mathcal{B}_d)$ is defined as $P_X := X_* \mathbf{P}$.
(b) An *I-marginal* of X is the distribution $X_I := (X_{k_1}, \ldots, X_{k_N})$, where $I = \{k_1 < \cdots < k_N\}$.

Note that an *I*-marginal of X according to Definition 3.9 is the same as an *I*-marginal of $\mathbf{P}_X$ according to Definition 3.8 (Exercise 3.15).

Lemma 3.5. *Suppose that $X : \Omega \to \mathbb{R}^d$ is a random variable and $f : \mathbb{R}^d \to \mathbb{R}$ a measurable function (with respect to the Borel algebra on both the domain and range). Further, suppose that $f \circ X$ is integrable. Then,*

$$\mathbb{E}(f \circ X) = \int_{\mathbb{R}^d} f(x) d\mathbf{P}_X(x).$$

Proof. This is essentially the transformation formula, see Exercise 3.14.

$\square$

Lemma 3.6. *Two random variables*

$$X, Y : \Omega \longleftrightarrow \mathbb{R}^d$$

have the same distribution if and only if they have the same I-marginals.

Proof. If d is finite, then the I-marginal for $I = \{1, \ldots, d\}$ is actually the distribution. If d is infinite, let $A = \{x \in \mathbb{R}^\infty; x_{n_k} \in A_k, n_k \in I\}$ be a rectangular cylinder for some $I = \{n_1, \ldots, n_k\} \subset \mathbb{N}$ and some $A_1, \ldots, A_k \in \mathcal{B}_1$. Then,

$$P_{X_I}(A_1 \times \cdots \times A_k) = \mathbf{P}(\{\omega; X_{n_k} \in A_k, n_k \in I\})$$
$$= \mathbf{P}(X \in A) = P_X(A) \tag{3.17}$$

and the same for Y. If the I-marginals agree, then Eq. (3.17) shows that P_X and P_Y agree on rectangular cylinders, so Theorem 3.10 gives $P_X = P_Y$. If on the other hand $P_X = P_Y$, then Eq. (3.17) (read from right to left) shows that the I-marginals agree. $\square$

3.6.1 *Independence*

This paragraph has only two definitions. Some facts about independent random variables will be explored in the exercises.

Definition 3.10. Let $X_1, X_2, \ldots$ random variables with values in $\mathbb{R}$. They are called independent if any I marginal is a product probability. This means that for any $N \in \mathbb{N}$, any index set $I = \{k_1 < \cdots < k_N\}$ and any selection of sets $B_1, \ldots, B_N$ in $\mathcal{B}(\mathbb{R})$ the relation

$$\mathbf{P}(X_{k_1} \in B_1, \ldots, X_{k_N} \in B_N) = \mathbf{P}(X_{k_1} \in B_1) \cdot \ldots \cdot \mathbf{P}(X_{k_N} \in B_N)$$

holds.

Definition 3.11. For any random variable $X : \Omega \to \mathbb{R}^d$, $d < \infty$ define the Covariance matrix

$$\text{Cov}(X) = \mathbb{E}([X_i - \mathbb{E}X_i][X_j - \mathbb{E}X_j])_{i,j}$$

and the variance

$$\mathbb{V}(X) = \text{tr}[\text{Cov}(X)] = \mathbb{E}([X - \mathbb{E}X]^2)$$

(both are finite if $\sum_{k=1}^{n} X_k^2$ is integrable). Finally, if $Y : \Omega \longrightarrow \mathbb{R}^{d'}$ is a random variable $(d' < \infty)$, then

$$\mathrm{Cov}(X,Y) = \mathbb{E}([X_i - \mathbb{E}X_i][Y_j - \mathbb{E}Y_j]) \quad \in \mathbb{R}^{d \times d'}.$$

It is easy to see that $\mathrm{Cov}(X,Y) = \mathrm{Cov}(Y,X)^T$. Note that $\mathrm{Cov}(X)$ is symmetric and non-negative definite, because

$$v^T \mathrm{Cov}(X)v = \mathbb{E}((v^T(X - \mathbb{E}(X)))^2) \geqslant 0.$$

We write $A \geqslant 0$ if $A \in \mathbb{R}^{d \times d}$ symmetric non-negative definite. Also, $A \geqslant B$ means A, B symmetric and $A - B \geqslant 0$. Similarly, ">" means positive definite.

3.6.2   *Modelling with random variables*

The following lemma might sound abstract, but its interpretation is very simple. Suppose we observe data from the real world, and we want to model them as random variables, say $X_1, \ldots, X_d$, each of them real valued. But "modelling" almost always means to merely specify the distribution of those random variables. The probability space $(\Omega, \mathcal{A}, \mathbf{P})$ on which these random variables live, and in fact the variables themselves are usually not specified. The following lemma simply says that this is not a problem, and in the proof you will find a canonical choice for these missing ingredients:

Lemma 3.7. *If μ is a probability on $(\mathbb{R}^d, \mathcal{B}_d)$, then there exists a probability space $(\Omega, \mathcal{A}, \mathbf{P})$ and a measurable random variable $X : \Omega \longrightarrow \mathbb{R}^d$ so that $\mu = P_X$.*

Proof. Take $\Omega = \mathbb{R}^d$, $\mathcal{A} = \mathcal{B}_d$, $\mathbf{P} = \mu$ and $X(\omega) = \omega$. □

Exercises for Sec. 3.6

Exercise 3.14. *In the setup of Lemma 3.5, show that $f \circ X$ is a random variable and prove the formula.*

Exercise 3.15. *For P_X the distribution of some $X : (\Omega, \mathcal{A}, \mathbf{P}) \to (\mathbb{R}^d, \mathcal{B}_d)$, we have defined the concept of I-marginals in Definition 3.9. Show that Definition 3.8 however is also applicable and gives the same concept of I-marginals. (Hint: this is used in the proof of Lemma 3.6).*

Exercise 3.16. *In this exercise, d is finite. Consider a random variable* $X : \Omega \to \mathbb{R}^d$ *with distribution* P_X *which has a density* $p : (\mathbb{R}^d, \mathcal{B}_d) \to (\mathbb{R}_{\geqslant 0}, \mathcal{B})$ *with respect to the n-dimensional Lebesgue measure.*

(a) *Let* $f : (\mathbb{R}^d, \mathcal{B}_d) \to (\mathbb{R}, \mathcal{B})$ *be integrable with respect to* P_X. *Show that*

$$\int f(x) dP_X(x) = \int f(x) p(x) dx.$$

Start with f *being a simple function and proceed as usual. (Note that this extends Lemma 3.5)*

(b) *Show that the marginals of* μ *have densities as well. Hint: For example* μ_1 *has the density*

$$p_1(x_1) = \int_{\mathbb{R}^{d-1}} p(x_1, x_2, \ldots, x_d) dx_2 \ldots dx_d.$$

Exercise 3.17. *Here are some properties of the notion of independence:*

(a) *Show that random variables* $X_1, X_2, \ldots$ *with values in* $\mathbb{R}$ *are independent if and only if for any* $n \in \mathbb{N}$ *and any selection* $f_1, \ldots, f_n$ *of bounded and measurable functions the relation*

$$\mathbb{E}(f_1(X_1) \cdot \ldots \cdot f_n(X_n)) = \mathbb{E}(f_1(X_1)) \cdot \ldots \cdot \mathbb{E}(f_n(X_n))$$

holds.

(b) *Suppose that random variables* $X_1, \ldots, X_d$ *with values in* $\mathbb{R}$ *are independent, and their distribution has a density* p *as in Exercise 3.16. Show that*

$$p(x) = p_1(x_1) \cdot \ldots \cdot p_d(x_d),$$

where p_k *is the density of the distribution of* X_k *for each* $k = 1, \ldots, d$.

(c) *Suppose that random variables* X_1, X_2 *with values in* $\mathbb{R}$ *are independent, and there are sets* B_1, B_2 *in* $\mathcal{B}(\mathbb{R})$ *so that*

$$\{\omega; X_1(\omega) \in B_1\} = \{\omega; X_2(\omega) \in B_2\}.$$

Then $\mathbf{P}(\{\omega; X_1(\omega) \in B_1\}) = 0$ *or* 1.

Exercise 3.18. *Let* $X = (X_1, \ldots, X_d)$ *random variables (d is finite). The distribution of* X *is said to be* normal *or* Gaussian *if it has a density* $p :$

$\mathbb{R}^d \to \mathbb{R}_{>0}$ *with respect to Lebesgue measure given by the formula*

$$p(x;\mu,\Gamma) = \frac{1}{\sqrt{\det(2\pi\Gamma)}} \exp\left(-\frac{1}{2}(x-\mu)^T \Gamma^{-1}(x-\mu) \right)$$

where $\mu \in \mathbb{R}^d$ and Γ is a positive definite $d \times d$–matrix.

(a) *Show that $\mathbb{E}(X_k) = \mu_k$ and $\text{Cov}(X) = \Gamma$.*
(b) *Show that the marginals of the distribution of X are normal as well, and determine the expectation value and covariance matrix.*
(c) *Let A be a surjective $m \times d$–matrix ($m \leqslant d$) and $b \in \mathbb{R}^m$. Show that $AX + b$ has again a normal distribution, and determine the expectation value and covariance matrix.*
(d) *Show that $X_1, \ldots, X_d$ are independent if and only if the covariance matrix is diagonal.*

3.7   Introduction to Statistics

One use of statistics is to "identify" the distribution P_X for some random variable $X : \Omega \longrightarrow \mathbb{R}^d$ given X assumes certain values $(x_1, \ldots, x_d)$. Here, d is finite, but one is also interested what happens if d becomes large, e.g., are we able to reconstruct P_X if $d \to \infty$?

Definition 3.12 (Parametric estimation problem). Let $\Theta \subset \mathbb{R}^p$ open ($p < \infty$). This is the *parameter space*. Let $\mathcal{H} := \{\mathbf{P}_\theta : \theta \in \Theta\}$ be a set of probability measures on $(\mathbb{R}^n, \mathcal{B}_n)$ (i.e., distributions). We call $\mathcal{H}$ the *hypothesis*. The quadruple $(\mathbb{R}^n, \mathcal{B}_n, \Theta, \mathcal{H})$ will be called a *parametric estimation problem*.

Since the parametric estimation problem is about identifying the *distribution* of $X = (X_1, \ldots, X_n)$, we are free to chose Ω, $\mathcal{A}$, and X as in Lemma 3.7 to ensure that X has the desired distribution. This is the choice we made in Definition 3.12. We will write $\mathbb{E}_\theta(\phi) := \int_{\mathbb{R}^n} \phi(x)d\mathbf{P}_\theta$ for any function ϕ integrable with respect to $\mathbf{P}_\theta$. Note that $\mathbf{P}_\theta$ is *not* the distribution of some random variable θ, but that θ is a parameter of that distribution. So our notation in the present section differs slightly from that of Sec. 3.6.

The aim in the parametric estimation problem is to find measurable functions ("estimators")

$$t : (\mathbb{R}^n, \mathcal{B}_n) \to (\Theta, \mathcal{B}_p)$$

so that the distribution of t under $\mathbf{P}_\theta$, that is $t_* \mathbf{P}_\theta$ "concentrates" around θ for all $\theta \in \Theta$. There are many ways to understand the word "concentrates". We will use the concept of *mean square error* to quantify this:

Definition 3.13. Fix $\mathcal{H}$ and estimator t.

(a) The bias of t is the function

$$b : \Theta \longrightarrow \mathbb{R}^p,$$

$$b(\theta) = \mathbb{E}_\theta(t) - \theta.$$

(b) t is unbiased if $b(\theta) = 0$ for all $\theta \in \Theta$.
(c) The mean-square-error (MSE) of t is

$$\mathrm{mse}(\theta) = \mathbb{E}_\theta([t - \theta]^2).$$

Lemma 3.8. *If* $\mathbb{E}_\theta(t^2) < \infty$ *for all* $\theta \in \Theta$, *then*

$$\mathrm{mse}(\theta) = \underbrace{\mathbb{E}_\theta([t - \mathbb{E}_\theta(t)]^2)}_{=: V_\theta(t)} + b(\theta)^2.$$

Proof.

$$\mathrm{mse}(\theta) = \mathbb{E}_\theta([t - \mathbb{E}_\theta(t)]^2) = \mathbb{E}_\theta([t - \mathbb{E}_\theta(t) + \mathbb{E}_\theta - \theta]^2),$$

$$= V_\theta(t) + (b(\theta))^2 - 2 \underbrace{\mathbb{E}_\theta \left([t - \mathbb{E}_\theta(t)] \underbrace{[\theta - \mathbb{E}_\theta(t)]}_{\text{const.}} \right)}_{0}. \qquad \square$$

Definition 3.14 (The standard setup). Let $(\mathbb{R}^n, \mathcal{B}_n, \Theta, \mathcal{H})$ be a parametric estimation problem. We will say that the parametric estimation problem has the *standard setup* if $\mathcal{H}$, the set of candidate distributions, is given by a family of product densities, that is

$$\mathbf{P}_\theta(A) = \int_{\mathbb{R}^n} 1_A \cdot p(x, \theta) dx,$$

where $p(x; \theta) = \prod_{k=1}^n f(x_k, \theta)$ and $f : \mathbb{R} \to \mathbb{R}_{\geq 0}$ is a density (i.e., $\int_{\mathbb{R}} f(x, \theta) dx = 1$).

It follows that if the parametric estimation problem is in the standard setup, the $X_1, \ldots, X_n$ are independent, with all X_k having the same distribution given by the density $f(\cdot, \theta)$.

Example 3.4. Assume the standard setup Definition 3.14 with $f(x,\theta) = g(x-\theta)$, where $g : \mathbb{R} \longrightarrow \mathbb{R}_{\geqslant 0}$ has properties

(1) $\int g(x)dx = 1$.
(2) $\int xg(x)dx = 0$.
(3) $\int x^2 g(x)dx = 1$.

Now item (1) ensures f is a density. From item (2) we get

$$\mathbb{E}_\theta(x_k) = \int x \cdot f(x,\theta)dx = \int xg(x-\theta)dx \overset{(2)}{=} \theta,$$

and $\mathrm{Cov}_\theta(x) = 1$. We will now try to estimate θ. We use $t : \mathbb{R}^n \to \mathbb{R}, t(x) = \frac{1}{n}\sum_{k=1}^n x_k$. First we calculate the bias:

$$\mathbb{E}_\theta(t) = \mathbb{E}_\theta\left(\frac{1}{n}\sum_{k=1}^n x_k\right) = \frac{1}{n}\sum_{k=1}^n \mathbb{E}_\theta x_k = \theta \Rightarrow \mathrm{Bias} = 0.$$

Variance:

$$\mathbb{V}_\theta(t) = \mathbb{E}_\theta\left([t-\theta]^2\right)$$

$$= \mathbb{E}_\theta\left(\left[\frac{1}{n}\sum_{k=1}^n x_k - \theta\right]^2\right)$$

$$= \mathbb{E}_\theta\left(\left[\frac{1}{n}\sum_{k=1}^n (x_k - \theta)\right]^2\right)$$

$$= \frac{1}{n^2}\sum_{k,j=1}^n \mathbb{E}\left([x_k-\theta][x_j-\theta]\right)$$

$$= \frac{n}{n^2} = \frac{1}{n},$$

so $\mathrm{mse}(\theta) = \mathbb{V}_\theta(t) = \frac{1}{n} \longrightarrow 0$.

Definition 3.15 (The Fisher Information regularity conditions). A parametric estimation problem $(\mathbb{R}^n, \mathcal{B}_n, \Theta, \mathcal{H})$ with d finite is said to satisfy the *Fisher Information (FI) regularity conditions* if the following is true:

(1) There is a measure ν on $(\mathbb{R}^d, \mathcal{B}_d)$ and a function $p : (\mathbb{R}^d, \Theta) \longrightarrow \mathbb{R}_{\geqslant 0}$ so that $p(\cdot, \theta)$ is a random variable $\forall \theta \in \Theta$, and

$$\mathbf{P}_\theta(A) = \int_A p(x, \theta)\, d\nu(x).$$

So p is a density with respect to ν (see Exercise 3.7).

(2) $\forall x \in \mathbb{R}^d, \theta \in \Theta : D_\theta p(x, \theta)$ exists.

(3) The following interchange is permitted:

$$\int D_\theta p(x, \theta) d\nu(x) = D_\theta \int p(x, \theta) d\nu(x) = D_\theta 1 = 0.$$

(4) $C(\theta) = \{x \in \mathbb{R}^d; p(x, \theta) > 0\}$ does not depend on θ.

Theorem 3.12 (The Cramér–Rao lower bound). *Suppose the FI-conditions in Definition 3.15 are met. Further, suppose that t is an unbiased estimator of θ. Then*

$$\mathrm{Cov}_\theta(t) = \mathbb{E}_\theta\left[(t - \theta)_i (t - \theta)_j\right] \geqslant \mathcal{I}^{-1},$$

where $\mathcal{I} = \mathrm{Cov}_\theta(D_\theta \log p(x, \theta))$ is the Fisher *information, provided that $\mathcal{I}$ is invertible.*

We will need the following

Lemma 3.9. *Let $(\Omega, \mathcal{A}, \mathbf{P})$ be a probability space. Suppose $X : \Omega \to \mathbb{R}^{d_1}$, $Y : \Omega \to \mathbb{R}^{d_2}$, have finite variances. If $\mathrm{Cov}(Y) > 0$, then*

$$\mathrm{Cov}(X) \geqslant \mathrm{Cov}(X, Y)\mathrm{Cov}(Y)^{-1}\mathrm{Cov}(Y, X).$$

Further, equality holds here if and only if there is $M \in \mathbb{R}^{d_1 \times d_2}$ so that $X = MY$.

Proof. Put

$$Z = \begin{pmatrix} X \\ Y \end{pmatrix}$$

and consider

$$\mathrm{Cov}(Z) = \begin{bmatrix} \mathrm{Cov}(X) & \mathrm{Cov}(X, Y) \\ \mathrm{Cov}(Y, X) & \mathrm{Cov}(Y) \end{bmatrix}.$$

Next, choose

$$W = \begin{bmatrix} 1 \\ -\operatorname{Cov}(Y)^{-1} \cdot \operatorname{Cov}(Y,X) \end{bmatrix} \in \mathbb{R}^{(d_1+d_2) \times d_1}.$$

Now,

$$0 \leqslant W^T \operatorname{Cov}(Z)W = \operatorname{Cov}(X) - \operatorname{Cov}(X,Y)\operatorname{Cov}(Y)^{-1}\operatorname{Cov}(Y,X).$$

If equality holds, then $W^T \operatorname{Cov}(Z)W = \operatorname{Cov}(W^T Z) = 0$. Checking the diagonal elements, we get $W^T Z = 0$ or $X = \underbrace{\operatorname{Cov}(X,Y)\operatorname{Cov}(Y)^{-1}}_{M} \cdot Y$.   □

Proof of Theorem 3.12. By taking the derivative of the identity

$$1 = \int p(x,\theta)d\nu(x),$$

we get

$$0 = \int \partial_{\theta_j} \log p(x,\theta) \cdot p(x,\theta)d\nu(x). \tag{3.18}$$

From unbiasedness, we get

$$0 = \int [t(x) - \theta]_i p(x,\theta)d\nu(x) = \mathbb{E}_\theta(t_i) - \theta_i,$$

and taking the derivative of this identity and using Eq. (3.18), we obtain

$$0 = -\delta_{ij} + \int [t(x) - \theta_i] \cdot \partial_{\theta_j} \log p(x,\theta) \cdot p(x,\theta)d\nu. \tag{3.19}$$

Equation (3.19) can be written as $1 = \operatorname{Cov}(t - \theta, D_\theta \log p(X,\theta))$. Now the lemma gives the result.   □

Lemma 3.10. *Under appropriate regularity conditions*

$$\mathcal{I} = -\mathbb{E}_\theta(\partial^2_{\theta_i \theta_j} \log p(X,\theta)).$$

Proof. Exercise 3.19.   □

Definition 3.16. Let $(\mathbb{R}^n, \mathcal{B}_n, \Theta, \mathcal{H})$ be a parametric estimation problem, and suppose that t is an unbiased estimator. If

$$\operatorname{Cov}_\theta(t) \leqslant \operatorname{Cov}_\theta(t') \quad \forall \theta \in \Theta$$

for any other unbiased estimator t', then t is called Uniformly Minimum Variance Unbiased (UMVU) estimator.

Example 3.5 (Another example). Standard setup in Definition 3.14, with $f(x,\theta) = \frac{1}{\theta}\exp(-\frac{x}{\theta})$, $\theta \in \mathbb{R}_+$, $x \in \mathbb{R}_{\geqslant 0}$. Consider the estimator $t = \frac{1}{n}\sum_{k=1}^{n} x_k$. For this estimator, we get $\mathbb{E}_\theta(t) = \theta$, $\mathbb{V}_\theta(t) = \frac{\theta^2}{n}$. The FI-regularity conditions are easily checked, and

$$\mathcal{I} = -\mathbb{E}_\theta\left(\partial_\theta^2 \log p(x,\theta)\right) = \frac{n}{\theta^2}$$

as the following calculations show:

$$-\log p(x,\theta) = \sum_{k=1}^{n}\log(\theta) + \frac{x_k}{\theta} = n\log(\theta) + \sum_{k=1}^{n}\frac{x_k}{\theta},$$

$$\partial_\theta^2(-\log p(x,\theta)) = -\frac{n}{\theta^2} + 2\frac{\sum_k x_k}{\theta^3},$$

$$\mathbb{E}_\theta\left(\partial_\theta^2(-\log p(X,\theta))\right) = -\frac{n}{\theta^2} + 2\frac{n\cdot\theta}{\theta^3} = \frac{n}{\theta^2}.$$

Hence, t has minimum variance!

Definition 3.17 (The maximum likelihood estimator). Suppose a parametric estimation problem is in the standard setup and satisfies the FI-regularity conditions in Definition 3.15. The *Likelihood* $L : \Theta \times \mathbb{R}^n \to \mathbb{R}$ is given by

$$L(\theta, x) = p(x, \theta).$$

Consider the set $C = \{x \in \mathbb{R}^d; f(x,\theta) > 0\}$ and suppose that for any $x \in C$ the likelihood as a function of θ, that is $\theta \to L(\theta, x)$, has a unique maximiser $t(x)$. In other words, for any $x \in C$ and $\theta \in \Theta$ it holds that

$$L(t(x), x) \geqslant L(\theta, x),$$

with equality here if and only if $\theta = t(x)$. Then t is called the *Maximum Likelihood estimator* (MLE).

Remark 3.2.

(a) It is often useful to maximise the log-likelihood function, $l := \log L(\cdot, x)$.

(b) In the standard setup, we have

$$p(x, \theta) = \prod_{k=1}^{n} f(x_k, \theta),$$

so $l = \log L(\theta, x) = \sum_{k=1}^{n}\log f(x_k, \theta)$.

Example 3.6.

(a) Assume the standard setup with $f(x,\theta) = \frac{1}{\theta}\exp\left(-\frac{x}{\theta}\right)$, $\theta > 0, x \geqslant 0$. Then

$$l(\theta,x) = -\frac{1}{\theta}\sum_{k=1}^{n} x_k - n\log\theta.$$

To find the MLE, we solve the normal equations:

$$\frac{\partial}{\partial\theta}l(\theta,x) = \frac{1}{\theta^2}\sum_{k=1}^{n} x_k - \frac{n}{\theta},$$

$$\frac{\partial}{\partial\theta}l(t,x) = 0 \Leftrightarrow 0 = \frac{1}{t^2}\sum_{k=1}^{n} x_k - \frac{n}{t},$$

$$\Leftrightarrow t(x_1,\ldots,x_n) = \frac{1}{n}\sum_{k=1}^{n} x_k \quad \text{as in Example 3.5.}$$

(b) Assume the standard setup with

$$f(x,\theta) = \frac{x^m \cdot \exp(-\theta x)\theta^{m+1}}{m!},$$

where $m \in \mathbb{N}$ is a fixed.

$$l(\theta,x) = \sum_{k=1}^{n}\left\{\log\frac{1}{m!} + m\log x_k - \theta x_k + (m+1)\log\theta\right\},$$

$$\frac{\partial}{\partial\theta}l(\theta,x) = -\sum_{k=1}^{n} x_k + n\cdot(m+1)/\theta,$$

$$\frac{\partial}{\partial\theta}l(t,x) = 0 \Leftrightarrow t = \frac{(m+1)}{\frac{1}{n}\sum_{k=1}^{n} x_k}.$$

Remark 3.3. It is not always true that the MLE is UMVU. But if the parametric estimation problem satisfies the standard setup, the FI-regularity conditions, and some more conditions, then t has, asymptotically for large n, a normal distribution with mean θ and variance $\mathcal{I}(\theta)^{-1}$. See Chapter 5 from van der Vaart (2000) for precise statements and proofs of this result.

Exercises for Sec. 3.7

Exercise 3.19. *Deduce the formula in Lemma 3.10 for the Fisher information and state the conditions under which this formula is correct.*

Exercise 3.20. *Assume the $\mathbb{R}$-valued random variables $X_1, \ldots, X_n$ (n is finite) are independent, non-negative, and for all $k = 1, \ldots, n$, the distribution of X_k has a density with respect to Lebesgue measure, given by*

$$p(x, \theta) = \begin{cases} \dfrac{1}{\theta} & \text{if } 0 \leqslant x \leqslant \theta, \\ 0 & \text{else,} \end{cases}$$

where $\theta \in \mathbb{R}_{>0}$. This is called a uniform distribution. *Given that X_k can never be larger than θ, we put $T(x) = \max\{x_1, \ldots, x_n\}$ as an estimator for θ.*

(a) *Show that*

$$\mathbf{P}(T \leqslant z) = \frac{z^n}{\theta^n}, \qquad 0 < z < \theta.$$

(b) *Conclude from the previous item that T has a density given by*

$$q(z) = \frac{n z^{n-1}}{\theta^n}, \qquad 0 < z < \theta.$$

(c) *Show that T is biased, and suggest an estimator for θ based on T which would be unbiased.*

Exercise 3.21. *Let $x_1, \ldots, x_n$ be independent and identically distributed random variables, and the distribution of each x_k has a density $f(x; \theta)$ with respect to Lebesgue measure of the form*

$$f(x; \theta) = a(x) \exp(\theta h(x) - \phi(\theta)),$$

where $a \geqslant 0$ and h are given random variables, and $x \in I$ with I some interval not depending on θ. Densities of this kind are referred to as exponential families *(not to be confused with the exponential distribution, which is a special case). You can assume that $f(x; \theta)$ is a well-defined density for all θ in some open interval Θ. Further, you can assume that differentiation under the integral sign is permitted in the following, that the function $\theta \to \mathbb{E}_\theta(h)$ is one-to-one and continuous in both directions, and that $0 < \mathbb{V}_\theta(h) < \infty$ for all $\theta \in \Theta$.*

(a) *By considering the condition $\int_{x \in I} f(x; \theta) dx = 1$, provide an expression for the function ϕ.*

(b) *By differentiating $\int_{x \in I} f(x; \theta) dx = 1$ with respect to θ, prove that*
$\mathbb{E}_\theta(h) = \frac{d\phi}{d\theta}$. *Differentiate again to show that* $\mathbb{V}_\theta(h) = \frac{d^2\phi}{d\theta^2}$.

(c) *Show that the maximum likelihood estimator $\hat{\theta}$ for θ is a solution of the equation* $\frac{1}{n} \sum_{k=1}^{n} h(x_k) = \frac{d\phi}{d\theta}(\hat{\theta})$.

(d) *Show that the maximum derived in the last item is actually a maximum at least locally (Hint: You might want to use results from item (b) for this proof.)*

(e) *Prove that the Fisher information (of the joint density) is given by* $n\frac{d^2\phi}{d\theta^2}$.

(f) *For this last item, you need to use the Law of large numbers: If $Y_1, \ldots, Y_n$ are independent and identically distributed random variables with finite variance, then $\frac{1}{n} \sum_{k=1}^{n} Y_k \to \mathbb{E}(Y_1)$ almost surely for $n \to \infty$. Using this, prove that if θ_0 corresponds to the true distribution of $x_1, \ldots, x_n$, then $\hat{\theta} \to \theta_0$.*

3.8   Conditional Probabilities and Conditional Expectations

Let $(\Omega, \mathcal{A}, \mathbf{P})$ probability space. Consider L_2, the space of all random variables $f : \Omega \to \mathbb{R}$ so that $\int f^2 d\mathbf{P} < \infty$. This is a Hilbert space with scalar product $\langle f, g \rangle := \int f g d\mathbf{P}$. Let f be an element of this Hilbert space and S be a closed subspace. Then there exists $\hat{f} \in S$ which is the "best approximation" f, which means

$$\|f - \hat{f}\|^2 = \langle f - \hat{f}, f - \hat{f} \rangle \leqslant \|f - g\|^2, \quad \forall g \in S,$$

and equality occurs here if and only if $g = \hat{f}$. We now claim that $f - \hat{f}$ (i.e., the approximation error) is perpendicular to S, that is $\langle f - \hat{f}, g \rangle = 0$ for any $g \in S$. To see this, note that for any $g \in S$ we have

$$\|f - (\hat{f} + g)\|^2 = \|f - \hat{f}\|^2 + \|g\|^2 - 2\langle f - \hat{f}, g \rangle.$$

Suppose there exists $g \in S$ with $\langle f - \hat{f}, g \rangle = m \neq 0$, then replace g in the relation above with $g' = \frac{m}{\|g\|^2} g$, which gives

$$\|f - (\hat{f} + g')\|^2 = \|f - \hat{f}\|^2 + \frac{m^2}{\|g\|^2} - 2\frac{m^2}{\|g\|^2}$$

$$= \|f - \hat{f}\|^2 - \frac{m^2}{\|g\|^2} < \|f - \hat{f}\|^2,$$

which means that $\hat{f} + g'$ is a better approximation than $\hat{f}$, which is a contradiction. Hence $\langle f - \hat{f}, g \rangle = 0$, or

$$\int f \cdot g d\mathbf{P} = \int \hat{f} \cdot g d\mathbf{P} \quad \text{for all } g \in S. \tag{3.20}$$

We want to use this with S being the space of all random variables g so that $\int g^2 d\mathbf{P} < \infty$ and g is $\mathcal{F}$-measurable, where $\mathcal{F}$ is some σ-algebra on Ω with $\mathcal{F} \subset \mathcal{A}$. It is clear that S is a subspace of L_2 and is in itself a Hilbert space. This implies that S is closed in L_2, hence we can find $\hat{f}$ so that Eq. (3.20) is correct. Note that for the special S we have chosen here, Eq. (3.20) would be true if

$$\int f \cdot 1_A d\mathbf{P} = \int \hat{f} \cdot 1_A d\mathbf{P}, \quad \forall A \in \mathcal{F} \tag{3.21}$$

by approximation. But Eq. (3.21) makes sense even if $\int |f| d\mathbf{P} < \infty$, which is weaker that $\int f^2 d\mathbf{P} < \infty$. This leads us to the following definition.

Definition 3.18. Let $\int |f| d\mathbf{P} < \infty$, $\mathcal{F} \subset \mathcal{A}$, $\mathcal{F}$ a sigma-algebra. Then the *conditional expectation of f given $\mathcal{F}$*, written as $\mathbb{E}(f|\mathcal{F})$, is any $\mathcal{F}$-measurable function $\hat{f}$ satisfying Eq. (3.21).

Theorem 3.13. *Let $\int |f| d\mathbf{P} < \infty, \mathcal{F} \subset \mathcal{A}, \mathcal{F}$ a sigma algebra.*

(a) *There exists conditional expectation $\mathbb{E}(f|\mathcal{F})$.*
(b) *Suppose $f^{(1)}, f^{(2)}$ are $\mathcal{F}$-measurable and satisfy Eq. (3.21), then*

$$f^{(1)}(\omega) = f^{(2)}(\omega)$$

for $\omega \in \Omega_1$, with $\mathbf{P}(\Omega_1) = 1$.

Proof. Suppose first $f \geqslant 0$. Put $c = \int f d\mathbf{P}$ and define

$$F(A) := \frac{1}{c} \int f \cdot 1_A d\mathbf{P} \tag{3.22}$$

for any $A \in \mathcal{F}$. Note that this is a probability on $\mathcal{F}$ (see Exercise 3.22). Using the Radon–Nykodym theorem, it can be shown that $\exists \mathcal{F}$-measurable random variable $\hat{f} \geqslant 0$ with

$$F(A) = \frac{1}{c} \int \hat{f} \cdot 1_A d\mathbf{P} \tag{3.23}$$

for any $A \in \mathcal{F}$. Combining Eqs. 3.22 and 3.23 is Eq. (3.21). The uniqueness is like the uniqueness for densities. For general f, consider f_+ and f_-. $\qquad\square$

Remark 3.4 (Defining properties of the conditional expectation).
Let us say you have some $\hat{f}$ and you suspect that

$$\hat{f} = \mathbb{E}(f|\mathcal{F}).$$

To verify this, you have to check that

(a) $\hat{f}$ is $\mathcal{F}$-measurable
(b)

$$\int f \cdot g d\mathbf{P} = \int \hat{f} \cdot g d\mathbf{P}$$

for any function g which is $\mathcal{F}$-measurable and bounded (in fact, it suffices to check this for all g of the form $\mathbf{1}_A$ with $A \in \mathcal{F}$).

Lemma 3.11 (Properties of $\mathbb{E}(f|\mathcal{F})$).

(a) *Linear in f.*
(b) $f \geqslant 0 \Rightarrow \mathbb{E}(f|\mathcal{F}) \geqslant 0$ *a.s.*
(c) *If $\mathcal{Y} \subset \mathcal{F} \subset \mathcal{A}$ are sigma-algebras, then*

$$\mathbb{E}(\mathbb{E}(f|\mathcal{F})|\mathcal{Y}) = \mathbb{E}(f|\mathcal{Y})$$

(*Law of the Iterated Expectations*).

Proof. Exercise 3.23. □

Definition 3.19.

(a) Let $g : (\Omega, \mathcal{A}) \to (\Omega', \mathcal{A}')$ measurable. The family of sets

$$\sigma(g) = \{g^{-1}(A) : A \in \mathcal{A}'\}$$

is a sigma algebra, called the *sigma algebra generated by g* ($\mathcal{A}'$ is fixed).
Measurability of g implies $\sigma(g) \subset \mathcal{A}$.
(b) $\mathbb{E}(f|g) := \mathbb{E}(f|\sigma(g))$. Note that this is a random variable on $(\Omega, \mathcal{A})$.
(c) The following is a slightly different concept of conditional expectation.
Let $X : (\Omega, \mathcal{A}) \to (\mathbb{R}^d, \mathcal{B}_d)$. Then $\mathbb{E}(f|X = x)$ is any random variable $\hat{f}$ satisfying

$$\int \mathbf{1}_B(x) \cdot \hat{f}(x) d\mathbf{P}_X(x) = \int \mathbf{1}_B \circ X(\omega) \cdot f(\omega) d\mathbf{P}(\omega)$$

for all $B \in \mathcal{B}_d$. Note that $\mathbb{E}(f|X = x)$ is a random variable on $(\mathbb{R}^d, \mathcal{B}_d)$.

Lemma 3.12. $\hat{f}(x) = \mathbb{E}(f|X = x) \Leftrightarrow \hat{f}(X(\omega)) = \mathbb{E}(f|\sigma(X))(\omega).$

Proof. Exercise 3.24. $\qquad\qquad\qquad\qquad\qquad\qquad\qquad\qquad\qquad\qquad\square$

Definition 3.20 (Conditional Probability). The conditional expectation of an indicator function has a special interpretation. Let $A \in \mathcal{A}$.

(a) $\mathbf{P}(A|\mathcal{F}) := \mathbb{E}(\mathbf{1}_A|\mathcal{F})(\omega)$ is called the *conditional probability* of A given $\mathcal{F}$.

(b) If X is a random variable, we define $\mathbf{P}(A|X) := \mathbb{E}(\mathbf{1}_A|X)$ and call it conditional probability of A given X.

(c) Using the alternative concept of conditional expectation in Definition 3.19, item (c) we define $\mathbf{P}(A|X = x) := \mathbb{E}(\mathbf{1}_A|X = x)$.

Note that for $B \in \mathcal{F}$ we have the formula

$$\int \mathbf{P}(A|\mathcal{F}) \cdot \mathbf{1}_B \, d\mathbf{P} = \int \mathbf{1}_A \cdot \mathbf{1}_B \cdot d\mathbf{P} = \mathbf{P}(A \cap B).$$

Lemma 3.13 (Bayes–Rule).

$$X : (\Omega, \mathcal{A}) \longrightarrow (\mathbb{R}^{d_1}, \mathcal{B}_{d_1}),$$

$$Y : (\Omega, \mathcal{A}) \longrightarrow (\mathbb{R}^{d_2}, \mathcal{B}_{d_2}),$$

$d_1 + d_2 < \infty$. *Suppose that* $Z = (X, Y)$ *has density* $p(z) = p(x, y)$. *Then*

$$\mathbf{P}(X \in B|Y = y) = \frac{\int_B p(x, y) dx}{\int_{\mathbb{R}^{d_1}} p(x, y) dx}$$

for all $B \in \mathcal{B}_{d_1}$.

Proof. Exercise 3.25. $\qquad\qquad\qquad\qquad\qquad\qquad\qquad\qquad\qquad\qquad\square$

3.8.1   *Regular conditional probabilities*

Consider a sigma-algebra $\mathcal{F} \subset \mathcal{A}$ and the conditional probability $\mathbf{P}(A|\mathcal{F})(\omega)$. We have a mapping

$$\mu : (\mathcal{A} \times \Omega) \longrightarrow [0, 1],$$

$$\mu(A, \omega) = \mathbf{P}(A|\mathcal{F})(\omega),$$

so that

(a) For all $A \in \mathcal{A}$, $\omega \to \mu(A, \omega)$ is $\mathcal{F}$-measurable random variable indexed by ω.

We would also like to have

(b) For all $\omega \in \Omega$; $A \longrightarrow \mu(A, \omega)$ is a probability on A.

But there is a problem: Note that the relation

$$\lim_{n \to \infty} \sum_{k=1}^{n} \mu(A_k, \omega) = \mu \left(\bigcup_{k=1}^{\infty} A_k, \omega \right) \tag{3.24}$$

for pairwise disjoint $A_1, A_2, \ldots, \in \mathcal{A}$ merely holds for $\omega \in \Omega_0$ with $\mathbf{P}(\Omega_0) = 1$. Although these are "almost all ω", the set Ω_0 where Eq. (3.24) holds depends on $A_1, A_2, \ldots$. Now μ would have to be modified on Ω_0^c in order to render Eq. (3.24) correct for *all* ω. We then have to repeat this for any sequence $(A_1, A_2, \ldots)$ of measurable and pairwise disjoint sets. There are uncountably many such sequences, hence uncountably many "problem sets" Ω_0^c, and their union might have non-zero measure.

Theorem 3.14. *Let* $X : (\Omega, \mathcal{A}) \to (\mathbb{R}^d, \mathcal{B}_d), d = \infty$ *permitted,* $\mathcal{F} \subset \mathcal{A}$ *sigma algebra. Then the conditional distribution*

$$P_X(B|\mathcal{F}) := \mathbf{P}(\{\omega; X(\omega) \in B | \mathcal{F})$$

has a regular *version* $\mu : (\mathcal{B}_d \times \Omega) \to [0, 1]$, *that is for any* $B \in \mathcal{B}_d$ *the equation* $P_X(B|\mathcal{F})(\omega) = \mu(B, \omega)$ *holds, provided* $\omega \in \Omega_B$, *where* $\mathbf{P}(\Omega_B) = 1$, *and* μ *satisfies conditions* (a) *and* (b) *at the beginning of this paragraph.*

Proof. See Theorem 4.34 from Breiman (1973). The structure of $\mathcal{B}_d$ enters in an essential way. □

Exercises for Sec. 3.8

Exercise 3.22. *Prove that the mapping F in the proof of Theorem 3.13 is a probability on $\mathcal{F}$.*

Exercise 3.23. *Prove Lemma 3.11.*

Exercise 3.24. *Prove Lemma 3.12.*

Exercise 3.25. *Prove Lemma 3.13.*

3.9   Convergence in Distribution and Characteristic Functions

Let $X_1, X_2, \ldots$ random variables $X_k : \Omega \longrightarrow \mathbb{R}$, which are independent with identical distribution (i.i.d.), $\mathbb{E}(X_k) = m$, $\mathbb{E}((X_k - m)^2) = \sigma^2 < \infty$. Then, $\mathbb{E}\left(\sum_{k=1}^{n}(X_k - m)\right) = 0$, $\mathbb{E}\left(\left(\sum_{k=1}^{n}(X_k - m)\right)^2\right) = n \cdot \sigma^2$, so $Z_n :=$ $\frac{1}{\sqrt{n}} \sum_{k=1}^{n} \frac{X_k - m}{\sigma}$ has mean zero and unit variance. Use your computer to try out specific examples and convince yourselves that Z_n does not converge pointwise (or any strong sense). On the other hand, if for example $X_k = 1$ or -1, both with probability $1/2$, it can be shown that

$$\mathbf{P}(Z_n \leqslant z) = P_{Z_n}((-\infty, z])$$

$$\longrightarrow \frac{1}{\sqrt{2\pi}} \int_{-\infty}^{z} \exp\left(-\frac{1}{2}x^2\right) dx \qquad \text{(standard normal)}.$$

Note that this does *not* involve the convergence of random variables but rather of *distributions*, that is probabilities on $(\mathbb{R}, \mathcal{B})$; write $\mathcal{D}$ for the family of all distributions on $(\mathbb{R}, \mathcal{B})$. For $\mu_1, \mu_2, \ldots \in \mathcal{D}$ we could define

$$\mu_n \xrightarrow{\mathcal{D}} \mu \in \mathcal{D} \quad \text{if } \mu_n(A) \longrightarrow \mu(A) \; \forall A \in \mathcal{B},$$

but this is too strong. Let for example $\mu_n = $ uniform measure of $[0, 1/n]$, then $\mu_n(\{0\}) = 0$ for all n but clearly we would like

$$\mu_n \xrightarrow{\mathcal{D}} \delta_0,$$

which according to our too strong definition would necessitate $\mu_n(\{0\}) \to 1$. The following definition of "$\xrightarrow{\mathcal{D}}$" is weaker.

Definition 3.21.

(a) If $\mu \in \mathcal{D}$, call $F_\mu(z) := \mu((-\infty, z])$ the *Cumulative Distribution Function* (CDF) of μ.

(b) We will say that $\mu_n \xrightarrow{\mathcal{D}} \mu$ (read "μ_n converges in distribution to μ") if one of the equivalent condition holds (we will not prove this):

   (i) $\int \phi d\mu_n \longrightarrow \int \phi d\mu$ for any bounded and continuous function ϕ on $\mathbb{R}$.
   (ii) $F_{\mu_n}(z) \longrightarrow F_\mu(z)$ at every z where F_μ continuous.

Note that the integral in item (b)(i) is always well defined because ϕ is assumed bounded. Further, it can be shown that two distributions μ_1 and

μ_2 agree if

$$\int \phi d\mu_1 = \int \phi d\mu_2$$

for any bounded and continuous function ϕ on $\mathbb{R}$. This implies that the limit in distribution is well defined.

Definition 3.22. The characteristic function of $\mu \in \mathcal{D}$ is

$$f(u) := \int e^{iux} d\mu(x) \quad u \in \mathbb{R}.$$

This is well defined since $|e^{iux}| = 1$.

Lemma 3.14. *The characteristic function f of some $\mu \in \mathcal{D}$ has the following properties:*

(a) *f is uniformly continuous.*
(b) *$f(0) = 1$.*
(c) *$|f(u)| \leqslant 1$.*
(d) *$f(-u) = \overline{f}(u)$.*

Proof. Only item (a) is non-trivial

$$|f(u+h) - f(u)| = \left| \int e^{i(u+h)x} - e^{iux}) d\mu(x) \right|,$$

$$= \left| \int (e^{ihx} - 1) e^{iux} d\mu(x) \right|,$$

$$\leqslant \int |(e^{ihx} - 1)| d\mu(x) =: \delta(h),$$

and $\delta(h) \to 0$ if $h \to 0$ by the bounded convergence theorem. $\qquad \square$

Theorem 3.15. *Let f_k be the characteristic function of μ_k, $k = 1, 2$. Then $f_1 = f_2$ implies $\mu_1 = \mu_2$.*

Proof. We will show that $\int \phi d\mu_1 = \int \phi d\mu_2$ for any bounded and continuous function ϕ on $\mathbb{R}$. If

$$\int \exp(iux) d\mu_1 = \int \exp(iux) d\mu_2,$$

then $\int g d\mu_1 = \int g d\mu_2$ for

$$g(x) = \sum_{\text{finite } k} \alpha_k e^{i u_k x}. \tag{3.25}$$

Now let $\epsilon_n \leqslant 1$, $\epsilon_n \to 0$. Any bounded and continuous function ϕ can be approximated uniformly on $[-n, n]$ by functions g_n of the form in Eq. (3.16), so

$$|\phi(x) - g_n(x)| \leqslant \epsilon_n$$

for $x \in [-n, n]$. But since $\phi(x) \leqslant M$ and $\epsilon_n \leqslant 1$, and since g is periodic, we have

$$|g_n| \leqslant M + 1.$$

So Eq. (3.25) gives

$$\int g_n d\mu_1 = \int g_n d\mu_2$$

and hence

$$\int \phi d\mu_1 = \int g_n d\mu_1 + \int \phi - g_n d\mu_1$$

$$= \int g_n d\mu_2 + \int \phi - g_n d\mu_1$$

$$= \int \phi d\mu_2 + \int \phi - g_n d\mu_1 + \int g_n - \phi d\mu_2.$$

It remains to be proven that by choosing n large enough, $\int \phi - g_n d\mu_1$ and $\int \phi - g_n d\mu_2$ can be made as small as we want. Fix $\epsilon > 0$. Note that for any $z > 0$ we have

$$\int |\phi - g_n| d\mu_1 = \int_{[-z, z]} |\phi - g_n| d\mu_1 + \int_{[-z, z]^c} |\phi - g_n| d\mu_1$$

$$\leqslant \int_{[-z, z]} |\phi - g_n| d\mu_1 + (M + 1)\mu_1([-z, z]^c).$$

Since $\mu_1([-z, z]^c) \to 0$ for $z \to \infty$, we can pick z so large that $(M + 1)\mu_1([-z, z]^c) \leqslant \frac{\epsilon}{2}$. Next, pick n_0 so large that $z \leqslant n$ and further $\epsilon_n \cdot 2z \leqslant \frac{\epsilon}{2}$ whenever $n \geqslant n_0$. Using this in the estimate above gives

$$\int |\phi - g_n| d\mu_1 \leqslant \epsilon,$$

whenever $n \geqslant n_0$. The reasoning for the integral $\int |\phi - g_n| d\mu_2$ is the same.

$\square$

Theorem 3.15 demonstrates that no two distributions can have the same characteristic function. The following theorem strengthens this and demonstrates why characteristic functions are so important.

Theorem 3.16. *Assume that $f_n, n \in \mathbb{N}$ are characteristic functions of some distributions μ_n, and further that $f_n \to f$ pointwise if $n \to \infty$, where f is also the characteristic function of some distribution μ. Then $\mu_n \xrightarrow{\mathcal{D}} \mu$.*

Proof. (*The proof will be incomplete.*) We know that $\int \phi d\mu_n \to \int \phi d\mu$ for every function ϕ of the form $\phi(x) = \exp(iux)$ for some u, so we can assume that this convergence takes place if ϕ is of the form in Eq. (3.25) in the proof of Theorem 3.15. We have to show that this implies $\int \phi d\mu_n \to \int \phi d\mu$ for every continuous and bounded function ϕ on $\mathbb{R}$. We will do this under the additional assumption that all μ_n are concentrated on a compact interval I. So let ϕ be a continuous and bounded function on I, and let $\epsilon > 0$. (All integrals that follow will be over I.) As in the proof of Theorem 3.15, we can find a function g of the form in Eq. (3.25) in the proof of Theorem 3.15 so that

$$\sup_{x \in I} |g(x) - \phi(x)| \leq \frac{\epsilon}{3}.$$

With g chosen, take n_0 so large that

$$\left| \int g d\mu_n - \int g d\mu \right| \leq \frac{\epsilon}{3}$$

for any $n \geq n_0$. This implies

$$\left| \int \phi d\mu_n - \int \phi d\mu \right|$$

$$\leq \left| \int (\phi - g) d\mu_n + \int g d\mu_n - \int g d\mu + \int (g - \phi) d\mu \right|$$

$$\leq \left| \int (\phi - g) d\mu_n \right| + \left| \int g d\mu_n - \int g d\mu \right| + \left| \int (g - \phi) d\mu \right|$$

$$\leq \epsilon. \qquad \square$$

Two remarks on Theorem 3.16:

(a) The additional assumption of compactness can be removed, because from the fact that $f_n \to f$ it is possible to show (with some further work) that for any $\epsilon > 0$ there exists a compact interval I_ϵ so that $\mu_n(I_\epsilon^c) < \epsilon$ for all n.

(b) Note that in our version of the theorem, we do not only assume that $f_n \to f$ but also that f is a characteristic function. In general, a pointwise limit of characteristic functions *need not be* a characteristic function! If we don't assume that f is a characteristic function, a compensating assumption has to be made.

We will use Theorem 3.16 to prove the Central Limit Theorem. The following Taylor expansion is the key:

Lemma 3.15. *Let* $\mathbb{E}(X^2) < \infty$, *then*

$$f(u) = 1 + iu \cdot \mathbb{E}(X) - \tfrac{1}{2}u^2 \mathbb{E}(X^2) + \delta(u)u^2,$$

with $\delta(u) \to 0$ *as* $u \to 0$.

Proof.

$$\exp(iux) = 1 + iux - \tfrac{1}{2}(ux)^2 + \tfrac{1}{2}(ux)^2 \cdot \varphi(u, x),$$

with $\varphi(u, x) = \cos(\theta_1(u, x)) + i\sin(\theta_2(u, x)) - 1$, $\theta_1, \theta_2 \in \mathbb{R}$, $|\theta_n| \leqslant 1$. Further $\mathbb{E}\left((uX)^2 \varphi(u, X)\right) = u^2 \cdot \mathbb{E}(X^2 \cdot \varphi(u, X))$ so we get that $(uX)^2 \varphi(u, X)$ is integrable for every u. Further $|X^2 \cdot \varphi(u, X)| \leqslant 3X^2$, and $\varphi(u, X) \to 0$ if $u \to 0$, so we get the statement by the dominated convergence theorem. $\qquad\square$

3.9.1 The Central Limit Theorem

Consider the situation at the beginning of this section: $X_1, X_2, \ldots$ are random variable, $X_k : \Omega \longrightarrow \mathbb{R}$, independent with identical distribution μ, and $\mathbb{E}(X_k) = m$, $\mathbb{E}((X_k - m)^2) = \sigma^2 < \infty$. We consider the characteristic function of $Z_n = \frac{1}{\sqrt{n}}\sum_{k=1}^{n}\frac{X_k - m}{\sigma}$:

$$f_n(u) = \mathbb{E}\left(\exp\left\{i \cdot u \cdot \frac{1}{\sqrt{n}}\sum_{k=1}^{n}\frac{X_k - m}{\sigma}\right\}\right)$$

$$= \mathbb{E}\left(\prod_{k=1}^{n}\exp\left\{i \cdot u \cdot \frac{X_k - m}{\sqrt{n} \cdot \sigma}\right\}\right)$$

$$= \left[\mathbb{E}\left(\exp\left\{iu\frac{X_k - m}{\sqrt{n}\sigma}\right\}\right)\right]^n$$

$$\overset{3.15}{=} \left[1 - \frac{1}{n} \left(\frac{1}{2} \cdot u^2 + \delta \left(\frac{u}{\sqrt{n}} \cdot u^2 \right) \right) \right]^n$$

$$\longrightarrow e^{-\frac{1}{2}u^2}$$

for $n \to \infty$. A simple calculation gives that this is the characteristic function of the standard normal distribution, i.e.,

$$\frac{1}{\sqrt{2\pi}} \int \exp(-iux) e^{-\frac{1}{2}x^2} \, dx = e^{-\frac{1}{2}u^2}.$$

Using Theorem 3.16, we arrive at the following:

Theorem 3.17. *If $X_1, X_2, \ldots$ are random variable, $X_k : \Omega \longrightarrow \mathbb{R}$, independent with identical distribution μ, $\mathbb{E}(X_k) = m$, and $\mathbb{E}((X_k - m)^2) = \sigma^2 < \infty$. Then the characteristic function of $Z_n = \frac{1}{\sqrt{n}} \sum_{k=1}^n \frac{X_k - m}{\sigma}$ converges in distribution to the standard normal distribution.*

3.10   Introduction to Bayesian Computation

In this part of the chapter, we will address the important and practical computational aspects of statistical inference. Now that we have a solid understanding of the measure theoretic underpinnings of probability, we can relax our notation slightly in favour of the more succinct, less explicit notation often found in the computational statistics literature. There are many excellent books on the topic of Bayesian computation and the interested reader will find Gelman *et al.* (2004); Sivia and Skilling (2006); Bolstad (2010); Albert (2007) worthwhile.

3.10.1   *Setting up the statistical framework*

We wish to model some underlying processes and we observe data y in the form of a vector $y = (y_1, \ldots, y_n)$. We will assume that the observations we make are drawn from some underlying distribution and will be observed with some degree of uncertainty, which may for example be due to natural variation of the population or measurement error. We therefore consider y to be the observed value of a random vector $Y = (Y_1, \ldots, Y_n)$, and we consider a parametric model for the distribution of Y, which is parameterised by Θ. We note that Θ may be a vector of parameters. If $\Theta = \theta$, then the data vector Y has a sampling distribution with the following density

$$y \to f_{Y|\Theta}(y|\theta). \tag{3.26}$$

Having observed the data $Y = y$, we can consider this as a function of θ, and we call this the likelihood function

$$\theta \to f_{Y|\Theta}(y|\theta). \tag{3.27}$$

In the frequentist setting, we assume there is a single true but unknown parameter $\theta \in \Theta$ and we consider the problem of finding a point estimate for it. In contrast, in Bayesian statistics we consider the parameter as a *random variable*, which expresses our belief in the range of values it may take conditioned on the observed data. Within the Bayesian inference framework, we need to construct not only a likelihood function defining the error model for our observed data, but also a prior distribution over any *a priori* unknown parameters. If the data and the parameter are jointly continuously distributed, we may write its density in the following form

$$(y, \theta) \to f_{Y, \Theta} = f_{Y|\Theta}(y|\Theta)f_{\Theta}(\theta), \tag{3.28}$$

where f_{Θ} is the density of the prior distribution over the parameter values. The prior distribution reflects our knowledge regarding the plausible values of the parameter Θ *before* we observe any data. This may be influenced by the physical constraints of the system we are modelling, or we might wish to take into account previous knowledge available from the scientific literature.

Bayes' theorem makes use of elementary probability theory to give us an expression for the posterior distribution. We may think of Bayesian inference as a mathematically self-consistent procedure for updating our existing prior knowledge regarding the possible values of θ, in light of new measurements available to us. The posterior is therefore the probability of θ conditioned on the data y and has the following density:

$$\theta \to f_{\Theta|Y}(\theta|y) = \frac{f_{Y,\Theta}(y,\theta)}{f_Y(y)} = \frac{f_{Y|\Theta}(y|\theta)f_{\Theta}(\theta)}{\int f_{Y|\Theta}(y|\theta^*)f_{\Theta}(\theta^*)d\theta^*}. \tag{3.29}$$

We note that the marginal distribution of Y in the denominator may be written as an integral over parameter θ of the likelihood times the prior, $f_{Y,\Theta}(y,\theta)$. This quantity is often referred to as the marginal likelihood, marginalised likelihood or the evidence, and it gives us a normalising constant which ensures that the posterior distribution is a well-defined probability density and integrates to 1.

Sometimes however we will only be able to compute the posterior density up to this constant of proportionality,

$$f_{\Theta|Y}(\theta|y) \propto f_{\Theta}(\theta) f_{Y|\Theta}(y|\theta), \tag{3.30}$$

and the computational challenge lies in drawing samples from this posterior and in estimating the (often) high dimensional integral that is the marginal likelihood, $f_Y(y)$. In the rest of this chapter, we shall investigate some of the computational methodology that is available to us to tackle this difficult problem.

In Bayesian statistics, we often use a more relaxed and concise notation than we have seen so far. In particular, it is usual to overload notation such that a random variable and its observed value are both represented by the same symbol. Often the terms *distribution* and *density* will be employed interchangeably and the same notation will be used for density functions of continuous distributions and probability mass functions of discrete distributions. In order to simplify the notation, we may also introduce a different symbol for each of the distributions of interest, rather than distinguishing between different densities with subscripts, for example,

$$h(y,\theta) = g(\theta) f(y|\theta) = m(y) p(\theta|y), \tag{3.31}$$

where h is what we previously denoted by $f_{Y,\Theta}$, g previously denoted by f_{Θ}, f previously denoted by $f_{Y|\Theta}$ and so forth. This is particularly convenient when considering more complex statistical models with many parameters. Finally, in cases where the meaning is clear, we may even go one step further and use $p(\cdot)$ generically to represent different densities, such that the argument of p indicates both the random quantity under consideration and the value it may assume, for example, $p(\theta)$ represents f_{Θ}, and $p(y)$ represents f_Y. We will now use this compact form of notation, although we will revert to the more explicitly stated form wherever notational misunderstandings might occur.

3.10.2   *Should we employ frequentist or Bayesian approaches?*

The Bayesian paradigm is, of course, not only the approach to statistics. Throughout most of the 20th century, the dominant approach to inference has been frequentist statistics, and it is only relatively recently, towards the turn of the century, that Bayesian statistics has become more widely adopted with the advent of widespread and easily accessible computational capabilities.

From a frequentist perspective, a model parameter is considered as a deterministic but unknown quantity, for which we may construct an estimator to obtain a point estimate. We note that we don't define any probability distributions over the parameter space and we work only with a likelihood function. The most commonly employed estimation technique involves the principle of maximum likelihood, which we saw previously in this chapter. The performance of such a statistical procedure is then evaluated by considering a large number of hypothetical repetitions of the observations under identical conditions. Whether this is a sensible approach really depends on the problem under consideration. For example, if we wish to estimate the bias in a coin, we might consider the frequentist approach appropriate, since we can easily obtain a large number of data points collected under roughly the same experimental conditions. In other words, a frequentist statistician is mostly interested in what happens when data is repeatedly drawn from the sampling distribution, $f_{Y|\Theta}(y|\theta_0)$.

Within the Bayesian framework on the other hand, the probability is conditioned the other way round; we always condition on the observed data. We therefore make probabilistic statements regarding the *parameter* given the observed data, rather than probabilistic statements about hypothetical repetitions of the data conditional on a single but unknown value of the parameter. This is a very important conceptual point that we should bear in mind when deciding to employ one approach or the other.

In practice, it is necessary to be familiar with both approaches, since even if we adopt a Bayesian perspective we will still often end up making use of frequentist arguments, for example, when estimating the resulting integrals using Monte Carlo, as we shall see. Reassuringly, if we base our inferences on a large amount of data, then the predictions using both approaches to inference are usually quite similar, albeit with slightly different mathematical interpretations. Bayesian approaches are particularly useful when we are dealing with small amounts of data from experiments that are difficult or impossible to repeat, and for which asymptotic approximations are not accurate.

3.11 Bayesian Inference

From one point of view, Bayesian inference is really quite straightforward. We derive the posterior distribution simply by combining a likelihood function with a prior distribution, then dividing by the appropriate normalising constant, which can be written as the integral of the likelihood times the prior with respect to the model parameters. For models with a very small number of parameters we can often calculate this quantity directly using numerical integration and quadrature methods. For more complex models however, Bayesian inference is generally not so easy, since standard quadrature methods for evaluating integrals become inefficient in higher dimensions to the point of becoming unusable. In some simple situations, we can avoid these problems by using conjugate priors in order to obtain analytical solutions for the posterior. We note that in contrast, frequentist statistics is often conceptually harder to understand, although the calculations may be computationally simpler. Indeed, it is this property that lead to their widespread use throughout the 20th century.

3.11.1   *Conjugate priors*

The use of conjugate priors can greatly simplify Bayesian computation. Some likelihood functions have the property that if the prior is selected from a certain family of distributions, $\mathcal{P}$, then the resulting posterior distribution will also belong to the same family, $\mathcal{P}$. For a given likelihood function, this collection of distributions is called a *conjugate family*. Conjugate families can be described as a parameterised family of distributions of the form

$$\mathcal{P} = \{\theta \to f(\theta|\phi) : \phi \in S\},$$

where S is a set in some Euclidean space, and $\theta \to f(\theta|\phi)$ is a density for each value of the hyperparameter vector $\phi \in S$.

If the likelihood $p(y|\theta)$ admits a conjugate prior in this family, such that $p(\theta)$ is $f(\theta|\phi_0)$ with some known value ϕ_0, then the posterior is also of this form and can be written as another member of the same family of distributions,

$$\theta \to p(\theta|y) = f(\theta|\phi_1),$$

where $\phi_1 \in S$. In order to find the posterior, we simply need to find the value of the updated hyperparameter vector $\phi_1 = \phi_1(y)$, which of course is conditioned on the data defined through the likelihood. Assuming we are satisfied with approximating our prior knowledge by some distribution $f(\theta|\phi_0)$ within a conjugate family, and also with using a likelihood to which this prior is conjugate, then Bayesian inference is straightforward and we can derive an analytic expression for updating hyperparameters $\phi_1(y)$.

We note that such conjugate families are only possible when the likelihood belongs to the exponential family of distributions (we refer the interested reader to Schervish (1995) for further mathematical details).

There are many research papers on how to select an appropriate prior given previous expert knowledge and this topic is known as prior elicitation. As an example, we may consider the case of statistically modelling a physical system consisting of multiple chemical reactions. The priors may be chosen according to physical constraints, since firstly we know that chemical reactions must have positive reaction rates and secondly that there is a physical upper limit on the speed at which such reactions may take place. In other cases, we might construct a prior based on the knowledge of domain experts and previous published research using summary statistics to describe the range within which a particular parameter value might be expected to lie. We can then choose a prior within a conjugate family which satisfies these statistical summaries.

3.11.1.1 *Conjugate prior example 1*

Let us now consider a concrete example of Bayesian computation using a conjugate prior. We can consider a binomial likelihood with sample size n and success probability θ. We will use the beta density as our prior, which we recall has strictly positive hyperparameters a, b and whose probability density function is given by

$$Be(\theta|a,b) = \frac{1}{B(a,b)}\theta^{a-1}(1-\theta)^{b-1},$$

where θ lies in the range $(0,1)$. If we employ a binomial likelihood and a beta prior $Be(a,b)$, we obtain an analytic form for the posterior distribution over θ, which we notice also belongs to the family of beta distributions,

$$p(\theta|y) \propto p(y|\theta)p(\theta)$$

$$\propto \theta^s(1-\theta)^{n-s}\theta^{a-1}(1-\theta)^{b-1}$$

$$\propto Be(\theta|a+s, b+n-s), \quad 0 < \theta < 1.$$

The posterior is therefore $Be(a+s, b+n-s)$, where s is the number of successes, and $n-s$ is the number of failures. When performing the calculation, we note that the posterior distribution is a function of the parameter θ, and so we can write the posterior up to a normalising constant by omitting any terms that do not depend on θ. We can simplify the calculation by keeping in mind which variables we are actually interested in and which variables we are not, i.e., those that we may therefore treat as constants. Once we have the posterior distribution written up to a normalising constant, we can then recognise whether it has a particular functional form, such that we can write down its normalising constant and hence its full distribution.

Indeed, we can find the normalising constant by integration since we know that the posterior has the unnormalised density $\theta^{a+s-1}(1-\theta)^{b+n-s-1}$, with $0 < \theta < 1$, such that

$$p(\theta|y) = \frac{1}{C(y)}\theta^{a+s-1}(1-\theta)^{b+n-s-1}, \quad 0 < \theta < 1,$$

with

$$C(y) = \int_0^1 \theta^{a+s-1}(1-\theta)^{b+n-s-1}d\theta = B(a+s,b+n-s),$$

where the last step is immediate, since the integral is the normalising constant of the beta density $\mathrm{Be}(\theta|a_1,b_1)$, where $a_1 = a+s$ and $b_1 = b+n-s$. Therefore,

$$p(\theta|y) = \mathrm{Be}(\theta|a+s,b+n-s).$$

In general, we can therefore arrive at a normalised posterior distribution simply by recognising its unnormalised functional form.

3.11.1.2   *Conjugate prior example 2*

Let us consider another example where the likelihood is defined such that the random variables $(Y_1,\ldots,Y_n)$ are independently exponentially distributed with rate θ,

$$p(y_i|\theta) = \theta e^{-\theta y_i}, \quad y_i > 0.$$

The full likelihood then follows as

$$p(y|\theta) = \prod_{i=1}^n p(y_i|\theta) = \theta^n e\left(-\theta\sum_{i=1}^n y_i\right).$$

Suppose that our prior is a gamma distribution, $\mathrm{Gam}(a,b)$, with known strictly positive hyperparameters a,b, such that

$$p(\theta) = \frac{b^a}{\Gamma(a)}\theta^{a-1}\exp^{-b\theta},$$

where $\theta > 0$. We can then write down the posterior distribution, which we note is a function of θ,

$$p(\theta|y) \propto p(y|\theta)p(\theta),$$

$$\propto \theta^{a-1}e^{-b\theta}\theta^n \exp\left(-\theta \sum_{i=1}^{n} y_i\right),$$

$$= \theta^{a+n-1} \exp\left(-\left(b + \sum_{i=1}^{n} y_i\right)\theta\right).$$

We have therefore shown that the posterior distribution is again a gamma distribution of the following form,

$$\text{Gam}\left(a+n, b+\sum_{i=1}^{n} y_i\right).$$

In this case, the prior distribution and posterior distribution belong to the same parametric family of distributions and we have derived an explicit formula for updating our prior information to obtain our posterior distribution in light of new data. We note that our posterior distribution can become our new prior for future calculations, which we can update once more in the same manner as more data is obtained. In this case, Bayesian inference reduces to simply finding updating formulas for the hyperparameters of the conjugate family being used.

Subsequently, we might want to make predictions regarding a possible state of future observable Y^*, whose distribution is also exponential with rate θ. We write this posterior predictive distribution as $p(y^*|y)$. We may assume that Y^* is conditionally independent of the previous observations $(y_1, \ldots, y_n)$ conditioned on θ, and therefore we can factorise the joint posterior of θ and y^* as

$$p(y^*, \theta|y) = p(y^*|\theta)p(\theta|y).$$

We can then marginalise by integrating out all possible values of θ, such that

$$p(y^*|y) = \int p(y^*, \theta|y)d\theta = \int p(y^*|\theta)p(\theta|y)d\theta,$$

$$= \int_0^\infty \theta e^{-\theta y^*} \frac{(b+\sum_1^n y_i)^{a+n}}{\Gamma(a+n)} \theta^{a+n-1} e^{-(b+\sum_1^n y_i)\theta} d\theta.$$

We can immediately see that the integral can once again be obtained explicitly and is expressed in terms of the gamma function.

Of course, the use of the above calculations assumes that the gamma distribution is an adequate representation of our prior knowledge. If it is not, then we should choose our prior from a different family of distributions, and accept that we will likely have to resort to numerical methods in order to evaluate the posterior distribution.

Exercise 3.26.

(a) *In a Bayesian model, suppose that $\theta \sim N(0,1)$, and that $X_1, \ldots, X_n \sim N(\theta, 1)$ are independent given θ. Show that this choice of Gaussian prior is conjugate to the likelihood.*

(b) *This question involves the Poisson distribution $P(x|\lambda)$, whose PDF is $f(x; \lambda) = \frac{\lambda^x e^{-\lambda}}{x!}$. Assuming a series of i.i.d. observations $X_1, \ldots, X_n$ are sampled from $P(x|\lambda)$, calculate the joint likelihood and show that we could use the gamma distribution as a conjugate prior.*

3.11.2   *Posterior summaries*

The posterior distribution completely describes our knowledge of the values of the model parameters *after* we have observed the data. We have seen that in the case of a conjugate prior and likelihood, we can summarise this posterior using only the hyperparameters of the distribution. For example, if the posterior is multivariate normal then we can fully describe the distribution by the mean and covariance matrix. In non-analytic cases, where we do not have a simple functional form of the posterior, we will need to summarise it in other ways.

In low dimensions, we may simply describe the posterior by plotting its density function. We might also want to summarise the distribution using the posterior mean together with some other statistics, for example the posterior variance, mode or median. If we can not plot the posterior distribution, but are able to draw samples from it, we could also plot the histogram and calculate summary statistics from the random realisations. In higher dimensions, it becomes trickier to plot the full distribution in all its complexity, for example, visualising all multidimensional correlations. In practice, we might simply want to summarise the posterior using one-dimensional marginal distributions of the individual parameters. Similarly, we might display all pairwise contour plots or correlation scatter plots.

Let us now explicitly consider how to do this. Suppose that $\theta = (\phi, \varphi)$, where ϕ is the scalar component of interest. Then the marginal posterior of ϕ is simply

$$p(\phi|y) = \int p(\phi, \varphi|y) d\varphi.$$

The integration above may be very difficult to perform analytically, however if we have samples available to us from the joint posterior distribution of $\theta = (\phi, \varphi)$,

$$(\phi_1, \varphi_1), (\phi_2, \varphi_2), \ldots, (\phi_N, \varphi_N),$$

then simply considering $\phi_1, \phi_2, \ldots, \phi_N$ gives us a sample from the marginal posterior of ϕ. We can therefore straightforwardly summarise the marginal posterior of ϕ using these samples. Later in this chapter, we will consider a number of approaches for obtaining such samples.

3.11.2.1 *Posterior intervals*

Credible intervals are also useful for summarising the posterior distribution. In the simplest case, a $[100 \times (1 - \alpha)]\%$ posterior credible interval of a single continuous parameter θ may be defined as any interval C in the parameter space such that

$$P(\theta \in C | Y = y) = \int_C p(\theta|y) d\theta = 1 - \alpha.$$

Such posterior intervals have a direct probabilistic interpretation in terms of straightforwardly describing the probability of believing that a parameter value will lie within the given range. In contrast, traditional confidence intervals in the frequentist statistical framework have a much more subtle and less straightforward interpretation, where we can only consider the calculated interval as a single realisation drawn from some underlying probability distribution and we must therefore consider this in the context of what would happen if we were to calculate multiple intervals based on multiple sets of data drawn from the assumed distribution.

Within the frequentist framework, we assume there exists a true but unknown parameter. A frequentist confidence interval either covers the true parameter or does not, and a frequentist confidence interval at significance level $[100 \times \alpha]\%$ is constructed in such a way that if it were possible to sample the data repeatedly under identical conditions and with the same parameter value, then the relative frequency of coverage in a long run of repetitions would be about $[100 \times (1 - \alpha)]\%$. However, for the single set of data that we actually have, the resulting frequentist confidence interval either covers the true parameter value or does not. We cannot interpret this single interval any further, and we should beware making the mistake of interpreting a confidence interval using the straightforward interpretation that we can indeed use with Bayesian posterior credible intervals.

There are many possible credible intervals that cover a given percentage of the posterior distribution, and so in practice we often use an equal-tailed interval, such that the end points of the interval are selected with half of the remaining posterior probability lying on each side of the interval. If a quantile function is not available directly, but instead we have a random sample $\theta_1, \ldots, \theta_N$ from the posterior, then we can simply calculate the empirical quantiles. We may equivalently define the highest posterior density (HPD) region as

$$C_t = \{\theta : f_{\Theta|Y}(\theta|y) > t\},$$

where we can choose the threshold t such that

$$P(\Theta \in C_t) = 1 - \alpha.$$

Finally, we note that when our model has multiple parameters, we often simply examine one parameter at a time.

3.12   Monte Carlo Integration

Let us suppose that f is a density, which we are able to simulate from, and that we are interested in calculating the expectation

$$I = \int h(x)f(x)dx = \mathbb{E}[h(X)].$$

Let us further suppose that we simulate $X_1, X_2, \ldots$ independently from the density f and set $Y_i = h(X_i)$. Then the sequence $Y_1, Y_2, \ldots$ is independently and identically distributed and

$$\mathbb{E}[Y_i] = \mathbb{E}[h(X_i)] = \int h(x)f(x)dx = I.$$

If we calculate the mean of the N values $h(X_1), \ldots, h(X_N)$, then we obtain the following estimate,

$$\hat{I}_N = \frac{1}{N} \sum_{i=1}^{N} h(X_i).$$

By the Strong Law of Large Numbers, $\hat{I}_N$ converges to I as N increases, provided that the condition $\mathbb{E}|h(X)| < \infty$ holds. In Monte Carlo simulations we are free to select N as large as our computational budget allows. We also note that

$$\mathbb{E}[\hat{I}_N] = \frac{1}{N} \sum_{i=1}^{N} \mathbb{E}[h(X_i)] = I,$$

and so our estimate $\hat{I}_N$ is unbiased. It is straightforward to derive the variance and the standard error of this estimator. If the variance of a single term $h(X)$ is finite, then the variance of the average follows as:

$$\mathbb{V}(\hat{I}_N) = \frac{1}{N} \mathbb{V}(h(X)).$$

This is called the sampling variance, simulation variance or Monte Carlo variance of the estimator $\hat{I}_N$.

A more meaningful quantity for measuring the accuracy of $\hat{I}_N$ is the square root of the variance (the standard error). In a similar manner, the standard error of a Monte Carlo estimate is called its sampling standard error, simulation standard error or Monte Carlo standard error. This Monte Carlo standard error is of order $1/\sqrt{N}$, which we can see directly by considering

$$\sqrt{\mathbb{V}\hat{I}_N} = \frac{1}{\sqrt{N}} \sqrt{\mathbb{V}(h(X))}.$$

The population variance $\mathbb{V}(h(X))$ appears in both of the above equations and usually needs to be estimated using the sample variance of the realisations of $h(X_i)$,

$$s^2 = \hat{\mathbb{V}}(h(X)) = \frac{1}{N-1} \sum_{i=1}^{N} (h(X_i) - \hat{I}_N)^2.$$

From the above, we can get an approximate $[100 \times (1 - \alpha)]\%$ confidence interval for I,

$$\hat{I}_N \pm z_{1-\frac{\alpha}{2}} \frac{s}{\sqrt{N}},$$

where $z_{1-\frac{\alpha}{2}}$ denotes the inverse CDF of a standard normal distribution evaluated at $1 - \frac{\alpha}{2}$.

We note that we are now using a frequentist approach here by defining a confidence interval, which makes sense in this context since we can indeed straightforwardly repeat the simulation as many times as we want. The accuracy of Monte Carlo integration therefore has a convergence rate of $1/\sqrt{N}$ as N increases. In other words, to get an extra decimal place of accuracy it is necessary to increase N by a factor of 100. In practice, one usually achieves moderate accuracy with a moderate simulation sample size N, and in order to achieve high accuracy, one usually needs an extremely large simulation sample size. The main advantage of Monte Carlo integration is that it works equally well regardless of the dimensionality. In contrast, classical quadrature approaches for estimating integrals become prohibitively expensive as dimensionality increases. We also note that we can estimate multiple expectations $\mathbb{E}[h_1(X)], \mathbb{E}[h_2(X)], \ldots, \mathbb{E}[h_k(X)]$ using a single sample $X_1, \ldots, X_N$ from the density f, such that

$$\mathbb{E}[h_j(X)] \approx \frac{1}{N} \sum_{i=1}^{N} h_j(X_i), \quad j = 1, \ldots, k$$

There are many very good books on Monte Carlo methodology and we refer the interested reader to Chen *et al.* (2000); Liu (2004); Robert and Casella (2004, 2010); Rubinstein and Kroese (2008) for further details.

3.12.1　*Variance reduction techniques*

Our aim will always be to estimate some non-analytically available integral by rewriting it in the form,

$$\int h(x) f(x) dx$$

There are many ways of writing an integral in this form, and if we choose carefully, we may obtain a Monte Carlo estimator with significantly lower variance than a naive choice. Particular approaches for this are known as variance reduction methods, and often the strategy will be to compute analytically as much as possible, using Monte Carlo integration as little as possible.

3.12.1.1　*Rao–Blackwellisation*

One approach for variance reduction involves conditioning the random variable of interest Z on another random variable Y, which may decrease the variance of the estimator, such that

$$\mathbb{V}\left(\mathbb{E}[Z|Y]\right) \leqslant \mathbb{V}(Z).$$

When performing a Monte Carlo integration, we should therefore seek to rewrite the integral in terms of a conditional expectation, whenever that is possible. In essence by conditioning we calculate part of the original integration analytically, with the remainder by Monte Carlo. This kind of conditioning is often called Rao–Blackwellisation. Let us consider the following example in which we want to estimate the integral

$$I = \mathbb{E}[h(X,Y)] = \mathbb{E}\left[\mathbb{E}[h(X,Y)|Y]\right].$$

Let us suppose that we are able to compute the conditional expectation analytically,

$$\mathbb{E}[h(X,Y)|Y = y] = m(y).$$

We can therefore estimate I either by simulating (X_i, Y_i), $i = 1, \ldots, N$ from the joint distribution of $p(X,Y)$ and by calculating

$$\hat{I}_N^{(1)} = \frac{1}{N} \sum_{i=1}^{N} h(X_i, Y_i),$$

or alternatively by calculating

$$\hat{I}_N^{(2)} = \frac{1}{N} \sum_{i=1}^{N} m(Y_i).$$

If the computational cost of evaluating $h(X_i, Y_i)$ or $m(Y_i)$ is roughly the same, then we should use the second method since it exhibits a lower variance, and hence we require a smaller number of samples compared to the first method to achieve a given accuracy.

This approach can be particularly useful for estimating posterior predictive expectations, where we wish make predictions about a future observation Y^* conditioned on the currently observed data. Often in such a case, the data Y and future observation Y^* are assumed to be conditionally independent given the model parameters Θ, which allows us to compute the conditional distribution $p(y^*|\theta)$ analytically. We begin by factorising the joint posterior of Θ and Y^* as

$$p(y^*, \theta|y) = p(\theta|y)p(y^*|\theta),$$

Naively we would simulate from the joint posterior distribution of Y^* and Θ by first sampling θ_i from the posterior distribution $p(\theta|y)$, and then sampling y_i^* conditioned on the simulated value θ_i. We can estimate the mean $\mathbb{E}[Y^*|Y=y]$ of the posterior predictive distribution by straightforward Monte Carlo as follows:

$$\hat{I}_N^{(1)} = \frac{1}{N}\sum_{i=1}^{N} y_i^*.$$

However, as already mentioned, we often know the mean of Y^* given the value of the parameter Θ,

$$m(\theta) = \mathbb{E}[Y^*|\Theta = \theta] = \int y^* p(y^*|\theta)dy^*,$$

and in such a scenario we can obtain a lower variance estimator for $\mathbb{E}[Y^*|Y]$ by rewriting the integral as

$$\hat{I}_N^{(2)} = \frac{1}{N}\sum_{i=1}^{N} m(\theta_i).$$

3.12.1.2   *Control variates*

We now consider an alternative approach for variance reduction based on the idea of introducing another random variable, whose expectation we know and which is correlated with the random variable of interest. Let us suppose we want to estimate the expectation $I = \mathbb{E}[h(X)]$ and further we know that

$$\mu = \mathbb{E}[m(X)],$$

where m is a known function and μ is a known constant. We can therefore define the following random variable,

$$W = h(X) - \beta\big(m(X) - \mu\big),$$

where β is a constant, and whose expectation is also I. We can immediately see why this construction is useful by considering its variance as follows,

$$\mathbb{V}(W) = \mathbb{V}(h(X)) - 2\beta\,\mathrm{Cov}(h(X), m(X)) + \beta^2\mathbb{V}(m(X)),$$

We can obtain the lowest possible variance for our new random variable W by selecting the following value for β,

$$\beta^* = \frac{\mathrm{Cov}(h(X), m(X))}{\mathbb{V}(m(X))}.$$

We can therefore substitute $\beta = \beta^*$ into the previous equation to see that,

$$\mathbb{V}(W) = \mathbb{V}(h(X)) - \frac{\text{Cov}^2(h(X), m(X))}{\mathbb{V}(m(X))},$$

In particular, we can observe that $\mathbb{V}(W) < \mathbb{V}(h(X))$, if the random variables $h(X)$ and $m(X)$ are correlated, such that $\text{Cov}(h(X), m(X)) \neq 0$, and that a stronger correlation results in a greater reduction in variance. If we can select the value β such that $\mathbb{V}(W) < \mathbb{V}(h(X))$, then we can employ a Monte Carlo estimate of the mean of W to estimate I as follows:

$$\hat{I}_N = \frac{1}{N} \sum_{i=1}^{N} [h(X_i) - \beta(m(X_i) - \mu)].$$

We can therefore estimate the integral of interest by drawing i.i.d. samples $X_1, \ldots, X_N$ from the distribution of X, and hence calculate samples from the control variate $m(X)$, whose expectation we know analytically. We can see intuitively why the variance of this control variate estimator is lower than simply averaging the values $h(X_i)$ by considering the case when $\text{Cov}(h(X), m(X))$ is positive. Whenever we obtain a large value for $\bar{h}$, the sample average of the $h(X_i)$ values, it tends to be associated with an unusually high outcome for $\bar{m}$ the sample average of the $m(X_i)$ values. In this case, we can select a positive value for β such that the control variate estimate adjusts the naive Monte Carlo estimate $\bar{h}$ of $\mathbb{E}[h(X)]$ downward,

$$\hat{I}_N = \frac{1}{N} \sum_{i=1}^{N} [h(X_i) - \beta(m(X_i) - \mu)] = \bar{h} - \beta(\bar{m} - \mu),$$

where

$$\bar{h} = \frac{1}{N} \sum_{i=1}^{N} h(X_i), \quad \bar{m} = \frac{1}{N} \sum_{i=1}^{N} m(X_i).$$

Similarly, whenever the correlation is negative we can choose a suitable value of β that adjusts the naive Monte Carlo estimate upwards. An optimal choice of β^* of course depends on the moments of random variables $h(X)$ and $m(X)$, which we can simply estimate using an additional set of samples drawn from X.

Exercise 3.27. *We wish to estimate the integral $\int_0^1 \exp(x) dx$.*

(a) *Rewrite this integral in terms of an expectation with respect to the standard uniform distribution.*

(b) *Write some computer code to estimate this expectation using* 1000 *samples. What is the variance of your estimator?*

(c) *Now construct an alternative estimator using the uniform random variable as a control variate and derive the optimal value of the scaling parameter β.*

(d) *Write some computer code to compare the mean and variance of your new estimator to that of the first one.*

3.13   Simulating Random Variables

In order to make use of the Monte Carlo methods developed so far, we need computational procedures for simulating random values from the distribution of interest. In general, we generate pseudo-random numbers using deterministic algorithms that exhibit particular properties, such as uniformity and independence, which allow us to use them as if they were the observed values of an i.i.d. sequence of random variables. We refer the interested reader to Devroye (1986) for further details.

In particular, we are interested in generating random numbers from a uniform target distribution $U(0,1)$ on the unit interval, since there is a variety of techniques available to us that use these samples to obtain realisations from other distributions. Nowadays random number generators are available for a variety of standard distributions in most programming languages and statistical packages. Let us now consider how to simulate i.i.d. random variables from a given non-uniform distribution.

3.13.1   *Transformation methods*

Let us consider a distribution of interest with cumulative distribution function F. The simplest approach for generating random numbers is simply to find some deterministic transformation T, such that $T(U) \sim F$. We remember that the cumulative density function (CDF) is defined as $F(x) = P(X < x)$, and the generalised inverse of the CDF is defined as $F^-(u) = \inf(x : F(x) \geqslant u)$. It is clear therefore that we can draw samples by drawing uniformly distributed random numbers in the set $[0,1]$ and mapping them through an appropriately chosen inverse CDF function.

3.13.1.1   *Transformation method example*

We may consider the exponential distribution as a concrete example. Its CDF follows as

$$F(x) = 1 - \exp(-\lambda x),$$

where $\lambda > 0$ is the rate parameter and $x > 0$. We can rearrange this formula to obtain the inverse CDF,

$$F^-(u) = -\frac{\log(1-u)}{\lambda}.$$

Such an approach works for many distributions and we refer the interested reader to the excellent Devroye (1986).

3.13.2 Rejection sampling

The rejection sampling method was first proposed by von Neumann (1951) and this is applicable for simulating from any density function that is given up to a constant of proportionality, for example $zf(x)$, for which we cannot analytically calculate the inverse CDF. In order to use this approach, we introduce an "envelope" distribution $h(x)$, from which we can straightforwardly sample, and a constant M such that $Mh(x) \geqslant f(x)$ for every x. We then draw samples using the following simple algorithm,

(1) Draw a sample $\tilde{x}$ from h and compute $R = \frac{f(\tilde{x})}{Mh(\tilde{x})}$.
(2) With probability R, accept the sample and return to step 1.

This procedure provides samples from the target distribution f. The efficiency of this method depends on both the constant M and on the envelope distribution h; if there is a large discrepancy between $Mh(x)$ and $f(x)$, then we will likely reject the majority of the proposed samples (Robert and Casella, 2004). Furthermore, it is usually challenging to find an appropriate envelope distribution in high dimensional problems.

Exercise 3.28.

(a) *Consider the Cauchy distribution $C(x|x_o, \gamma)$, whose PDF is $f(x; x_0, \gamma) = \frac{1}{\pi\gamma} \left[\frac{\gamma^2}{(x-x_0)^2 + \gamma^2} \right]$ and whose CDF is $F(x; x_0, \gamma) = \frac{1}{\pi} \arctan \left(\frac{x-x_0}{\gamma} \right) + \frac{1}{2}$. Derive the inversion method for drawing samples from a standard Cauchy distribution, $C(x|0, 1)$.*
(b) *Rejection sampling from an unnormalised density $\tilde{p}(x)$ proceeds by proposing with probability $q(x)$ and accepting with probability $\frac{\tilde{p}(x)}{Mq(x)}$. Show that the efficiency of a rejection sampling method is dependent on the constant M and explain how M should therefore be chosen.*
(c) *Consider the random variable Z with pdf $f(z) = 6z(1-z)$ on $[0,1]$. Propose a rejection sampling approach for drawing samples from Z, calculate the optimal scaling factor M, and the optimal acceptance probability.*

(d) *Consider scalar random variables X, Y with PDFs f_X, f_Y respectively. Let us assume, we can sample from Y directly and we know some constant M such that $f_X \leqslant M f_Y$.*

(e) *Show that a rejection sampling algorithm does indeed produce samples from f_X.*

Hint: Consider a random variable that defines all draws of Y that are accepted, i.e., $f_Z \equiv f_{Y|A}$, then use Bayes' theorem.

3.13.3   *Importance sampling*

We now consider the case where we wish to estimate an integral, where the density f might be difficult to sample from directly using the methods we have seen so far. We consider the integral

$$I = \mathbb{E}_f[h(X)] = \int h(x) f(x) dx.$$

We can introduce an auxiliary distribution and rewrite this integral as

$$I = \int h(x) \frac{f(x)}{g(x)} g(x) dx = \mathbb{E}_g \left[h(X) \frac{f(X)}{g(X)} \right],$$

where the subscript denotes the auxiliary distribution, with respect to which the expectation is calculated. In the literature this is sometimes called the fundamental identity of importance sampling (Robert and Casella, 2004). The main idea behind importance sampling is therefore quite straightforward; we use samples from g together with the above fundamental identity to obtain a Monte Carlo estimate of the integral.

We have a lot of freedom regarding the choice of the auxiliary density g, although we must ensure that it has a support equal to or greater than the product of functions appearing in the integral of interest. We therefore require that

$$g(x) = 0 \implies h(x) f(x) = 0.$$

The fundamental identity works for discrete distributions too. For example, let us suppose X has a discrete distribution with probability mass function f. We may then estimate the integral according the summation

$$I = \mathbb{E}_f[h(X)] = \sum_x h(x) \frac{f(x)}{g(x)} g(x)$$

by employing samples drawn from another pmf g, and using a Monte Carlo estimator. Once again, we must ensure that the support of the function hf is equal to or included within the support of the function g.

In its most general form, the fundamental identity of importance sampling uses the concept of the Radon–Nikodym derivative of measures. Let μ be the probability distribution of X. We assume that μ, when restricted to the set

$$\{h \neq 0\} = \{x : h(x) \neq 0\},$$

is absolutely continuous relative to another probability measure ν, which means that

$$\nu(B \cap \{h \neq 0\}) = 0 \Rightarrow \mu(B \cap \{h \neq 0\}) = 0,$$

for all sets B. Then

$$\mathbb{E}[h(X)] = \int h(x)\mu(dx) = \int_{\{h \neq 0\}} h(x)\mu(dx)$$

$$= \int_{\{h \neq 0\}} h(x)\frac{d\mu}{d\nu}(x)\nu(dx),$$

where the function $d\mu/d\nu$ is the Radon–Nikodym derivative of μ relative to ν. When the distribution μ has pdf f and the probability measure ν has the pdf g, then the Radon–Nikodym derivative $d\mu/d\nu$ is simply the ratio $f(x)/g(x)$.

3.13.3.1 *Unbiased importance sampling*

We consider f to be the pdf of a continuous distribution, which we assume we know completely, including its normalising constant. We select a density g that is easy to sample from. Then, we generate a sample $X_1, \dots, X_N$ from g and calculate

$$\hat{I}_N = \frac{1}{N}\sum_{i=1}^{N} h(X_i)\frac{f(X_i)}{g(X_i)}.$$

Let us call the ratio of the densities

$$w_i = w(X_i) = \frac{f(X_i)}{g(X_i)}, \quad i = 1, \dots, N,$$

the importance weights. Then the importance sampling estimate can be written as

$$\hat{I}_N = \frac{1}{N} \sum_{i=1}^{N} w_i h(X_i).$$

Importance sampling gives more weight for those sample points X_i for which $f(X_i) > g(X_i)$ and down-weights the other sample points, in order to form an unbiased estimate of $I = \mathbb{E}_f[h(X)]$, given a sample $X_1, \ldots, X_N$ from g.

Different authors use different names for g, such as the importance sampling density, the approximation density, and the proposal density. Following Robert and Casella (2004), we call g the *instrumental density*. We can interpret the procedure as producing a weighted sample

$$(w_1, X_1), \ldots, (w_N, X_N),$$

where the weights are needed in order to correct for the fact that the sample is produced from the wrong density. Since the estimator is the arithmetic mean of terms $w_i h(X_i)$ each with mean I,

$$\mathbb{E}_g[w_i h(X_i)] = \mathbb{E}_g \left[\frac{f(X_i)}{g(X_i)} h(X_i) \right] = \int h(x) f(x) dx = I,$$

the estimator is unbiased. Its variance can be estimated in the same way as the variance of the basic Monte Carlo estimator.

In importance sampling we should strive for low variance. In particular, the variance should be finite. This is the case, if and only if the following expression holds,

$$\mathbb{E}_g \left[h^2(X) \frac{f^2(X)}{g^2(X)} \right] = \int h^2(x) \frac{f^2(x)}{g(x)} dx < \infty.$$

If this condition is not satisfied, then the estimator behaves erratically. One pair of conditions which guarantees finite variance is

$$\mathbb{V}_f[h(X)] < \infty, \quad \text{and} \quad \frac{f(x)}{g(x)} \leqslant M, \forall x,$$

for some $M > 0$. The second of these conditions means that the tails of g should be at least as heavy as the tails of f. In order to achieve minimal variance, one can show that it is optimal to choose the instrumental density g proportional to $|h|f$. Then the variance of the importance sampling estimator is smaller (or equal to) the variance of the naive Monte Carlo estimator, which uses samples from f. While the optimal choice $g \propto |h|f$ can hardly ever be used in practice, it can still provide some guidance in choosing the form of g, since the shape of the instrumental density should resemble the product $|h|f$ as closely as possible. One should focus sampling on the regions of interest where $|h|f$ is large in order to save computational resources. On the other hand, if the integrand h is not fixed in advance (e.g., one wants to estimate expectations for many functions h) then the instrumental density g should be selected so that $f(x)/g(x) = w(x)$ is nearly constant. If g is a good approximation to f, then all the importance weights will be roughly equal. If, on the other hand, g is a poor approximation to f, then most of the weights will be close to zero; a few of the X_is will therefore dominate the sum, resulting in inaccurate estimates. This problem can easily be identified by inspecting the importance weights, for example by examining their histogram. Finally, we can note that the importance weights can be utilised to form a control variate, since their expectation is known to be one.

$$\mathbb{E}_g[w_i] = \int \frac{f(x)}{g(x)} g(x) dx = 1.$$

3.13.3.2 *Self-normalised importance sampling*

It is also possible to apply importance sampling in situations where we want to estimate $I = \mathbb{E}_f[h(X)]$, but only know an unnormalised version of the density f, which we will denote f^*, such that

$$f(x) = \frac{1}{c} f^*(x)$$

with the normalising constant c unknown. Of course, c can easily be expressed as the integral

$$c = \int f^*(x) dx.$$

Such a situation is common in Bayesian statistics and in these cases we cannot calculate the importance sampling approximation directly. However, we can express the integral of interest in the following way

$$I = \int h(x) f(x) dx = \frac{\int h(x) f^*(x) dx}{\int f^*(x) dx},$$

and then estimate the numerator and denominator separately using importance sampling. We can proceed by sampling $X_1, \ldots, X_N$ from an instrumental density g, and then estimating the denominator by

$$\int f^*(x)dx = \int \frac{f^*(x)}{g(x)} g(x)dx \approx \frac{1}{N} \sum_{i=1}^{N} \frac{f^*(X_i)}{g(X_i)} = \frac{1}{N} \sum_{i=1}^{N} w_i,$$

where now we use the importance weights w_i corresponding to the unnormalised density f^*, given by

$$w_i = \frac{f^*(X_i)}{g(X_i)}.$$

Our estimate of the numerator may also be calculated using these weights,

$$\int h(x)f^*(x)dx \approx \frac{1}{N} \sum_{i=1}^{N} h(X_i) \frac{f^*(X_i)}{g(X_i)} = \frac{1}{N} \sum_{i=1}^{N} w_i h(X_i).$$

Cancelling the common factor $1/N$ appearing in both expressions, we obtain the following recipe for a self-normalised importance sampling estimator.

(1) Generate $X_1, X_2, \ldots, X_N$ from density g.
(2) Calculate the importance weights

$$w_i = \frac{f^*(X_i)}{g(X_i)}.$$

(3) Estimate I by the weighted average

$$\hat{I} = \frac{\sum_{i=1}^{N} w_i h(X_i)}{\sum_{j=1}^{N} w_j}.$$

The same method can be implemented using normalised importance weights,

$$\tilde{w}_i = \frac{w_i}{s}, \qquad \text{where } s = \sum_{j=1}^{N} w_j,$$

where we have divided the original weights by their sum. We may then alternatively calculate the self-normalised importance sampling estimate using the following simpler expression

$$\bar{I} = \sum_{i=1}^{N} \tilde{w}_i h(X_i).$$

Unlike the unbiased estimator, the self-normalised estimator is not unbiased. Its bias is however negligible when N is large. In self-normalised importance sampling, we can also examine the histogram of the importance weights to ascertain how accurate the estimate is likely to be; if there are only a few large weights, with the others all much smaller, then the estimate is unlikely to be very accurate.

3.13.3.3 *Estimating the variance for self-normalised importance sampling*

We have seen that the self-normalised estimator is the ratio of two averages,

$$\hat{I} = \frac{\frac{1}{N}\sum_{i=1}^{N} w_i h(X_i)}{\frac{1}{N}\sum_{j=1}^{N} w_j} = \frac{\bar{U}}{\bar{w}}$$

where $\bar{U}$ is the average of N random variables $U_i = h(X_i)w_i$, $\bar{w}$ is the average of N raw importance weights w_i. The pairs (U_i, w_i) are therefore i.i.d. random vectors. An approximation of the variance of this ratio can be computed by making use of the Delta Method (Liu, 2004; Owen, 2013), which is based on the Central Limit Theorem and a Taylor expansion of the random function of interest. Making use of this approach, an approximation of the variance follows as,

$$\hat{\mathbb{V}}(\hat{I}) = \frac{1}{N(N-1)} \frac{1}{\bar{w}^2} \sum_{i=1}^{N} (h(X_i) - \hat{I})^2 w_i^2$$

$$= \frac{N}{N-1} \frac{\sum_{i=1}^{N}(h(X_i) - \hat{I})^2 w_i^2}{(\sum_{j=1}^{N} w_j)^2},$$

such that the $(1 - \alpha) \times 100$ confidence interval for I is given by

$$\hat{I} + z_{1-\alpha/2}\sqrt{\hat{\mathbb{V}}(\hat{I})}.$$

3.14 Markov Chain Monte Carlo

The use of Markov chains to produce samples from an arbitrary probability distribution was first suggested in the 1950s by Metropolis *et al.* (1953). This approach allows us to simulate a Markov chain such that its stationary distribution is in fact the target distribution we are interested in; in other words, the samples obtained from our Markov chain are equivalent to correlated samples drawn from the target distribution, and this may be implemented even when the normalising constant is unknown. The amount of correlation impacts the variance of the Monte Carlo estimator, and we therefore wish to use a Markov Chain Monte Carlo (MCMC) method that reduces this as much as possible.

We begin by defining a transition density that dictates how the chain moves around and explores the parameter space. The aim is to find a transition density that proposes new minimally correlated parameter values that are accepted with high probability. The Metropolis–Hastings algorithm proceeds as follows:

(1) Given current state θ_i, draw proposed state θ^* from transition density $T(\theta^*|\theta_i)$.
(2) Calculate the acceptance ratio $R(\theta^*|\theta_i) = \min[1, \frac{p(\theta^*)T(\theta_i|\theta^*)}{p(\theta_i)T(\theta^*|\theta_i)}]$.
(3) Draw $U \sim \text{Uniform}[0,1]$.
(4) Let

$$\theta_{i+1} = \begin{cases} \theta^* & \text{if } U < R(\theta^*|\theta), \\ \theta_i & \text{otherwise.} \end{cases}$$

Thus, the new set of parameters θ^* are accepted with probability R. We note that this form is due to the generalisation by Hastings (1970) that allows T to be any normalised probability distribution, and in the original method by Metropolis the transition T was required to be symmetric. Indeed, it was Hastings who first described this algorithm in its more general form in terms of Markov chains sampling from an arbitrary target distribution $\pi(\theta)$; before this it was described purely in physical terms using the motivating example of statistical mechanics. Despite this publication, it was almost another 20 years until its utility within Bayesian statistics began to be fully realised.

We can show that such a Markov chain does indeed converge to the required stationary distribution by considering the following transition function, which we denote $A(\theta^*|\theta)$. This is the total probability of firstly a proposed point being sampled from T, and secondly this proposed point actually being accepted with probability R,

$$A(\theta^*|\theta) = T(\theta^*|\theta)R(\theta^*|\theta). \tag{3.32}$$

Generally speaking, a Markov chain will converge if

$$\int p(\theta)A(\theta^*|\theta)d\theta = p(\theta^*). \tag{3.33}$$

In other words, the average probability of moving from any point in parameter space, denoted by θ, to a particular point θ^* is equal to the probability of θ^* itself. This must hold for all points θ^*.

A Markov chain will also converge under the following, more restrictive, condition known as detailed balance

$$p(\theta)A(\theta^*|\theta) = p(\theta^*)A(\theta|\theta^*) \tag{3.34}$$

and chains that satisfy this symmetry constraint are known as reversible, since the probability of moving from θ to θ^* is the same as moving from θ^* to θ, for all values of θ and θ^*. Assuming detailed balance, we can see straightforwardly that Eq. (3.33) holds

$$\int p(\theta)A(\theta^*|\theta)d\theta = \int p(\theta^*)A(\theta|\theta^*)d\theta \tag{3.35}$$

$$= p(\theta^*) \tag{3.36}$$

since A is a normalised probability distribution that by definition integrates to 1. In order to show that using the acceptance ratio R satisfies detailed balance, we again see straightforwardly that

$$p(\theta)A(\theta^*|\theta) = p(\theta)T(\theta^*|\theta)\min\left\{1, \frac{p(\theta^*)T(\theta|\theta^*)}{p(\theta)T(\theta^*|\theta)}\right\} \tag{3.37}$$

$$= \min\{p(\theta)T(\theta^*|\theta), p(\theta^*)T(\theta, \theta^*)\} \tag{3.38}$$

$$= p(\theta^*)A(\theta|\theta^*). \tag{3.39}$$

Alternative acceptance rules are explored in Barker (1965), however Peskun, a PhD student of Hastings, showed that the form given by Hastings was asymptotically optimal (Peskun, 1973). There are also a couple of technical conditions on a Markov chain that are required for it to converge; a chain must be *irreducible* and *aperiodic*, which state that a chain has a non-zero probability of reaching any point in parameter space from any other point within a finite number of steps, and that the chain does not get stuck in any loops such that it repeatedly visits the same location with some fixed regularity.

There are many excellent expositions on the Metropolis–Hastings algorithm, and this is no doubt linked to the fact that it is considered one of the most important of all Monte Carlo algorithms (Geyer, 1992; Chib and Greenberg, 1996; Beichl and Sullivan, 2000; Gamerman and Lopes, 2006; Liu, 2004).

Exercise 3.29.

(a) *Show that Metropolis–Hastings reduces to regular simulation when using the proposal $q(\theta^*|\theta) = \pi(\theta^*)$.*

(b) *Give the simplified version of Metropolis–Hastings when the proposal distribution is symmetric, i.e., $q(\theta^*|\theta) = q(\theta|\theta^*)$.*

3.15 Literature on Measure Theory and Integration

The following books cover measure theory and integration, mostly somewhat more generally than in this chapter. Jacod and Protter (2000) is nice and brief, strongly recommended. Some proofs are omitted. Dudley (1989) is unusual in that it covers analysis and probability alongside each other, including aspects of functional analysis, measure theory, and advanced aspects of probability theory. The presentation is superb. Halmos (1974) is an absolute classic. Halmos' fame as a mathematical expositor began with this book. Focusses on measures on locally compact spaces which is somewhat outdated. Doob (1994) is a very concise text which nonetheless covers everything that is important.

Concerning probability theory, we recommend the following. Klenke (2014) is a modern account of measure theory which touches upon many aspects of probability theory as well. For an introductory text it is often somewhat too concise. Breiman (1973) is a classic in probability theory. Written in Breiman's very personal but highly readable style, it gives a wonderful introduction to the subject, and whoever thinks it "too theoretical" should look at Breiman's later career. This book does not cover measure theory and integration in detail though. Feller's two books on probability theory (Feller, 1966, 1970) are even more classic in probability theory than Breiman (1973). Again, they do not cover measure theory and integration in detail.

For data assimilation, Jazwinski (1970) is a good introduction, albeit not a rigorous account, and written by an engineer rather than an atmospheric scientist. It is a must have for everyone working in data assimilation, however.

Finally, there is a growing amount of very decent lecture notes available on the internet, for instance,

Daniel Ocone's homepage:
`http://www.math.rutgers.edu/~ocone`
Stefan Grossinsky's homepage:
`http://homepages.warwick.ac.uk/~masgav`

Pavel Chigansky's homepage:
`http://pluto.huji.ac.il/~pchiga/teaching.html`

3.16   Solutions to Selected Exercises

Exercise 3.1

(a) For a set Ω the power set 2^{Ω} of Ω is the set of all of its subsets. Following Definition 3.1, we need to verify three properties for the power set to be a sigma algebra. First, since $\varnothing$ is a subset of Ω, $\varnothing \in 2^{\Omega}$. Second, if $A \in 2^{\Omega}$ then $A \subset \Omega$ which implies that $A^{c} \subset \Omega$. This means that $A^{c} \in 2^{\Omega}$. Finally for countably many elements $A_1, A_2, \ldots \in 2^{\Omega}$, we have $A_1, A_2, \ldots \subset \Omega$, hence $\cup_{k=1}^{\infty} A_k \subset \Omega$. This means that $\cup_{k=1}^{\infty} A_k \in 2^{\Omega}$ as required.

(b) Suppose $\mathcal{S}_1, \mathcal{S}_2$ are sigma algebras. Following Definition 3.1, we need to verify three properties for $\mathcal{S}_1 \cap \mathcal{S}_2$ to be a sigma algebra. First, since $\varnothing \in \mathcal{S}_1$ and $\varnothing \in \mathcal{S}_2$ we have that $\varnothing \in \mathcal{S}_1 \cap \mathcal{S}_2$. Second, if $A \in \mathcal{S}_1 \cap \mathcal{S}_2$ then $A \in \mathcal{S}_k$, $k = 1, 2$. Since $\mathcal{S}_k$, $k = 1, 2$ are sigma algebras, we deduce that $A^{c} \in \mathcal{S}_k$, $k = 1, 2$ which further implies that $A^{c} \in \mathcal{S}_1 \cap \mathcal{S}_2$. Finally, let $A_1, A_2, \ldots \in \mathcal{S}_1 \cap \mathcal{S}_2$. Then, $A_1, A_2, \ldots \in \mathcal{S}_k$, $k = 1, 2$. Since $\mathcal{S}_k$, $k = 1, 2$ are sigma algebras we deduce that $\cup_{j=1}^{\infty} A_j \in \mathcal{S}_k$, $k = 1, 2$ which further implies that $\cup_{j=1}^{\infty} A_j \in \mathcal{S}_1 \cap \mathcal{S}_2$. In conclusion, the three required properties for a sigma algebra are satisfied.

(c) For a set Ω, let $\mathcal{A}$ be an arbitrary family of subsets of Ω. We define $\mathfrak{F}$ to be the family of all sigma algebras on Ω that contain the family $\mathcal{A}$ of subsets of Ω. The power set 2^{Ω} by definition contains $\mathcal{A}$ and from the previous item it is a sigma algebra on Ω. Hence, $2^{\Omega} \in \mathfrak{F}$. So $\mathfrak{F}$ contains at least one sigma algebra (and maybe more). We take the intersection of all these sigma algebras and call the result $\bar{\mathcal{A}}$. For sure, $\bar{\mathcal{A}} \supset \mathcal{A}$. But since the intersection of sigma algebras is also a sigma algebra we deduce that $\bar{\mathcal{A}}$ is a sigma algebra on Ω containing $\mathcal{A}$. Further, $\bar{\mathcal{A}}$ is contained in any other sigma algebra in $\mathfrak{F}$ and is therefore the smallest sigma algebra containing $\mathcal{A}$.

Exercise 3.2

(a) Since $\varnothing$ and Ω are disjoint, the additivity property implies:

$$P(\varnothing) + P(\Omega) = P(\varnothing \cup \Omega) = P(\Omega),$$

and since $P(\Omega) = 1 < \infty$ we deduce that $P(\varnothing) = 0$.

(b) ($\Rightarrow$) Consider countably many pairwise disjoint sets $A_n, n = 1, 2, \ldots$ in $\mathcal{A}$ so that $\bigcup A_n$ is in $\mathcal{A}$ as well. We want to show that $\sum_{n=1}^{\infty} P(A_n) = P(\bigcup A_n)$. We define:

$$B_n = \bigcup A_k \setminus (A_1 \cup \cdots \cup A_n)$$

for $n \in \mathbb{N}$. We then have that $B_1 \supset B_2 \supset \cdots$ and $\cap B_n = \varnothing$. Hence, since we assume that continuity at the empty set holds we deduce:

$$\lim_{n \to \infty} P(B_n) = P\left(\bigcap B_n\right) = 0.$$

Furthermore, due to the disjointness of the A_i's, we have

$$0 = \lim_{n \to \infty} P(B_n) = \lim_{n \to \infty} \left(P\left(\bigcup A_k\right) - \sum_{k=1}^{n} P(A_k) \right)$$

$$= P\left(\bigcup A_k\right) - \sum_{k=1}^{\infty} P(A_k),$$

which gives the required equality.

($\Leftarrow$) For countably many sets $A_n, n = 1, 2, \ldots$ in $\mathcal{A}$ such that $A_1 \supset A_2 \supset \cdots$ and $\cap A_k = \varnothing$, we need to show that $\lim_{n \to \infty} P(A_n) = 0$. We define:

$$B_n = A_n \cap A_{n+1}^c$$

for $n \in \mathbb{N}$. We then have that for $i \neq j$, $B_i \cap B_j = \varnothing$. Moreover, $\bigcup B_k = \bigcup A_k = A_1 \in \mathcal{A}$ and from the sigma additivity of the family of subsets $\{B_n\}_{n \in \mathbb{N}}$ we also have that $\sum_{k=1}^{\infty} P(B_k) = P(\bigcup B_k) = P(A_1)$. So we have

$$0 = \lim_{n \to \infty} \left(P\left(\bigcup B_k\right) - \sum_{k=1}^{n} P(B_k) \right)$$

$$= \lim_{n \to \infty} P\left(\bigcup B_k \setminus (B_1 \cup \cdots \cup B_n)\right)$$

$$= \lim_{n \to \infty} P(A_1 \setminus (A_1 \cap A_{n+1}^c))$$

$$= \lim_{n \to \infty} P(A_{n+1}),$$

namely, $\lim_{n \to \infty} P(A_n) = 0$ as required.

(c) Using the previous item we will show that sigma additivity is equivalent to continuity from above.

($\Rightarrow$) Consider countably many sets A_n, $n \in \mathbb{N}$ with $A_1 \supset A_2 \supset \ldots$. We define

$$B_n = A_n \setminus A_{n+1}$$

for all $n \in \mathbb{N}$. We have then that for $i \neq j$, $B_i \cap B_j = \varnothing$ and $\bigcup B_k = \bigcup A_k = A_1 \in \mathcal{A}$. Moreover, using the sigma additivity of the family $\{B_n\}_{n \in \mathbb{N}}$, we have $\sum P(B_k) = P(\bigcup B_k)$. Furthermore,

$$0 = \lim_{n \to \infty} \left(P\left(\bigcup B_k\right) - \sum_{k=1}^{n} P(B_k) \right)$$

$$= \lim_{n \to \infty} P\left(\bigcup B_k \setminus (B_1 \cup \cdots \cup B_n)\right)$$

$$= \lim_{n \to \infty} P(A_1 \setminus (A_1 \cap A_{n+1}^c))$$

$$= \lim_{n \to \infty} P(A_{n+1}).$$

($\Leftarrow$) Consider sets $\{A_n\}_{n \in \mathbb{N}}$ with $A_i \cap A_j = \varnothing$ for $i \neq j$. We define:

$$B_n = \bigcup_{k=n}^{\infty} A_k$$

for all $n \in \mathbb{N}$. Then we have that $B_1 \supset B_2 \supset \cdots$ and $\bigcup B_k = \bigcup A_k$. Moreover, it is not difficult to verify that

$$\bigcap B_n = \bigcap_{n=1}^{\infty} \left(\bigcup_{k=n}^{\infty} A_k \right) = \varnothing.$$

From the continuity from above property on the family $\{B_n\}_{n \in \mathbb{N}}$ we have that $\lim_{n \to \infty} P(B_n) = P(\bigcap B_n) = 0$. Furthermore,

$$\bigcup A_k = \left(\bigcup_{k=1}^{n} A_k \right) \cup \left(\bigcup_{k=n+1}^{\infty} A_k \right) = \bigcup_{k=1}^{n} \bigcup B_{n+1},$$

and therefore

$$P\left(\bigcup A_k\right) = \sum_{k=1}^{n} P(A_k) + P(B_{n+1}), \quad \forall n \in \mathbb{N}.$$

By taking the limit $n \to \infty$, we get

$$P\left(\bigcup A_k\right) = \sum_{k=1}^{\infty} P(A_k) + P\left(\bigcap B_k\right) = \sum_{k=1}^{\infty} P(A_k).$$

(d) Using a previous item, we will show that sigma additivity is equivalent to continuity from below.

($\Rightarrow$) Consider countably many sets with $A_1 \subset A_2 \subset \ldots$ and $\bigcup A_k \in \mathcal{A}$. We define:

$$B_n = A_n \setminus (A_1 \cup \cdots \cup A_{n-1})$$

for all $n \in \mathbb{N}$. Then we have that for $i \neq j$, $B_i \cap B_j = \varnothing$ and $\bigcup B_k = \bigcup A_k$ and $\bigcup_{j=1}^k B_j = A_k$. So we have that $P(A_k) = \sum_{j=1}^k P(B_j)$ which implies that:

$$\lim_{k \to \infty} P(A_k) = \sum_{k=1}^{\infty} P(B_k) = P\left(\bigcup B_k\right) = P\left(\bigcup A_k\right),$$

where for the second equality we have used the sigma additivity for the family of sets $\{B_n\}_{n \in \mathbb{N}}$.

($\Leftarrow$) Consider countably many pairwise disjoint sets A_n, $n \in \mathbb{N}$. We define:

$$B_n = A_1 \cup \cdots \cup A_n$$

for all $n \in \mathbb{N}$. Then we have that $B_1 \subset B_2 \subset \ldots$ and $\bigcup B_k = \bigcup A_k$. Furthermore:

$$\sum_{n=1}^{\infty} P(A_n) = \lim_{n \to \infty} \sum_{k=1}^{n} P(A_k) = \lim_{n \to \infty} P(A_1 \cup \cdots \cup A_n)$$

$$= \lim_{n \to \infty} P(B_n) = P\left(\bigcup B_k\right) = P\left(\bigcup A_k\right),$$

where the fourth equality follows from the continuity from below property of the family $\{B_i\}_{i \in \mathbb{N}}$ of subsets.

(e) For $A_1, A_2, \ldots$ countably many disjoint sets in $\mathcal{A}$. Then from the additivity property we have for all $n \in \mathbb{N}$:

$$\sum_{k=1}^{n} P(A_k) = P\left(\bigcup_{k=1}^{n} A_k\right) \leqslant P(\Omega) = 1.$$

Since for all $n \in \mathbb{N}$, $\sum_{k=1}^{n} P(A_k) \leqslant 1$ by taking the limit as $n \to \infty$ we have:

$$\sum_{n=1}^{\infty} P(A_n) < \infty,$$

i.e., the series converges. This implies that $P(A_n) \to 0$ as $n \to \infty$.

Exercise 3.3

(a) We check Definition 3.1.

   (i) $\varnothing \in \mathcal{B}$, and $f^{-1}(\varnothing) = \varnothing$, so $\varnothing \in \mathcal{A}_0$.
   (ii) Let $A \in \mathcal{A}_0$. Then $\exists B \in \mathcal{B} : f^{-1}(B) = A$. Now $B^c \in \mathcal{B}$ and $f^{-1}(B^c) = f^{-1}(B)^c = A^c$, hence $A^c \in \mathcal{A}_0$.
   (iii) If $A_1, A_2, \ldots, \in \mathcal{A}_0$, then $\exists B_1, B_2, \ldots, \in \mathcal{B}$ so that $f^{-1}(B_k) = A_k$. Therefore $\bigcup_k A_k = \bigcup_k f^{-1}(B_k) = f^{-1}\underbrace{\left(\bigcup_k B_k\right)}_{\in \mathcal{B}} \in \mathcal{A}_0$.

(b) We check Definition 3.1.

   (i) Since $f^{-1}(\varnothing) = \varnothing \in \mathcal{A}$, $\varnothing \in \mathcal{B}_0$.
   (ii) If $f^{-1}(B) \in \mathcal{A}$, then $f^{-1}(B^c) = f^{-1}(B)^c \in \mathcal{A}$. This shows $B \in \mathcal{B}_0 \implies B^c \in \mathcal{B}_0$.
   (iii) If $f^{-1}(B_k) \in \mathcal{A}$ $\forall k \in \mathbb{N}$, then $f^{-1}(\bigcup_k B_k) = \bigcup_k f^{-1}(B_k) \in \mathcal{A}$. This shows $B_1, B_2, \ldots \in \mathcal{B}_0 \implies \bigcup_k B_k \in \mathcal{B}_0$.

(c) If $\mathcal{B}_0$ contains $\mathcal{B}$, then $f^{-1}(B) \in \mathcal{A}$ for all sets $B \in \mathcal{B}$. Hence, f is a random variable.

(d) Let $\mathcal{D}$ be the sets of the form $\{x \in \mathbb{R}, x > a\}$, and $\mathcal{B}_0$ as in item (b). We know by assumption $\mathcal{D} \subset \mathcal{B}_0$. Since $\mathcal{B}_0$ is a σ-algebra by (b), we have $\sigma(\mathcal{D}) \subset \mathcal{B}_0$. But by (4.2), $\sigma(\mathcal{D}) = \mathcal{B}$. Hence, $\mathcal{B} \subset \mathcal{B}_0$. It follows from (c) that f is a random variable.

Exercise 3.5

(a) You can find simple $\tilde{g} \leqslant f$ so that $\int \tilde{g} d\mathbb{P}$ is arbitrarily close to $\sup \int g d\mathbb{P}$ in the theorem. Hence, if $c < \sup \int g d\mathbb{P}$, you could find g so that $c < \int g d\mathbb{P}$, violating the statement, i.e., the statement implies $c \geqslant \sup \int g d\mathbb{P}$. On the other hand, since all f_n are simple and no greater than f, we must have $c \leqslant \sup \int g d\mathbb{P}$.

(b) $f_n - g$ is measurable, so $M_n = \{f_n - g > -\epsilon\}$ are measurable sets. $M_1 \subset M_2 \subset \cdots$ follows because f_n is monotone increasing. Suppose that ω were in none of the M_n, then

$$f_n(\omega) \leqslant g(\omega) - \epsilon \leqslant f(\omega) - \epsilon$$

$\forall n$, so $f_n(\omega) \nrightarrow f(\omega)$, which is a contradiction. Hence, $\bigcap_n M_n = \varnothing \implies \bigcup_n M_n = \Omega$.

(c) We know that f_n, g and $\mathbf{1}_{M_n}$ are simple, so $f_n \cdot \mathbf{1}_{M_n}, g \cdot \mathbf{1}_{M_n}, \sum \mathbf{1}_{M_n}$ are too. Now $f_n \geqslant f_n \cdot \mathbf{1}_{M_n} \geqslant (g - \epsilon) \cdot \mathbf{1}_{M_n}$ due to the definition of M_n. Now the relation in Eq. (3.13) follows from monotonicity.

(d) $\mathbb{P}(M_n) = \mathbb{P}\big(\bigcup_{k=1}^{n} M_k\big) \to \mathbb{P}\big(\bigcup_{k}^{\infty} M_k\big) = 1$. Now let $g\mathbf{1}_{M_n} = \sum_{k=1}^{m} g_k \cdot \mathbf{1}_{B_k \cap M_n}$. By the same argument as above we get

$$\mathbb{P}(B_k \cap M_n) = \mathbb{P}\left(\bigcup_{l=1}^{n}(B_k \cap M_k)\right) \xrightarrow{n \to \infty} \mathbb{P}(B_k).$$

Hence, $\int \mathbf{1}_{M_n} g d\mathbb{P} = \sum_{k=1}^{m} g_k \cdot \mathbb{P}(B_k \cap M_n) \xrightarrow{n \to \infty} \sum_{k=1}^{m} g_k \cdot \mathbb{P}(B_k) = \int g d\mathbb{P}$.

Exercise 3.6

Put $A_n = \{\omega; f(\omega) > \frac{1}{n}\}$, then for $\omega \in A_n$; $n \cdot f(\omega) \geqslant 1$, and if $\omega \notin A_n$, $n \cdot f(\omega) \geqslant 0$, so $n \cdot f(\omega) \geqslant \mathbf{1}_{A_n}(\omega) \ \forall \omega \in \Omega$. This gives $0 = n \int f d\mathbb{P} \geqslant n \cdot \int \mathbf{1}_{A_n} d\mathbb{P} = \mathbb{P}(A_n)$, so

$$\mathbb{P}\left(\bigcup_{n=1}^{\infty} A_n\right) = \lim_{m \to \infty} \mathbb{P}\left(\bigcup_{n=1}^{m} A_n\right) \leqslant \sum_{n=1}^{m} \mathbb{P}(A_n) = 0.$$

But if $f(\omega) > 0$ for some ω, then $f(\omega) > \frac{1}{n}$ for some n, hence $\omega \in A_n$ for some n, hence $\omega \in \bigcup_{n \in \mathbb{N}} A_n$, but this set has probability zero.

Exercise 3.8

(a) We shall check the three defining properties of probability one by one.

(i) Note that $T^{-1}(\Omega_2) = \Omega_1$. Then $T_\star \mathbb{P}(\Omega_2) = \mathbb{P}(T^{-1}(\Omega_2)) = \mathbb{P}(\Omega_1) = 1$, using the assumption that $\mathbb{P}$ is a probability on $(\Omega_1, \mathcal{A}_1)$.

(ii) If $A, B \in \mathcal{A}_2$ and $A \cap B = \phi$ then $T^{-1}(A) \cap T^{-1}(B) = T^{-1}(A \cap B) = T^{-1}(\phi) = \phi$, hence $T^{-1}(A)$ and $T^{-1}(B)$ are disjoint. Therefore by additivity of $\mathbb{P}$, $T_\star \mathbb{P}(A \cup B) = \mathbb{P}(T^{-1}(A \cup B)) = \mathbb{P}(T^{-1}(A) \cup T^{-1}(B)) = \mathbb{P}(T^{-1}(A)) + \mathbb{P}(T^{-1}(B)) = T_\star \mathbb{P}(A) + T_\star \mathbb{P}(B)$, i.e., $T_\star \mathbb{P}$ is additive.

(iii) Suppose $A_k \in \mathcal{A}_2$ for all $k \in \mathbb{N}$ and $A_1 \supseteq A_2 \supseteq \cdots$ with $\bigcap_{k \in \mathbb{N}} A_k = \phi$. It follows that $T^{-1}(A_1) \supseteq T^{-1}(A_2) \supseteq \cdots$ and $\bigcap_{k \in \mathbb{N}} T^{-1}(A_k) = T^{-1}(\bigcap_{k \in \mathbb{N}} A_k) = \phi$. Hence by continuity of $\mathbb{P}$ at ϕ, $T_\star \mathbb{P}(A_k) = \mathbb{P}(T^{-1}(A_k)) \to 0$, so $T_\star \mathbb{P}$ is continuous at ϕ as well.

(b) If $f : (\Omega_2, \mathcal{A}_2) \to (\mathbb{R}, \mathcal{B})$ is a random variable, then $f^{-1}(B) \in \mathcal{A}_2$ for all $B \in \mathcal{B}$. Further, if $T : (\Omega_1, \mathcal{A}_1) \to (\Omega_2, \mathcal{A}_2)$ is measurable, then

$T^{-1}(A_2) \in \mathcal{A}_1$ for all $A_2 \in \mathcal{A}_2$, in particular if we take $A_2 = f^{-1}(B)$. Hence $(f \circ T)^{-1}(B) = T^{-1}(f^{-1}(B)) \in \mathcal{A}_1$ for all $B \in \mathcal{B}$, implying that $f \circ T$ is a random variable.

Exercise 3.9

(a) Suppose $f : (\Omega_2, \mathcal{A}_2) \to (\mathbb{R}, \mathcal{B})$ is measurable and non-negative. Take $(f_n)_{n \in \mathbb{N}}$ a sequence of simple functions with $f_n \uparrow f$ (e.g., as in step 4 of the integral construction). We have, from Theorem 5.1,

$$\int_{\Omega_2} f_n \, d(T_\star \mathbb{P}) = \int_{\Omega_1} f_n \circ T \, d\mathbb{P}.$$

By monotone convergence, the left-hand side converges to $\int_{\Omega_2} f \, d(T_\star \mathbb{P})$. Further, since $f_n \circ T \uparrow f \circ T$, the right-hand side converges to $\int_{\Omega_1} f \circ T \, d\mathbb{P}$, again by monotone convergence. By uniqueness of limits we get

$$\int_{\Omega_2} f \, d(T_\star \mathbb{P}) = \int_{\Omega_1} f \circ T \, d\mathbb{P}.$$

(b) If $f = f_+ - f_-$ is integrable with respect to $T_\star \mathbb{P}$, then

$$\infty > \int_{\Omega_2} f_+ \, d(T_\star \mathbb{P}) = \int_{\Omega_1} f_+ \circ T \, d\mathbb{P}$$

and

$$\infty > \int_{\Omega_2} f_- d(T_\star \mathbb{P}) = \int_{\Omega_1} f_- \circ T d\mathbb{P}$$

using part (a). Subtracting the second expression from the first and observing $f_+ \circ T - f_- \circ T = (f_+ - f_-) \circ T = f \circ T$ gives the result.

Exercise 3.10

We first prove the statement "f is a random variable implies f_k are random variables for all $k \in \mathbb{N}$." Fix $k \in \mathbb{N}$ and $B \in \mathcal{B}$, and consider the rectangular cylinder $C := \{x \in \mathbb{R}^\infty : x_k \in B\}$. Then $C \in \mathcal{B}_\infty$ and hence by our assumption $f^{-1}(C) \in \mathcal{A}$. But

$$f^{-1}(C) = \{\omega \in \Omega : f_k(\omega) \in B\}$$
$$= f_k^{-1}(B).$$

Hence $f_k^{-1}(B) \in \mathcal{A}$, implying that f_k is a random variable.

For the converse statement, fix a non-negative integer L, the indices $k_1, \ldots, k_L \in \mathbb{N}$, the Borel sets $B_1, \ldots, B_L \in \mathcal{B}$, and the rectangular cylinder

$$C = \{x \in \mathbb{R}^\infty : x_{k_1} \in B_1, \ldots, x_{k_L} \in B_L\}.$$

Then

$$f^{-1}(C) = \{\omega \in \Omega : f_{k_1}(\omega) \in B_1, \ldots, f_{k_L}(\omega) \in B_L\}$$

$$= \bigcap_{m=1}^{L} \{\omega \in \Omega : f_{k_m}(\omega) \in B_m\}$$

$$= \bigcap_{m=1}^{L} f_{k_m}^{-1}(B_m).$$

Since we have assumed that f_k are random variables for all $k \in \mathbb{N}$ and $\mathcal{A}$ is a sigma-algebra (which is closed under finite and countable intersections), the right-hand side is in $\mathcal{A}$. So we have shown $f^{-1}(C) \in \mathcal{A}$ for any rectangular cylinder. The conclusion now follows as in Exercise 4.1: $\mathcal{B}_0$, the family of all sets $B \subseteq \mathbb{R}^\infty$ such that $f^{-1}(B) \in \mathcal{A}$, is a sigma-algebra. Since we have shown that $\mathcal{B}_0$ contains all rectangular cylinders, we have $\mathcal{B}_\infty = \sigma(\{C : C \text{ is a rectangular cylinder}\}) \subseteq \mathcal{B}_0$ and in particular the pre-image of any rectangular cylinder is in $\mathcal{A}$.

Exercise 3.20

Following the assumptions we have:

(a) For $z \in [0, \theta]$ we have:

$$P(T \leqslant z) = \prod_{k=1}^{n} P(x_k \leqslant z) = \prod_{k=1}^{n} \int_0^z \frac{1}{\theta} \, dx = \prod_{k=1}^{n} \frac{z}{\theta} = \frac{z^n}{\theta^n}.$$

(b) From the previous item, we have that:

$$P(T \leqslant z) = \frac{z^n}{\theta^n} = \int_0^z \frac{n x^{n-1}}{\theta^n} \, dx$$

for all $z \in [0, \theta]$. Hence, we have that the distribution of T has a density function $q(z) = \frac{n z^{n-1}}{\theta^n}$, $z \in [0, \theta]$.

(c) Using the density, we find:

$$E_\theta(T) = \int_0^\theta x \frac{nx^{n-1}}{\theta^n}\, dx \frac{n}{n+1}\theta$$

so T is biased. From the linearity of E_θ we deduce that putting $\hat{T} = \frac{n+1}{n}T$, we have that $E_\theta(\hat{T}) = \frac{n+1}{n}E_\theta(T) = \theta$, so $\hat{T}$ is unbiased.

Exercise 3.21

(a) We have

$$1 = \int_I f(x;\theta)\, dx = \int_I a(x)e^{\theta h(x)-\phi(\theta)}\, dx = e^{-\phi(\theta)}\int_I a(x)e^{\theta h(x)}\, dx.$$

Hence,

$$e^{\phi(\theta)} = \int_I a(x)e^{\theta h(x)}\, dx \Leftrightarrow \phi(\theta) = \log\left(\int_I a(x)e^{\theta h(x)}\, dx\right).$$

(b) We differentiate with respect to θ and then use differentiation under the integral sign which is permitted by assumption. We have

$$0 = \int_I a(x)(h(x) - \phi'(\theta))e^{\theta h(x)-\phi(\theta)}\, dx = \int_I (h(x) - \phi(\theta))f(x;\theta)\, dx,$$

which then give us:

$$\int_I h(x)f(x;\theta)\, dx = \phi'(\theta)\int_I f(x;\theta)\, dx = \phi'(\theta) = \frac{d\phi}{d\theta}(\theta).$$

But the left-hand side of the above equation is by definition equal to $E_\theta(h)$ i.e., $E_\theta(h) = \frac{d\phi}{d\theta}$ as required. Now we differentiate again with respect to θ and by using again differentiation under the integral sign we obtain:

$$\frac{d^2\phi}{d\theta^2} = \int_I h(x)\frac{d}{d\theta}f(x;\theta)\, dx$$

$$= \int_I h(x)\left(h(x) - \frac{d\phi}{d\theta}\right)f(x;\theta)\, dx$$

$$= \int_I h(x)^2 f(x;\theta)\, dx - \frac{d\phi}{d\theta}\int_I h(x)f(x;\theta)\, dx$$

$$= E_\theta(h^2) - E_\theta(h)E_\theta(h)$$

$$= V_\theta(h).$$

Here, we have used the definition of variance and the properties of E_θ.

(c) Let $\hat{\theta}$ the MLE for θ. Then $\hat{\theta}$ maximises the log-likelihood

$$l(\theta) = \log L(\theta, x) = \sum_{k=1}^{n} \log f(x_k; \theta).$$

(Here, we have the standard setup in Definition 3.14 so we use the log-likelihood.) Hence, we have that

$$
\begin{aligned}
0 = \frac{dl}{d\theta}(\hat{\theta}) &= \frac{d}{d\theta}\Big|_{\theta=\hat{\theta}} \sum_{k=1}^{n} \log f(x_k; \theta) \\
&= \sum_{k=1}^{n} \frac{d}{d\theta}\Big|_{\theta=\hat{\theta}} \left(\log a(x_k) + \theta h(x_k) - \phi(\theta) \right) \\
&= \sum_{k=1}^{n} h(x_k) - n\frac{d\phi}{d\theta}(\hat{\theta}) \\
&\Rightarrow \frac{1}{n} \sum_{k=1}^{n} h(x_k) = \frac{d\phi}{d\theta}(\hat{\theta}).
\end{aligned}
$$

(d) From the previous item we have that the derivative of log-likelihood with respect to θ is:

$$\frac{dl}{d\theta} = \sum_{k=1}^{n} h(x_k) - n\frac{d\phi}{d\theta}.$$

We differentiate again with respect to θ and have:

$$\frac{d^2 l}{d\theta^2} = -n\frac{d^2\phi}{d\theta^2} = -nV_\theta(h),$$

where the last equality follows from the second item. Now $\hat{\theta}$ derived in the previous item (i.e., as a critical point of the log-likelihood) satisfies:

$$\frac{dl}{d\theta}(\hat{\theta}) = 0, \quad \frac{d^2 l}{d\theta^2}(\hat{\theta}) = -nV_{\hat{\theta}}(h) < 0.$$

This implies that the critical point of the log-likelihood is at least a local maximum for the log-likelihood.

(e) We calculate:

$$\log f(x_k, \theta) = \sum_{k=1}^{n} \{\log a(x_k) + \theta h(x_k) - \phi(\theta)\},$$

$$\partial_\theta (\log f(x_k, \theta)) = \sum_{k=1}^{n} \{h(x_k) - \phi'(\theta)\},$$

$$\frac{\partial^2}{\partial \theta^2} (\log f(x_k, \theta)) = -\sum_{k=1}^{n} \phi''(\theta) = -n\phi''(\theta).$$

Then using Lemma 3.10, we have that the Fisher information is given by:

$$I = -E_\theta(-n\phi''(\theta)) = n\phi''(\theta).$$

(f) We will be using the law of large numbers which says that if $Y_1, \ldots, Y_n$ independent, identically distributed RV with finite variance then

$$\frac{1}{n} \sum_{k=1}^{n} Y_k \to E(Y_1)$$

almost surely for $n \to \infty$. We note that $\hat{\theta}$ is uniquely determined due to the one-to-one property of $E_\theta(h)$. By assumption $x_1, \ldots, x_n$ are independent and identically distributed and it is easy to see that the same holds for $Y_k = h(x_k)$. Since by assumption $V_\theta(h)$ is finite the law of large numbers implies $\frac{1}{n} \sum_{k=1}^{n} h(x_k) \to E_\theta(h(x_1))$ almost surely for $n \to \infty$. Using items (b) and (c) we then deduce that $\frac{d\phi}{d\theta}(\hat{\theta}) \to E_\theta(h(x_1))$ almost surely as n goes to infinity. Moreover, by assumption $\frac{d\phi}{d\theta}$ is one-to-one and continuous, thus $\hat{\theta}$ converges to θ as $n \to \infty$.

Exercise 3.23

We can assume that f is integrable (otherwise $\mathbb{E}[f|\mathcal{F}]$ is not well defined). We use Remark 3.4 throughout.

(a) Let f, g be integrable, $\lambda, \mu \in \mathbb{R}$. Then $\lambda f + \mu g$ is integrable. Put $h := \lambda \mathbb{E}[f|\mathcal{F}] + \mu \mathbb{E}[g|\mathcal{F}]$ then h is $\mathcal{F}$-measurable, and for any $A \in \mathcal{F}$ we have

$$\int 1_A \cdot h d\mathbb{P} = \int 1_A \{\lambda \mathbb{E}[f|\mathcal{F}] + \mu \mathbb{E}[g|\mathcal{F}]\} d\mathbb{P}$$

$$= \lambda \int 1_A \mathbb{E}[f|\mathcal{F}] d\mathbb{P} + \mu \int 1_A \mathbb{E}[g|f] d\mathbf{P}$$

$$\stackrel{(3.4,2)}{=} \lambda \int 1_A f d\mathbf{P} + \mu \int 1_A g d\mathbf{P}$$

$$= \int 1_A \{\lambda f + \mu g\} d\mathbf{P}.$$

This implies $\mathbb{E}[\lambda f + \mu g|\mathcal{F}] = \lambda \mathbb{E}[f|\mathcal{F}] + \mu \mathbb{E}[g|\mathcal{F}]$.

(b) Suppose not. Then there is $\epsilon > 0$ so that $A := \{\omega \in \Omega, \mathbb{E}[f|\mathcal{F}] \leqslant -\epsilon\}$ has positive probability. Further, $A \in \mathcal{F}$ by the first property in Remark 3.4. By the second property in Remark 3.4 and since $f \geqslant 0$ we get:

$$0 = \int f \cdot 1_A d\mathbb{P} = \int \mathbb{E}[f|\mathcal{F}] 1_A d\mathbb{P} \leqslant -\epsilon \mathbb{P}(A).$$

which is a contradiction.

(c) Put $h := \mathbb{E}[\mathbb{E}[f|\mathcal{F}]|\mathcal{Y}]$. Then by Remark 3.4, h is $\mathcal{Y}$-measurable and $\forall A \in \mathcal{Y}$,

$$\int 1_A \mathbb{E}[f|\mathcal{F}] d\mathbb{P} = \int 1_A h d\mathbb{P}.$$

But since $A \in \mathcal{F}$ as well, the left-hand side is equal to $\int 1_A f d\mathbb{P}$. By Remark 3.4, $h = \mathbb{E}[f|\mathcal{Y}]$.

Chapter 4

Dynamical Systems

Davoud Cheraghi[*] and Tobias Kuna[†]

Imperial College London, UK
†University of Reading, UK

4.1 Introduction

The area of dynamical systems is the branch of mathematics that studies the time evolution of systems. The evolution is given by a transition law, for example a recursion relation, a (partial) differential equation, an integral equation or even a random mechanism. The theory of dynamical systems takes a global and more qualitative viewpoint trying to work out properties of systems which are intrinsic and have features independent of the details of the considered dynamics; long-time behaviour, its sensitivity with respect to when the transition law is modified and how to classify complicated systems with highly complex dynamics by modelling them through simpler dynamics which have nevertheless the same complex dynamical structure.

Mathematically, a dynamical system consists of a phase space (or state space) X representing all possible states of the system, and a map from the phase space to itself that represents the evolution law of the system, that is, if the systems at time s is in the state x then the function $\Phi_{s,t}(x)$ gives the state that the system has evolved to at time t. Examples include the mathematical models that describe the swinging of a clock pendulum, the temperature at each point on Earth on 1st of January each year as well as complicated systems, such as the Earth's climate system as a whole. Other examples include fluid flows, numerical algorithms and stationary processes.

A dynamical system may evolve in *discrete* or *continuous* time. In the former case, the system is often described as $f : X \to X$, where the state at time $n+1$ is obtained from applying f to the state at time n. In the latter case, the system may be described e.g., by a differential equation, where the state at time $t > s$ is obtained from the state at time s by running along the flow of the equation for the time $t - s$. Any such continuous system can be considered as a discrete systems restricting attention to integer times. These systems are *deterministic*, i.e., the current state evolves to a unique future state, but one may consider stochastic systems as well, i.e., the future states are randomly selected, e.g., by a random choice from a collection of maps f. In general, the law may depend explicitly on the time variable.

Given a system $f : X \to X$ the *orbit* (path for continuous time) of x, is defined as the sequence of states x, $f(x)$, $f \circ f(x)$, $f \circ f \circ f(x), \dots$. One imaginable goal in the theory of dynamical systems is to describe the behaviour of individual orbits (or trajectories, respectively). While it is possible to know the orbits for some (rare) simple dynamical systems, most dynamical systems are too complicated to be understood in terms of individual orbits, however one can give an effective stochastic description.

In Secs. 4.2 and 4.3, two explicit examples are studied in detail qualitatively and explicitly, which are diametral in their properties. In Sec. 4.2, we look at the "rotations of the circle", where if we understand one orbit then we understand any other orbits and in Sec. 4.3, we investigate the *doubling map*, an example of a *chaotic* system where nearby points move apart at an exponential rate (and come back together). Although X is low-dimensional, that is one-dimensional, the doubling map shows surprisingly complex behaviour such as sensitive dependence on initial conditions, the existence of a space filling orbit and that the periodic orbits are dense. The rest of the section is dedicated to study these counter-intuitive properties in more detail. These complicated systems are typical in the sense that a randomly chosen f will exhibit the same complicated behaviour. In Sec. 4.4, it is shown that chaotic systems can be completely described by a simple model where the dynamic is just a shift in an infinite word. A surprising consequence of this description is that chaotic systems form large classes which are qualitatively similar. By qualitatively similar we mean that the full dynamical picture of two dynamical systems differ only by a coordinate transformation. More importantly, they are structurally stable, in the sense that a small change of the dynamics will lead to a qualitatively similar system. The latter is not true for the rotation on the circle, for example, which is the paradigmatic example of an integrable system. Hence, concerning long-time behaviour, chaotic systems exhibit more regularity and stability than integrable systems. In Sec. 4.5, the topological entropy is introduced. Topological entropy measures the complexity of a dynamic. Qualitatively similar dynamics have the same complexity.

The chaotic nature of a system restricts our ability to make deterministic predictions long into the future, for example, in weather forecasting. However, probabilistic predictions will get more reliable the further we look into the future and the more chaotic the system is, as for example, in the prediction of future climate. Some basic techniques of probabilistic descriptions of dynamical systems are given in Sec. 4.6, and show how they can be used to predict the behaviour in the distant future of dynamical systems. In Sec. 4.7, we demonstrate that expanding dynamical systems (which are badly predictable deterministically) fulfil all the properties required in Sec. 4.6 and even give much stronger regularity results for the probabilistic description. We show that the time average can be effectively described by a suitable unique space average and that the influence of the starting point decays exponentially, cf. Sec. 4.7.4. The space average is described by a probability on the set of all possible states of the system, called the "statistics of the dynamical system". The fluctuations of the temporal average can also be related to the aforementioned probability on states, see Sec. 4.7.5. Another aspect is that the aforementioned probability on states is not only stable under the changes of dynamics but also differentiable, see Sec. 4.7.6. Finally, the techniques are applied to derive the convergence and consistency of the simple stochastic numerical algorithm. Section 4.6 is based on Barreira and Valls (2013b); Halmos (1974); Krengel (1985). Section 4.7 follows mainly Baladi (2000); Viana (1998).

4.2 Homeomorphisms of the Circle

4.2.1 *Rigid rotations*

It is convenient to define the circle as a subset of the complex plane as

$$S^1 = \{z \in \mathbb{C} \mid |z| = 1\}.$$

For $\alpha \in [0, 2\pi)$, the rotation of angle α is defined as

$$R_\alpha : S^1 \to S^1, \quad R_\alpha(z) = e^{i\alpha} \cdot z, \quad \text{for all } z \in S^1.$$

That is, the point $e^{i\theta}$ on S^1 is mapped to the point $e^{i\alpha} \cdot e^{i\theta} = e^{i(\alpha+\theta)}$ on S^1.

However, sometimes it is convenient to use an alternative notation for the rotation of the circle. We may identify S^1 with the quotient space

$\mathbb{R}/\mathbb{Z}$, which is the same as the interval $[0,1]$ with 0 and 1 identified. The identification is given by the explicit map $x \in [0,1] \mapsto e^{2\pi i x} \in S^1$. Then, for $a \in [0,1)$, the rotation of angle $2\pi a$ becomes

$$T_a : [0,1) \to [0,1), \quad T_a(x) = x + a \pmod{1},$$

i.e.,

$$T_a(x) = \begin{cases} x + a & \text{if } 0 \leqslant x + a < 1, \\ x + a - 1 & \text{if } 1 \leqslant x + a < 2. \end{cases}$$

Exercise 4.1. *Show that the above definition provides a well-defined homeomorphism of the circle $\mathbb{R}/\mathbb{Z}$.*

Recall that the orbit of a point $z \in S^1$ under the rotation R_α is defined as the sequence

$$z, R_\alpha(z), R_\alpha \circ R_\alpha(z), R_\alpha \circ R_\alpha \circ R_\alpha(z), \dots.$$

To simplify the notations, we use the expression $f^{\circ n}$ to denote the map obtained from composing f with itself n times. For example, $f^{\circ 1} = f$, $f^{\circ 2} = f \circ f$, $f^{\circ 3} = f \circ f \circ f$, etc. Following the standard conventions, $f^{\circ 0}$ denotes the identity map.

Due to the basic algebraic form of the rigid rotations, we are able to obtain a simple formula for the orbits. However, this is very exceptional in the study of dynamical systems. Let us first consider the case that $\alpha = 2\pi \cdot \frac{p}{q}$, where p/q is a rational number. We assume that $q \neq 0$, $p \in \mathbb{Z} - \{0\}$, and p/q is in the reduced form, that is, p and q are relatively prime. Then, the orbit of z under R_α becomes

$$z, e^{2\pi i \frac{p}{q}} \cdot z, e^{2\pi i \frac{2p}{q}} \cdot z, \dots, e^{2\pi i \frac{qp}{q}} \cdot z, \dots$$

$$= z, e^{2\pi i \frac{p}{q}} \cdot z, e^{2\pi i \frac{2p}{q}} \cdot z, \dots, e^{2\pi i \frac{(q-1)p}{q}} \cdot z,$$

$$z, e^{2\pi i \frac{p}{q}} \cdot z, e^{2\pi i \frac{2p}{q}} \cdot z, \dots, e^{2\pi i \frac{(q-1)p}{q}} \cdot z, z, \dots.$$

This is a periodic sequence of q points on the circle.

In contrast, when α is irrational the situation is very different. Before we discuss that, we recall some basic definitions. A metric on a set X is a function $d : X \times X \to \mathbb{R}$ such that for all x_1, x_2, and x_3 in X we have

(i) $d(x_1, x_2) = d(x_2, x_1)$;
(ii) $d(x_1, x_2) \geqslant 0$, where the equality occurs only if $x_1 = x_2$;
(iii) $d(x_1, x_2) \leqslant d(x_1, x_3) + d(x_3, x_2)$.

The Euclidean metric, defined as $d(x, y) = |x - y|$ on $\mathbb{R}$ (or on any Euclidean space $\mathbb{R}^n$), is a prominent example of a metric. On S^1 (and any n-dimensional sphere) we may defined the function $d(x, y)$ as the length of the shortest arc on S^1 connecting x to y.

The notion of a metric allows one to talk about convergence of sequences on X. We say that a sequence x_n, $n \geqslant 1$, on X *converges* to some point $x \in X$ with respect to some metric d defined on X if the sequence of real numbers $d(x_n, x)$ tends to 0 as n tends to infinity. Let X be a set equipped with the metric d. An orbit x_n, $n \geqslant 1$, is said to be *dense* on X if for every x in X there is a sub-sequence of the sequence x_n that converges to x. This is equivalent to saying that for every $x \in X$ and every $\epsilon > 0$ there is $n \in \mathbb{N}$ such that $d(x_n, x) \leqslant \epsilon$.

Proposition 4.1. *If a is an irrational number, then for each $z \in S^1$, the orbit $\{R_{2\pi a}^{\circ n}(z) : n \in \mathbb{Z}\}$ is infinite and dense on S^1.*

Proof. Let z be an arbitrary point on S^1, and let $\alpha = 2\pi a$. If $R_\alpha^{\circ m}(z) = R_\alpha^{\circ n}(z)$ for some integers m and n, we must have $e^{(m-n)\alpha i} \cdot z = z$. As $z \neq 0$, and $(m - n)\alpha$ cannot be an integer multiple of 2π, we must have $m = n$. In other words, the orbit of z is an infinite sequence.

Fix an arbitrary $w \in S^1$ and an $\epsilon > 0$. We aim to find $n \in \mathbb{Z}$ with $d(R_\alpha^{\circ n}(z), w) \leqslant \epsilon$.

Choose $n_0 > 0$ with $(2\pi)/n_0 < \epsilon$. Consider the $n_0 + 1$ points on the circle $R_\alpha^i(z)$, for $0 \leqslant i \leqslant n_0$. There must be integers $0 \leqslant l < k \leqslant n_0$ such that $d(R_\alpha^{\circ k}(z), R_\alpha^{\circ l}(z)) < 2\pi/n_0$, where d is the arc length metric on S^1. Since R_α is an isometry, that is, it preserves distances, we must have $d(z, R_\alpha^{\circ(k-l)}(z)) < \epsilon$.

As $R_\alpha^{\circ(k-l)}$ is an isometry, by the above paragraph the sequence z, $R_\alpha^{\circ(k-l)}(z)$, $R_\alpha^{\circ 2(k-l)}(z)$, $R_\alpha^{\circ 3(k-l)}(z)$ consists of points on the circle that are at most ϵ apart. In particular, there is $j \in \mathbb{N}$ such that $d(R_\alpha^{\circ j(k-l)}(z), w) < \epsilon$. $\square$

Recall that a metric space (X, d) is *compact* if any sequence in X has a sub-sequence converging to some point in X. For example, the interval

$(0,1]$ is not compact since the sequence $1/n$, $n \geqslant 1$, does not converge to some point in $(0,1]$. On the other hand, each interval $[a,b]$ (with respect to the Euclidean metric), the circle S^1 with respect to the arc length, and the two-dimensional sphere as a subset of $\mathbb{R}^3$ are compact spaces.

Definition 4.1. Let X be a compact metric space and $T : X \to X$ be a continuous map. We say that $T : X \to X$ is topologically transitive if there exists $x \in X$ such that the orbit $\{T^{\circ n}(x) : n \in \mathbb{Z}\}$ is dense in X. We say that $T : X \to X$ is minimal if for every $x \in X$, the orbit $\{T^{\circ n}(x), n \in \mathbb{Z}\}$ is dense in X.

A topologically transitive dynamical system cannot be decomposed into two disjoint sets with non-empty interiors which do not interact under the transformation.

Exercise 4.2. *Give an example of a metric space X and a continuous map $T : X \to X$ that is topologically transitive, but not minimal.*

4.2.2   Distribution of orbits

We now look at the problem of quantifying the time of visiting an interval. If α is irrational then the proportion of the orbit $z, R_\alpha(z), R_\alpha^{\circ 2}(z), \ldots$, which lies inside a given arc becomes the length of the arc divided by 2π. This is made precise in the next theorem.

Theorem 4.1. *If a is irrational and $\phi : [0,1] \to \mathbb{R}$ is a continuous function with $\phi(0) = \phi(1)$, then for any $x \in [0,1)$,*

$$\lim_{n \to \infty} \left(\frac{1}{n} \sum_{k=0}^{n-1} \phi(T_a^{\circ k}(x)) \right) = \int_{[0,1]} \phi(y)\, dy.$$

Proof. Let us first consider the functions

$$e_m(x) = e^{2\pi i m x} = \cos(2\pi m x) + i\sin(2\pi m x), \quad m \in \mathbb{Z}.$$

We have $e_m(T_a^{\circ k}(x)) = e^{2\pi i m(x+ka)} = e^{2\pi i m k a} e_m(x)$. Thus, for $m \neq 0$,

$$\left| \frac{1}{n} \sum_{k=0}^{n-1} e_m(T_a^{\circ k}(x)) \right| = \frac{1}{n} \cdot |e^{2\pi i m x}| \cdot \left| \sum_{k=0}^{n-1} e^{2\pi i m k a} \right|$$

$$= \frac{1}{n} \cdot 1 \cdot \left| \frac{1 - e^{2\pi i n m a}}{1 - e^{2\pi i m a}} \right| \leqslant \frac{1}{n} \cdot \frac{2}{|1 - e^{2\pi i m a}|} \to 0$$

as n tends to infinity. Thus, if $\phi(x) = \sum_{m=-N}^{N} a_m e_m(x)$, with $a_{-N}, a_{-N+1}, \ldots, a_N \in \mathbb{C}$, then

$$\lim_{n \to \infty} \frac{1}{n} \sum_{k=0}^{n-1} \phi(T_a^{\circ k}(x)) = a_0 = \int_0^1 \phi(y) \, dy.$$

Since trigonometric polynomials are dense in the space of all periodic continuous functions, we obtain the result in the theorem. $\qquad \square$

Exercise 4.3. *By an example show that the continuity assumption in Theorem 4.1 is necessary.*

As an application of the above theorem, we look at the distribution of the first digits of 2^n, $n \geqslant 1$.

Proposition 4.2. *Fix $p \in \{1, 2, \ldots, 9\}$. The frequency of those n for which the first digit of 2^n is equal to p, that is,*

$$\lim_{N \to \infty} \frac{\{1 \leqslant n \leqslant N : \text{first digit of } 2^n \text{ is equal to } p\}}{N} = \log_{10}\left(1 + \frac{1}{p}\right).$$

Proof. The first digit of 2^n is equal to p if and only if for some $k \geqslant 1$,

$$p \times 10^k \leqslant 2^n < (p+1)10^k.$$

This is equivalent to $\log_{10} p + k \leqslant n \log_{10} 2 < \log_{10}(p+1) + k$, which is also equivalent to

$$n \log_{10} 2 \pmod 1 \in [\log_{10} p, \log_{10}(p+1)).$$

Let us define

$$\chi(x) = \begin{cases} 1 & \text{if } x \in [\log_{10} p, \log_{10}(p+1)], \\ 0 & \text{otherwise.} \end{cases}$$

Let $a = \log_{10} 2 \in \mathbb{R} \setminus \mathbb{Q}$, and $x = \log_{10} 2$. We claim that

$$\lim_{N \to \infty} \frac{1}{N} \sum_{n=0}^{N-1} \chi(T_a^{\circ n}(x)) = \int_{[0,1]} \chi(y) \, dy = \log_{10}\left(1 + \frac{1}{p}\right). \tag{4.1}$$

If χ was continuous, this would directly follows from the theorem. However, we need a bit more work. Given $\delta > 0$, there are continuous functions

$\chi_1 \leqslant \chi \leqslant \chi_2$ defined on $[0,1]$ such that $\int_{[0,1]} |\chi_1 - \chi_2| \, dy < \delta$. Then,

$$\int \chi_2 \, dy = \lim_{N\to\infty} \frac{1}{N} \sum_{n=0}^{N-1} \chi_2(T_a^{\circ n}(x)) \geqslant \overline{\lim_{N\to\infty}} \frac{1}{N} \sum_{n=0}^{N-1} \chi(T_a^{\circ n}(x))$$

$$\geqslant \underline{\lim_{N\to\infty}} \frac{1}{N} \sum_{n=0}^{N-1} \chi(T_a^{\circ n}(x)) \geqslant \lim_{N\to\infty} \frac{1}{N} \sum_{n=0}^{N-1} \chi_1(T_a^{\circ n}(x)) = \int \chi_1 \, dy.$$

The notation $\overline{\lim}$ denotes the lim sup (supremum limit) of a given sequence. When a sequence is convergent, $\overline{\lim}$ gives the same limit, but when there are many convergent sub-sequences, it gives the maximum of all the limits of all convergent sub-sequences. The advantage is that $\overline{\lim}$ always exists, although it may be infinite.

Since $\delta > 0$ was arbitrary, the above inequalities imply that the limit in Eq. (4.1) exists. On the other hand, as $\delta \to 0$, $\int \chi_2 \, dy \to \int \chi \, dy$ and $\int \chi_1 \, dy \to \int \chi \, dy$. Hence, the limit must be equal to $\int \chi(y) \, dy$.

It is clear that Eq. (4.1) implies the equality in the theorem. □

4.2.3   *Homeomorphisms of the circle*

Consider the natural projection

$$\pi : \mathbb{R} \to \mathbb{R}/\mathbb{Z}, \quad \pi(x) = x \pmod 1.$$

For all $i \in \mathbb{Z}$ and $x \in \mathbb{R}$, $\pi(x+i) = \pi(x)$.

Proposition 4.3. *Let $f : S^1 \to S^1$ be a homeomorphism of the circle. Then there exists a homeomorphism $F : \mathbb{R} \to \mathbb{R}$, called a lift of f, such that $f \circ \pi = \pi \circ F$ on $\mathbb{R}$.*

Moreover, F is unique up to adding an integer. That is, if F and G are lifts of f then there is $n \in \mathbb{Z}$ such that for all $x \in \mathbb{R}$, $F(x) = G(x) + n$ (see Fig. 4.1).

Proof. Given x and $y = f(x) \in S^1$, choose x', and $y' \in \mathbb{R}$ with $\pi(x') = x$, $\pi(y') = y$. Define $F(x') = y'$. Now one can use the functional equation $f \circ \pi = \pi \circ F$ to extend F to a continuous map from $\mathbb{R}$ to $\mathbb{R}$.

If G is another lift, we must have $G(x') = y' + n'$, for some integer n'. This implies that for all $x \in \mathbb{R}$, $G(x) = F(x) + n'$. □

Exercise 4.4. *Show that any lift F of a circle homeomorphism under π satisfies $F(x+i) = F(x) + i$, for all i in $\mathbb{Z}$ and $x \in \mathbb{R}$. On the other hand,*

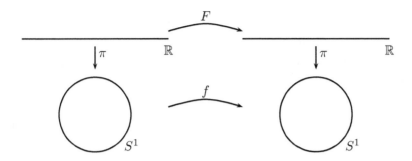

Figure 4.1.   Illustration of the maps in Proposition 4.3.

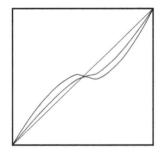

Figure 4.2.   The graph of the function f_ϵ for three different values of ϵ.

prove that any homeomorphism $F : \mathbb{R} \to \mathbb{R}$ that satisfies $F(x+c) = F(x)+c$, for some positive constant c, induces a homeomorphism of the circle.

Example 4.1. Let $f : S^1 \to S^1$ be a rotation by $\alpha = 2\pi a$. The lifts of f are given by the formulas

$$F : \mathbb{R} \to \mathbb{R}, \quad F(x) = x+a+n,$$

where $n \in \mathbb{Z}$.

Exercise 4.5. Let $f_\epsilon : \mathbb{R}/\mathbb{Z} \to \mathbb{R}/\mathbb{Z}$ be defined as $f_\epsilon(x) = x + \epsilon \sin(2\pi x)$ (mod 1), for $|\epsilon| < \frac{1}{2\pi}$. Then, the lifts of f_ϵ are defined by

$$F_\epsilon(x) = x + \epsilon \sin(2\pi x) + n,$$

for $n \in \mathbb{Z}$. Show that if f_ϵ is a homeomorphism, we must have $|\epsilon| < \frac{1}{2\pi}$ (see Fig. 4.2).

Exercise 4.6. *Is* $F(x) = x + \frac{1}{2}\sin(x)$ *the lift of a circle homeomorphism? What if* $F(x) = x + \frac{1}{4x}\sin(2\pi x)$?

Remark 4.1. We always assume that $f : S^1 \to S^1$ is orientation preserving, that is, the graph of f is strictly increasing.

It is possible to assign a rotation number to a homeomorphism of the circle that records the "combinatorial rotation" of the map on the circle. Note that individual points may be rotated by different values.

Proposition 4.4. *Let* $f : S^1 \to S^1$ *be an orientation preserving homeomorphism, and let* $F : \mathbb{R} \to \mathbb{R}$ *be a lift of* F. *Then, for each* $x \in \mathbb{R}$, *the limit*

$$\rho(f) = \lim_{n \to \infty} \frac{F^{\circ n}(x)}{n} \quad (\mathrm{mod}\ 1)$$

exists.[1] *Moreover, the limit is independent of* $x \in \mathbb{R}$ *and the choice of the lift* F.

Proof. We present the proof of the above proposition in several steps.

Step 1. If the limit exists, it is independent of the choice of the lift.

If G is another lift of f, by Proposition 4.3, there is $k \in \mathbb{Z}$ such that $G(x) = F(x) + k$. By Example 4.4, for all $x \in \mathbb{R}$ and all $i \in \mathbb{Z}$ we have $G(x+i) = G(x) + i$. Hence,

$$G^{\circ 2}(x) = G(G(x)) = G(F(x) + k) = G(F(x)) + k$$
$$= F(F(x)) + k + k = F^{\circ 2}(x) + 2k.$$

In general, one can see that for all $n \in \mathbb{N}$, $G^{\circ n}(x) = F^{\circ n}(x) + nk$. Therefore,

$$\lim_{n \to \infty} \frac{G^{\circ n}(x)}{n} = \lim_{n \to \infty} \frac{F^{\circ n}(x) + nk}{n} = \lim_{n \to \infty} \frac{F^{\circ n}(x)}{n} + k.$$

Therefore, if the limit exists, we obtain the same values modulo 1.

Step 2. The limit is independent of the choice of $x \in \mathbb{R}$.

Let $y \in \mathbb{R}$ be another choice that satisfies $|x - y| < 1$. Note that by the definition of the lift, for each x and y in $\mathbb{R}$ with $|x - y| < 1$ we have $|F(x) - F(y)| < 1$. Repeating this property inductively, we conclude that for

---

[1] $\alpha = \beta$ mod 1 if and only if there exists a $k \in \mathbb{Z}$ with $\alpha = \beta + k$. For the reader familiar with basic abstract algebra, the latter essentially means that we consider the quotient ring $\mathbb{R}/\mathbb{Z}$.

all $n \geqslant 1$, $|F^{\circ n}(x) - F^{\circ n}(y)| < 1$. Hence, $|F^{\circ n}(x)/n - F^{\circ n}(y)/n| < \frac{1}{n} \to 0$, as $n \to \infty$.

By the above paragraph, when $|x - y| < 1$, $\lim_{n \to \infty} F^{\circ n}(x)/n$ is the same as $\lim_{n \to \infty} F^{\circ n}(y)/n$, provided they exist. For arbitrary x and y in $\mathbb{R}$, there is a finite sequence of points $x = t_0 < t_1 < t_2 < \ldots < t_n = y$ with all $|t_{i+1} - t_i| < 1$. Then, provided the limits exist, we must have

$$\lim_{n \to \infty} \frac{F^{\circ n}(x)}{n} = \lim_{n \to \infty} \frac{F^{\circ n}(t_1)}{n} = \cdots = \lim_{n \to \infty} \frac{F^{\circ n}(y)}{n}.$$

Step 3. The limit exists.

For each $n \geqslant 1$, there is an integer k_n with $k_n \leqslant F^{\circ n}(0) < k_n + 1$. Then,

$$\left| \frac{F^{\circ n}(0)}{n} - \frac{k_n}{n} \right| \leqslant \frac{1}{n}.$$

Since each iterate $F^{\circ n}$ is a monotone map, $t_1 \leqslant t_2$ implies $F^{\circ n}(t_1) \leqslant F^{\circ n}(t_2)$. Thus,

$$2k_n \leqslant k_n + F^{\circ n}(0) \leqslant F^{\circ n}(k_n) \leqslant F^{\circ n}(F^{\circ n}(0)) = F^{\circ 2n}(0),$$

and

$$F^{\circ 2n}(0) = F^{\circ n}(F^{\circ n}(0)) \leqslant F^{\circ n}(k_n + 1) = k_n + 1 + F^{\circ n}(0) \leqslant 2(k_n + 1).$$

In general, repeating the above argument several times one concludes that for $m \geqslant 1$,

$$m k_n \leqslant F^{\circ(nm)}(0) \leqslant m(k_n + 1).$$

Thus,

$$\left| \frac{F^{\circ(nm)}(0)}{nm} - \frac{k_n}{n} \right| \leqslant \frac{1}{n},$$

and so

$$\left| \frac{F^{\circ m}(0)}{m} - \frac{F^{\circ n}(0)}{n} \right| \leqslant \left| \frac{F^{\circ m}(0)}{m} - \frac{k_m}{m} \right| + \left| \frac{k_m}{m} - \frac{F^{\circ(nm)}(0)}{nm} \right|$$

$$+ \left| \frac{F^{\circ(nm)}(0)}{nm} - \frac{k_n}{n} \right| + \left| \frac{k_n}{n} - \frac{F^{\circ n}(0)}{n} \right|$$

$$\leqslant \frac{1}{m} + \frac{1}{m} + \frac{1}{n} + \frac{1}{n}.$$

In particular, $F^{\circ n}(0)/n$ forms a Cauchy sequence, and hence it converges. $\qquad \square$

Exercise 4.7. *Let $f : S^1 \to S^1$ be a homeomorphism of S^1. Show that $\rho(f^{\circ m}) = m\rho(f) \mod 1$, where $\rho(f)$ denotes the rotation number of f.*

Exercise 4.8. *Let f and g be orientation preserving homeomorphisms of S^1. Prove that $\rho(f) = \rho(g^{-1}fg)$, where ρ denotes the rotation number.*

The notion of rotation number defined in Proposition 4.4 is quite informative, as illustrated in the next two lemmas.

Lemma 4.1. *If a homeomorphism $f : S^1 \to S^1$ has a periodic point $f^{\circ N}(z) = z \in S^1$, then, $\rho(f)$ is a rational number.*

Proof. Let F be a lift of f and choose x with $\pi(x) = z$. By the definition of the lift, we have $\pi \circ F^{\circ N}(x) = f^{\circ N} \circ \pi(x) = f^{\circ N}(z) = z$. Thus, there is $l \in \mathbb{Z}$ such that $F^{\circ N}(x) = x + l$.

For each $n \geq 1$ there are $k \geq 0$ and r with $0 \leq r \leq N - 1$ such that $n = kN + r$. Then,

$$\lim_{n \to \infty} \frac{F^{\circ n}(x)}{n} = \lim_{n \to \infty} \frac{F^{\circ(kN+r)}(x)}{n} = \lim_{n \to \infty} \frac{F^{\circ r}(F^{\circ(kN)}(x))}{n}$$

$$= \lim_{n \to \infty} \frac{F^{\circ r}(x+kl)}{n} = \lim_{n \to \infty} \frac{F^{\circ r}(x) + kl}{n}$$

$$= \lim_{n \to \infty} \frac{kl}{kN+r} = \frac{l}{N}. \qquad \square$$

Exercise 4.9. *Let $F(x) = x + c + b\sin(2\pi x)$. Show that if $|2\pi b| < 1$ then this is an orientation preserving homeomorphism from $\mathbb{R}$ to $\mathbb{R}$. If $|c| < |b|$ show that $\rho(f) = 0$ for the induced map $f : S^1 \to S^1$.*

Lemma 4.2. *If $f : S^1 \to S^1$ is a homeomorphism of the circle with a rational rotation number, then f has a periodic point.*

Proof. Let F be a lift of f with $\lim_{n \to \infty} \frac{F^{\circ n}(x)}{n} \pmod 1 = \frac{p}{q} \in \mathbb{Q}$. Note that $F^{\circ q}$ is a lift of $f^{\circ q}$, and we have

$$\lim_{n \to \infty} \frac{(F^{\circ q})^{\circ n}(x)}{n} = \lim_{n \to \infty} \frac{F^{\circ qn}(x)}{n} = q \lim_{n \to \infty} \frac{F^{\circ qn}(x)}{qn} = q\frac{p}{q} = p = 0 \mod 1.$$

Thus, $\rho(f^{\circ q}) = 0$. The map $G = F - p$ is also a lift of $f^{\circ q}$ and we have

$$\lim_{n \to \infty} \frac{G^{\circ n}(x)}{n} = 0. \tag{4.2}$$

We claim that $G : \mathbb{R} \to \mathbb{R}$ must have a fixed point. Assuming this for a moment, the fixed point projects to a fixed point for $f^{\circ q}$, which must be a periodic point for f.

If in the contrary G has no fixed point, then either (i) for all $y \in \mathbb{R}$ we have $G(y) > y$, or (ii) for all $y \in \mathbb{R}$ we have $G(y) < y$. If (i) occurs, since $G(y) - y$ is continuous on the closed interval $[0,1]$, and strictly positive, there is $\delta > 0$ such that $G(y) - y \geqslant \delta$. As $G : \mathbb{R} \to \mathbb{R}$ is a lift, for all $x \in \mathbb{R}$ and all $i \in \mathbb{Z}$, we have $G(x+i) = G(x)+i$. These imply that for all $y \in \mathbb{R}$, we have $G(y) - y \geqslant \delta$. Repeating this inequality inductively, we have $G^{\circ n}(0) \geqslant 0 + n\delta = n\delta$. Thus,

$$\lim_{n \to \infty} \frac{G^{\circ n}(0)}{n} \geqslant \frac{n\delta}{n} = \delta.$$

This contradicts Eq. (4.2).

The proof in case (ii) is similar to the above one where one shows that $\lim_{n \to \infty} G^{\circ n}(0)/n \leqslant -\delta$. $\square$

The following is a classical result on the homeomorphisms of the circle. See Theorem 11.2.7 in Katok and Hasselblatt (1995) for a proof.

Theorem 4.2 (Poincaré). *Assume that $f : S^1 \to S^1$ is a homeomorphism that is minimal and $\rho(f)$ is irrational. Then there is a homeomorphism $\phi : S^1 \to S^1$ such that, $R_{2\pi\rho(f)} \circ \phi = \phi \circ f$.*

The statement of the above theorem may be illustrated by the commutative diagram

$$
\begin{array}{ccc}
S^1 & \xrightarrow{\;\;f\;\;} & S^1 \\
\downarrow{\scriptstyle\phi} & & \downarrow{\scriptstyle\phi} \\
S^1 & \xrightarrow{R_{2\pi\rho(f)}} & S^1
\end{array}
$$

The homeomorphism ϕ in the above proposition is called *topological conjugacy*. It motivates the following definition.

Definition 4.2. Let $f : X \to X$ and $g : Y \to Y$ be continuous maps. We say that f is topologically conjugate to g if there is a homeomorphism $\phi : X \to Y$ such that $g \circ \phi = \phi \circ f$ holds on X. The conjugacy is called C^1, or smooth, or analytic, if we further require that ϕ is C^1, or C^∞, or analytic, respectively.

Exercise 4.10. *Let $Q_c(x) = x^2 + c$. Prove that if $c < \frac{1}{4}$ there is a unique $\mu > 1$ such that Q_c is topologically conjugate to $f_\mu(x) = \mu x(1-x)$ via a map of the form $h(x) = \alpha x + \beta$.*

When two dynamical systems are topologically conjugate, the two systems behave the same in terms of topological properties. For example, if some sub-sequence $f^{\circ n_k}(x)$ converges to some point $x' \in X$ the corresponding sub-sequence $g^{\circ n_k}(\phi(x))$ converges to $\phi(x')$. However, $f^{\circ n_k}(x)$ may converge to x' exponentially fast, but the latter convergence may be very slow. In general, higher regularity of the conjugacy is required to have similar fine properties for the two systems.

Exercise 4.11. *Let $f : \mathbb{R} \to \mathbb{R}$ be a C^1 map, and $x \in \mathbb{R}$ be a periodic point of f of minimal period q. That is, q is the smallest positive integer with $f^{\circ q}(x) = x$. The quantity $(f^{\circ q})'(x)$ is called the multiplier of f at x. Show that all points in the orbit of x have the same multipliers, i.e., the notion of multiplier is well defined for a periodic orbit.*

Definition 4.3. We say that a continuous function $w : [0,1] \to \mathbb{R}$ has bounded variation if

$$\sup\left\{ \sum_{i=0}^{n-1} |w(x_{i+1}) - w(x_i)| : 0 \leqslant x_1 < x_2 < \cdots < x_n = 1 \right\} < \infty.$$

Exercise 4.12. *For $n = 1, 2$, define the function $w_n : [0,1] \to \mathbb{R}$ as*

$$\begin{cases} 0, & \text{if } x = 0, \\ x^n \sin\left(\frac{1}{x}\right), & \text{if } x \neq 0. \end{cases}$$

Show that w_1 is not a function of bounded variation, but w_2 is a function of bounded variation.

Theorem 4.3 (Denjoy). *Let $f : S^1 \to S^1$ be an orientation-preserving homeomorphism of the circle with irrational rotation $\rho(f) = \rho$. Moreover, assume that $f : S^1 \to S^1$ is continuously differentiable and that $w(x) = \log|f'(x)|$ has bounded variation. Then $f : S^1 \to S^1$ is minimal.*

Exercise 4.13. *Assume $f : \mathbb{R} \to \mathbb{R}$ and $g : \mathbb{R} \to \mathbb{R}$ be smooth maps that are conjugate by a C^1 map ϕ. Prove that the map ϕ preserve the multipliers of periodic points. That is, if x is a periodic point of f, then x and $\phi(x)$ have the same multipliers. By giving an example, show that if the conjugacy is not smooth but only a homeomorphism, the multipliers are not necessarily* *preserved. Hint: build topologically conjugate maps that have fixed points with distinct multipliers.*

We do not give a proof of the above theorem here, see for instance, Theorem 12.1.1 in Katok and Hasselblatt (1995), but instead we present an example that shows the necessity of the assumption. This construction is known as "surgery", and is widely used in constructions of examples in dynamics and other areas of mathematics.

Example 4.2 (Denjoy's Example). For each irrational ρ, there is a C^1 diffeomorphism $f : S^1 \to S^1$ with rotation number $\rho(f) = \rho$, which is not minimal (see Fig. 4.3).

Let us introduce the positive numbers

$$l_n = \frac{1}{(|n|+3)^2}, \quad n \in \mathbb{Z}.$$

We have

$$\sum_{n\in\mathbb{Z}} l_n \leqslant 2 \sum_{n=3}^{\infty} \frac{1}{n^2} \leqslant 2 \int_2^{\infty} \frac{1}{x^2}\,dx = 1.$$

Fix $x \in [0,1)$, and note that since ρ is irrational, the points in the orbit $x_n = T_\rho^{on}(x)$, $n \in \mathbb{Z}$, are distinct. For each $n \in \mathbb{Z}$, we remove the point x_n from the segment $[0,1)$ and replace it by a closed interval I_n of length l_n. After repeating this process for all points, we end up with an interval of length $1 + \sum_{n\in\mathbb{Z}} l_n$.

On the complement of the intervals $\bigcup_n I_n$, we define f as the map induced from the rotation T_ρ. On the intervals I_n, we want to arrange the map so that $f(I_n) = I_{n+1}$, for each $n \in \mathbb{Z}$. It is enough to specify f' in

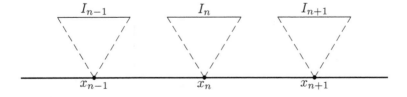

Figure 4.3.   Schematic presentation of the Denjoy surgery.

the intervals I_n so that f' is equal to 0 at the end of the intervals. Let $I_n = [a_n, a_n + l_n]$ and set

$$f'(x) = \begin{cases} 1, & x \notin \bigcup_{n \in \mathbb{Z}} I_n \\ 1 + c_n - \frac{c_n}{l_n} |2(x - a_n) - l_n|, & x \in I_n, \text{ for some } n \in \mathbb{Z}, \end{cases}$$

where $c_n = 2(\frac{l_{n+1}}{l_n} - 1)$. We have chosen c_n such that

$$\int_{I_n} f'(x) = \int_{a_n}^{a_n + l_n} \left(1 + c_n - \frac{c_n}{l_n} |2(x - a_n) - l_n| \right) dx$$

$$= l_n + c_n l_n - \frac{c_n}{l_n} \frac{l_n^2}{2} = l_{n+1}.$$

Exercise 4.14. *Show that the map f introduced above is not transitive. Hint: look at the orbit of x when $x \in \bigcup_n I_n$ and when $x \in S^1 \backslash \bigcup_n I_n$.*

4.3　Expanding Maps of the Circle

In this section, we consider a different class of dynamical systems on the unit circle S^1.

Definition 4.4 (Expanding). A C^1 map $f : S^1 \to S^1$ is called expanding if for all $x \in S^1$, $|f'(x)| > 1$.

Example 4.3. Let $m \geqslant 2$ be an integer, and define $f : [0, 1) \to [0, 1)$ as $f(x) = mx \pmod 1$. If we regard the circle as $S^1 = \{z \in \mathbb{C} : |z| = 1\}$, then f can be written as $f(z) = z^m$. Each of these maps is expanding (see Fig. 4.4).

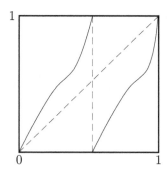

Figure 4.4.　The graph of an expanding degree two map of the circle.

An expanding map $f : S^1 \to S^1$ cannot be a homeomorphism. Also, since f' is continuous and S^1 is compact, there is $\beta > 1$ such that for all $x \in S^1$, $|f'(x)| \geqslant \beta$.

Definition 4.5. The degree of an expanding map $f : S^1 \to S^1$, denoted by $\deg(f)$, is defined as the number of points in the set $f^{-1}(x)$, for $x \in S^1$. One can see that the notion of degree is independent of the choice of x.

Lemma 4.3. *If f and $g : S^1 \to S^1$ are expanding maps, then we have* $\deg(f \circ g) = \deg(f) \cdot \deg(g)$. *In particular,* $\deg(f^{\circ n}) = (\deg(f))^n$, *for $n \geqslant 1$.*

Proof. Since for each $y \in f^{-1}(x)$ the set $g^{-1}(y)$ has $\deg(g)$ elements, $(f \circ g)^{-1}(x)$ has $\deg(g)\deg(f)$ elements. □

Proposition 4.5. *If $f : S^1 \to S^1$ is an expanding map with $\deg(f) = d \geqslant 2$, the number of periodic points of period n is $(d^n - 1)$.*

Proof. First assume $n = 1$, the number of fixed points of f is equal to the number of points on the intersection of the diagonal with the graph of f, which is $d - 1$. For arbitrary $n \geqslant 2$, we consider $f^{\circ n}$ with $\deg(f^{\circ n}) = d^n$. Note that the number of periodic points of period n is equal to the number of the fixed points of $f^{\circ n}$, that is, $d^n - 1$. □

Proposition 4.6. *Let X be a compact metric space, $f : X \to X$ be continuous. The following are equivalent*

(i) *f is topologically transitive;*
(ii) *for all non-empty and open sets U and V in X, there is $n \in \mathbb{N}$, with $f^{-n}(V) \cap U \neq \varnothing$.*

Proof. First we prove that (i) implies (ii). Let $\{f^{\circ n}(x)\}_{n=1}^{\infty}$ be a dense orbit in X. Choose integers $m > n$ with $f^{\circ n}(x) \in U$ and $f^{\circ m}(x) \in V$. Then,

$$f^{\circ n}(x) \in U \cap f^{-(m-n)}(V).$$

Thus, the intersection is non-empty.

Now we prove that (ii) implies (i). Let $Y = \{y_i\}_{i=1}^{\infty}$ be a countable dense set in X. Any compact metric space has a countable dense set. (For instance, when $X = S^1$, one can take all the points with rational angles.) Let U_i denote the ball of radius $1/i$ about y_i. We aim to find a point $x \in X$ whose orbit visits every U_i.

Choose $N_1 \geqslant 0$ such that $f^{-N_1}(U_2) \cap U_1 \neq \varnothing$. Then choose an open disk V_1 of radius less than $1/2$ such that

$$V_1 \subseteq \overline{V}_1 \subseteq U_1 \cap f^{-N_1}(U_2).$$

Above, $\overline{V}_1$ denotes the closed disk obtained from adding the boundary of V_1 to V_1. Then choose N_2 such that $f^{-N_2}(U_3) \cap V_1 \neq \varnothing$. Choose an open disk V_2 of radius less than $1/2^2$ such that

$$V_2 \subseteq \overline{V}_2 \subseteq V_1 \cap f^{-N_2}(U_3).$$

Inductively repeating the above process, we obtain disks $V_1 \supseteq V_2 \supseteq V_3 \supseteq \cdots$ with radius $V_n \leqslant \frac{1}{2^n}$ and

$$\overline{V}_{n+1} \subseteq V_n \cap f^{-N_{n+1}}(U_{n+2}).$$

Now define x as the unique point in the intersection $\bigcap_{n=1}^{\infty} \overline{V}_n$. It easily follows that $f^{\circ N_n - 1}(x) \in U_n$, for $n \geqslant 1$. This implies that $\{f^{\circ n}(x)\}_{n=1}^{\infty}$ is dense in X. $\qquad\square$

Exercise 4.15. *Let $f : X \to X$ be a continuous map of a compact metric space. A point $p \in X$ is called topologically recurrent if for any open set V containing p, there exists $n \geqslant 1$ with $f^{\circ n}(p) \in V$. Clearly every periodic point is recurrent.*

(i) *Give an example of a map $f : X \to X$ with a non-periodic recurrent point.*

(ii) *Give an example of a map $f : X \to X$ with a non-periodic recurrent point p whose orbit is not dense in X.*

Hint: Look at the map in Example 4.2.

Definition 4.6. Let X be a compact metric space, and $f : X \to X$ be continuous. We say that f is topologically mixing if for any two non-empty open sets U and V in X, there exists $N \geqslant 0$ such that for all $n \geqslant N$, $U \cap f^{-n}(V) \neq \varnothing$.

By Proposition 4.6, any mixing transformation is topologically transitive. But, the notion of topological mixing is stronger than the notion of topological transitivity.

Exercise 4.16. *Show that an irrational rotation $R_\alpha : S^1 \to S^1$ is transitive, but is not topologically mixing.*

Exercise 4.17. *Let X be a compact metric space with more than one point and $f : X \to X$ be an isometry. Show that f cannot be topologically mixing.*

Exercise 4.18. *Let X be a compact metric space with at least three distinct points and let $f : X \to X$ be an isometry.*

(i) *Show that f is not mixing.*
(ii) *What if X has only two points?*

Proposition 4.7. *An expanding map $f : S^1 \to S^1$ is mixing.*

Proof. Because f is expanding, there is $\beta > 1$ such that for all $z \in S^1$, $|f'(z)| \geqslant \beta$. Let $d = \deg(f) \geqslant 2$. There is a lift of f to a homeomorphisms $F : \mathbb{R} \to \mathbb{R}$ that satisfies $\pi \circ F(x) = f \circ \pi(x)$ and $F(x+1) = F(x)+d$, for all $x \in \mathbb{R}$. The proof of this is similar to the one for Proposition 4.3.

It follows that $\forall x \in \mathbb{R}$, $|F'(x)| \geqslant \beta$. For an open set U in $\mathbb{R}$, choose an interval $(a,b) \subseteq U$. Since F is C^1 and one-to-one,

$$|F(b) - F(a)| = \int_a^b F'(t)\, dt \geqslant \beta(b-a).$$

That is, F increases the length of intervals by a factor of β. Similarly, $F^{\circ n}$ increases the length of intervals by a factor of at least β^n. Choose N large enough so that $\beta^N > \frac{1}{b-a}$. Then for $n \geqslant N$, the length of $F^{\circ n}(a,b)$ is at least 1 (see Fig. 4.5).

On the other hand, the relation $\pi \circ F = f \circ \pi$ implies that for all $n \geqslant N$, $\pi \circ F^{\circ N}(a,b) = S^1$, and hence $f^{\circ n}(U) \supseteq f^{\circ n}(a,b) = S^1$. In particular, for all open sets V, and all $n \geqslant N$, $f^{\circ N}(U) \cap V \neq \varnothing$, which implies,

$$U \cap f^{-n}(V) \neq \varnothing. \qquad \square$$

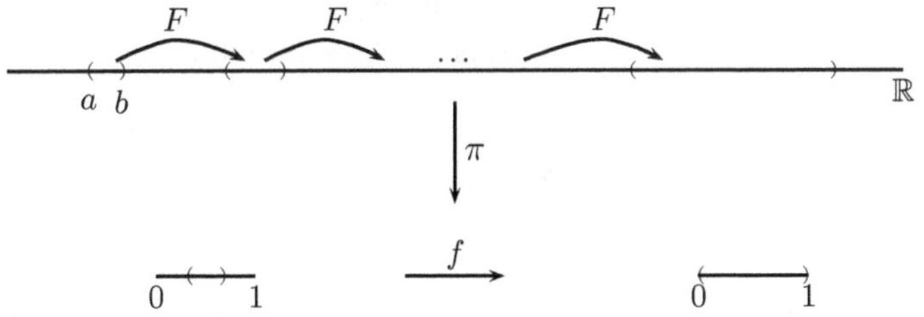

Figure 4.5. The iterates of the maps F and f on an open set.

As a consequence of the above proof, any expanding map of the circle is topologically transitive.

Definition 4.7 (Choatic). A continuous map $f : X \to X$ of a compact metric space is called chaotic if,

(i) f is topologically transitive; and
(ii) the set of periodic points of f is dense in X.

The notion of chaotic behaviour is invariant under topological conjugacy. That is, if two maps are topologically conjugate and one of them is chaotic, the other one is also chaotic.

Example 4.4. Consider the linear expanding map $f : S^1 \to S^1$, defined as $f(x) = mx$ (mod 1), $m \geqslant 2$. The periodic points of f take the form $x = \frac{j}{m^n-1}, 0 \leqslant j < m^n - 1$. That is because

$$f^{on}(x) = m^n \left(\frac{j}{m^n - 1} \right) = j \left(\frac{m^n - 1}{m^n - 1} \right) + \frac{j}{m^n - 1} = x \ (\text{mod } 1).$$

Such points form a dense subset of $[0,1)$. In Theorem 4.4, we shall show that any C^1 expanding map of the circle is chaotic.

Exercise 4.19. *Consider the linear map $A : \mathbb{R}^n \to \mathbb{R}^n$ defined as $A(x) = 2x$, observe that A induces a map $f : T^n \to T^n$, where $T^n = S^1 \times S^1 \times \cdots \times S^1$ (n times) is the n-dimensional torus.*

(i) *Prove that the periodic points of f are dense in T^n.*
(ii) *Prove that eventual fixed points, i.e., the points $x \in T^n$ with $f(f^{om}(x)) = f^{om}(x)$, for some m, are dense in T^n.*
(iii) *Prove that $f : T^n \to T^n$ is chaotic.*

Definition 4.8. A continuous map $f : X \to X$ on a compact metric space is said to have sensitive dependence on initial conditions if there is $\delta > 0$ such that for all $x \in X$ and all $\epsilon > 0$, there are $y \in X$ and a positive integer $n \geqslant 0$ with $d(x,y) < \epsilon$ and $d(f^{on}(x), f^{on}(y)) \geqslant \delta$.

Proposition 4.8. *Expanding circle maps have sensitive dependence on initial conditions.*

Proof. By the expansion property, continuity of f', and the compactness of S^1, there is $\beta > 1$ such that $|f'(z)| \geqslant \beta$, for all $z \in S^1$. For the same reason, there is $\alpha > 1$ such that $|f'(z)| \leqslant \alpha$, for all $z \in S^1$. Note that $\alpha \geqslant \deg(f)$.

For x and y in S^1 with $d(x,y) < 1/2$, let $I_{x,y}$ denote the arc of smallest length connecting x to y. By the above paragraph, if $d(x,y) < 1/(2\alpha)$, then f is monotone on $I_{x,y}$ and $d(f(x), f(y)) < 1/2$. In particular, $d(f(x), f(y)) \geqslant \beta d(x,y)$. We claim that $\delta = 1/(2\alpha)$ satisfies the definition of sensitive dependence on initial conditions.

Let x and $\epsilon > 0$ be given. If $\epsilon > \delta$ we choose $y \in S^1$ with $d(x,y) = \delta$ and $n = 0$. If $\epsilon < \delta$, we may choose any $y \in S^1$ with $0 < d(x,y) < \epsilon$. By the expansion of f and the above paragraph, there is an integer n such that $d(f^{\circ n}(x), f^{\circ n}(y)) \geqslant \delta$. $\qquad \square$

Proposition 4.9. *A chaotic map* $f : X \to X$ *of a compact metric space* X *is either a single periodic orbit, or has sensitive dependence on initial conditions.*

Proof. Since the set of periodic points of f are dense, if X is not a single periodic set, we must have two distinct sets

$$A = \{x, f(x), f^{\circ 2}(x), \ldots, f^{\circ n-1}(x) = x\},$$
$$B = \{y, f(y), f^{\circ 2}(y), \ldots, f^{\circ m-1}(y) = y\}.$$

Let

$$\delta = \tfrac{1}{8} \min\{d(f^{\circ i}(x), f^{\circ j}(y)) : 0 \leqslant i \leqslant n-1, 0 \leqslant j \leqslant (m-1)\} > 0.$$

We aim to show that f satisfies the sensitive dependence on initial conditions with respect to the constant δ. Fix an arbitrary $z \in X$, and $\epsilon > 0$. We may assume that $\epsilon < \delta$, otherwise, we may make ϵ smaller so that this condition holds.

We must have one of the following:

(i) $d(z, A) = \min\{d(z,w) : w \in A\} \geqslant 4\delta$;
(ii) $d(z, B) = \min\{d(z,w) : w \in B\} \geqslant 4\delta$.

We write the proof when (i) occurs. For (ii), one only needs to replace A with B in the following argument. Since periodic points are dense in X, there is a periodic point p with $d(p, z) \leqslant \epsilon$. Let N be the smallest positive integer with $p = f^{\circ N}(p)$.

Define

$$V = \{w \in X : d(f^{\circ i}(w), f^{\circ i}(x)) < \delta, \text{ for all } i \text{ with } 0 \leqslant i \leqslant N\}.$$

Since f has a dense orbit, there is $a \in X$ with $d(z,a) < \epsilon$ and $f^{\circ k}(a) \in V$. There exists an integer k' with $0 \leqslant k' \leqslant N-1$ and $k+k' = jN$, for some

$j \in \mathbb{N}$. Now $f^{\circ jN}(p) = p$, and

$$
\begin{aligned}
d(f^{\circ jN}(p), f^{\circ jN}(a)) &= d(p, f^{\circ jN}(a)) \\
&\geqslant d(z, f^{\circ jN}(a)) - d(p, z) \\
&\geqslant d(z, f^{\circ k'}(x)) - d(f^{\circ k'}(x), f^{\circ jN}(a)) - d(p, z) \\
&\geqslant 4\delta - \delta - \delta = 2\delta.
\end{aligned}
$$

However, a and p belong to $B(z, \epsilon)$ and $d(f^{\circ jN}(p), f^{\circ jN}(a)) \geqslant 2\delta$. Therefore, by the triangle inequality, at least one of $d(f^{\circ jN}(p), f^{\circ jN}(z))$ and $d(f^{\circ jN}(a), f^{\circ jN}(z))$ must be bigger than δ. This finishes the proof of the proposition. $\qquad\square$

4.4   Symbolic Dynamics

In this section, we introduce an approach to build a symbolic model for a dynamical system. We shall focus on two examples, but the method is far reaching.

4.4.1   *Coding expanding maps of the circle*

For an integer $n \geqslant 2$, define the set

$$
\Sigma_n = \{(w_0, w_1, w_2, \dots) \mid \forall i \geqslant 0, w_i \in \{1, 2, \dots, n\}\}.
$$

We define a metric on Σ_n as

$$
d((w_i)_{i=0}^{\infty}, (w_i')_{i=0}^{\infty}) = \sum_{i=0}^{\infty} \frac{|w_i - w_i'|}{2^i}.
$$

The shift map $\sigma : \Sigma_n \to \Sigma_n$ is defined as

$$
\sigma(w_0, w_1, w_2, \dots) = (w_1, w_2, w_3, \dots).
$$

Let $f : S^1 \to S^1$ be an expanding map of degree 2, where S^1 denotes the unit circle. There is a unique fixed point $p \in S^1$. Let $q \neq p$ be the other pre-image of p, i.e., $f(q) = p$. Let Δ_1 and Δ_2 denote the closed arcs on S^1 bounded by p and q, so that $S^1 = \Delta_1 \cup \Delta_2$.

Given $x \in S^1$, we want to associate a $w = (w_i)_{i=0}^{\infty} \in \Sigma_2$ such that,

$$
f^{\circ n}(x) \in \Delta_{w_n}, \quad \forall n \geqslant 0.
$$

However, if $f^{\circ n}(x) \in \Delta_1 \cap \Delta_2 = \{p, q\}$ then there are ambiguities. In this case, we can finish the sequence $w_n, w_{n+1}, \dots = $ with either, $1, 1, 1, 1, \dots$ or

$2,2,2,2,\ldots$ if $f^{\circ n}(x) = p$, and either $2,1,1,1,\ldots$ or $1,2,2,2,\ldots$ if $f^{\circ n}(x) = q$. To illustrate the situation we look at a familiar example.

Example 4.5. Let $T : S^1 \to S^1$ be defined as $T(x) = 2x \pmod 1$. Then $p = 0$ and $q = 1/2$, $\Delta_1 = [0, 1/2]$, $\Delta_2 = [1/2, 1]$. Here, the sequence $w = (w_n)_{n=0}^{\infty}$ associated to x corresponds to a dyadic expansion

$$x = \sum_{n=0}^{\infty} \frac{w_n - 1}{2^{n+1}}.$$

The coding is similar to the decimal expansion, with similar ambiguities.

Definition 4.9. Let X and Y be metric spaces, and $f : X \to X$ and $g : Y \to Y$. We say that g is a factor of f if there is a continuous and surjective map $\pi : X \to Y$ such that $g \circ \pi = \pi \circ f$ on X.

Proposition 4.10. *If $f : S^1 \to S^1$ is an expanding map of degree 2, then f is a factor of $\sigma : \Sigma_2 \to \Sigma_2$.*

Proof. We define the map $\pi : \Sigma_2 \to S^1$ as follows: For $w_0, w_1, \ldots, w_{n-1} \in \{1, 2\}$, define

$$\Delta_{w_0,\ldots,w_{n-1}} = \Delta_{w_0} \cap f^{-1}(\Delta_{w_1}) \cap f^{-2}(\Delta_{w_2}) \cap \cdots \cap f^{-(n-1)}(\Delta_{w_{n-1}}).$$

Given $w = (w_n)_{n=0}^{\infty}$, we obtain a nest of closed intervals

$$\Delta_{w_0} \supset \Delta_{w_0 w_1} \supset \Delta_{w_0 w_1 w_2} \supset \cdots.$$

That is because, $f^{\circ n} : \Delta_{w_0 \cdots w_{n-1}} \to S^1$ is monotone and

$$2\pi = \int_{\Delta_{w_0 w_1 \cdots w_{n-1}}} |(f^{\circ n})'(x)| \, |dx| \geqslant \beta^n \cdot \text{length}(\Delta_{w_0 w_1 \cdots w_{n-1}}),$$

where β is the expansion constant of f (see Fig. 4.6). This implies that length $(\Delta_{w_0 w_1 \cdots w_{n-1}})$ tends to 0 as n tends to infinity. In particular, the nest $\bigcap_{n=1}^{\infty} \Delta_{w_0 w_1 \cdots w_{n-1}}$ shrinks to a single point, which we defined it as $\pi(w)$.

The map π is surjective. That is because, for $x \in S^1$ define $w = (w_n)_{n=0}^{\infty}$ according to $f^{\circ n}(x) \in \Delta_{w_n}$. This gives us $\pi(w) = x$.

The map π is continuous. That is because, if $w = (w_i)_{i=0}^{\infty}$ is close to $w' = (w_i')_{i=0}^{\infty}$, then there is a large N such that $w_n = w_n'$ for all $n \leqslant N$.

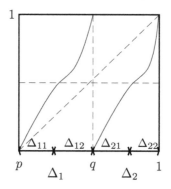

Figure 4.6.   The first two generations of the partitions.

Then, $\pi(w)$ and $\pi(w')$ belong to $\Delta_{w_0\cdots w_{N-1}}$ and we have

$$\left|\pi(w) - \pi(w')\right| \leqslant \text{length}(\Delta_{w_0\cdots w_{N-1}}) \leqslant \frac{2\pi}{\beta^N}.$$

Finally, the relation $f \circ \pi = \pi \circ \sigma$ follows immediately from the definition. That is,

$$\pi(\sigma(w)) = \bigcap_{n=1}^{\infty} \Delta_{w_1\cdots w_n} = f\left(\bigcap_{n=0}^{\infty} \Delta_{w_0\cdots w_{n-1}}\right) = f(\pi(w)). \qquad \square$$

Exercise 4.20. *If distinct points w and w' in Σ_2 satisfy $\pi(w) = \pi(w') = x$. Then, there is $n \geqslant 0$ such that $f^{\circ n}(x) = p$.*

Exercise 4.21. *Show that $\pi : \Sigma_2 \to S^1$ cannot be a homeomorphism. Hint: Σ_2 is a union of two disjoint and closed sets, but S^1 may not be decomposed as a union of two non-empty, disjoint, and closed sets.*

Corollary 4.1. *Let $f : S^1 \to S^1$ be an expanding map of the unit circle. Then,*

(i) *the periodic points of f are dense in S^1,*
(ii) *the map $f : S^1 \to S^1$ is topologically mixing.*

Proof. Let $d = \deg(f) \geqslant 2$. Part (i) follows from the corresponding statement for $\sigma : \Sigma_d \to \Sigma_d$. That is because $\sigma^{\circ p}(w) = w$ implies that $\pi(\sigma^{\circ p}(w)) = f^{\circ p}(\pi(w)) = \pi(w)$. In other words, the image of a σ-periodic point is an f-periodic point. As σ-periodic points are dense in Σ_d, and π is continuous and surjective, the result follows.

For part (ii), we already proved that every expanding map of the circle is mixing (and hence is transitive). Here, we give an alternative proof.

For non-empty and open sets U and V in S^1 we can choose $w_0, w_1, \ldots, w_{m-1}$ and $w_0', w_1', \ldots, w_{m-1}'$ in Σ_d such that

$$\Delta_{w_0 w_1 \ldots w_{m-1}} \subseteq U, \quad \Delta_{w_0' w_1' \ldots w_{m-1}'} \subseteq V.$$

Let $[a_0, a_1, \ldots, a_k] = \{(w_i)_{i=0}^\infty \in \Sigma_d \mid \forall i = 0, 1, \ldots, k, w_i = a_i\}$. Then, for all $n \geqslant m$,

$$[w_0, w_1, \ldots, w_{m-1}] \cap \sigma^{-n}[w_0', w_1', \ldots, w_{m-1}'] \neq \varnothing.$$

For w in the above intersection,

$$x = \pi(w) \in \Delta_{w_0 \ldots w_{m-1}} \cap f^{-n} \Delta_{w_0' \ldots w_{m-1}'}.$$

This finishes the proof of part (ii). □

We now come to the main classification result for expanding maps $f : S^1 \to S^1$ of degree 2.

Theorem 4.4. *If $f : S^1 \to S^1$ and $g : S^1 \to S^1$ are two expanding maps of degree 2, then f and g are topologically conjugate. That is, there exists a homeomorphism $\pi : S^1 \to S^1$ such that $g \circ \pi = \pi \circ f$.*

By the above theorem, every expanding map of the circle of degree 2 is topologically conjugate to the linear one in Example 4.4.

Proof. Consider the conjugacies $\pi_f : \Sigma_2 \to S^1$ and $\pi_g : \Sigma_2 \to S^1$ associated to the two expanding maps f and g in Proposition 4.10. We claim that $\pi(x) = \pi_g(\pi_f^{-1}(x))$ induces a well-defined map from S^1 to S^1. That is because,

(i) if $\pi_f^{-1}(x)$ is a single point then $\pi(x)$ is well defined;
(ii) if $\pi_f^{-1}(x)$ is two points, then the sequences end with infinitely many 1's or 2's. But, then $\pi_g \circ (\pi_f^{-1}(x))$ is again a single point.

It easily follows that the map $x \mapsto \pi_g(\pi_f^{-1}(x))$ is one-to-one and onto. The map π is continuous, since

$$\pi : \bigcap_{k=0}^{n-1} f^{-k}(\Delta_{w_k}) \to \bigcap_{k=0}^{n-1} g^{-k}(\Delta_{w_k}).$$

Finally,

$$g(\pi(x)) = g(\pi_g(\pi_f^{-1}(x))) = \pi_g(\sigma(\pi_f^{-1}(x)))$$
$$= \pi_g(\pi_f^{-1}(f(x))) = \pi(f(x)). \qquad \square$$

Exercise 4.22. *Show that in the above theorem, even if $f : S^1 \to S^1$ and $g : S^1 \to S^1$ are real analytic, π may not even be C^1. Hint: Give examples of f and g whose fixed points have different multipliers.*

Theorem 4.4 extends in an obvious manner to expanding maps of degree n.

4.4.2   Coding horseshoe maps

For an integer $n \geqslant 2$, define the set

$$\Sigma_n' = \{(\ldots, w_{-2}, w_{-1}, w_0, w_1, w_2, \ldots) \mid \forall i \in \mathbb{Z}, w_i \in \{1, 2, \ldots, n\}\}.$$

We define a metric on Σ_n' as

$$d((w_i)_{i \in \mathbb{Z}}, (w_i')_{i \in \mathbb{Z}}) = \sum_{i \in \mathbb{Z}} \frac{|w_i - w_i'|}{2^{|i|}}.$$

The shift map $\sigma : \Sigma_n' \to \Sigma_n'$ is defined as $\sigma(\ldots, w_{-2}, w_{-1}, w_0, w_1, w_2, \ldots) = w' \in \Sigma_n'$, where the entry in the ith coordinate of w is w_{i+1}. Note that $\sigma : \Sigma_n' \to \Sigma_n'$ is continuous, one-to-one, and onto.

Definition 4.10 (Linear Horseshoes). Let $\Delta = [0,1] \times [0,1] \subset \mathbb{R}^2$ be a rectangle. Assume $f : \Delta \to f(\Delta) \subset \mathbb{R}^2$ is a diffeomorphism onto its image such that

(i) $\Delta \cap f(\Delta)$ is a disjoint union of two (horizontal) sub-rectangles Δ_1 and Δ_2 with heights $\leqslant 1/2$;

(ii) the restriction of f to the components of $\Delta \cap f^{-1}(\Delta)$ are linear maps (see Fig. 4.7).

One can write $\Delta \cap f^{-1}(\Delta) = \Delta^1 \cup \Delta^2$, where Δ^1 and Δ^2 are (vertical) sub-rectangles of width $\leqslant \frac{1}{2}$.

The set of points in Δ that can be iterated infinitely many times forward and backward under f is

$$\Lambda = \bigcap_{n \in \mathbb{Z}} f^{-n}(\Delta).$$

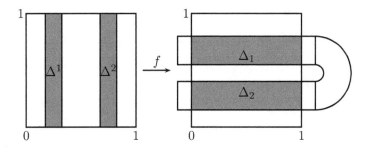

Figure 4.7. A schematic presentation of the horseshoe map. The grey rectangles are mapped to the grey rectangles by some linear maps.

Proposition 4.11. *The map* $f : \Lambda \to \Lambda$ *is topologically conjugate to* $\sigma \colon \Sigma'_2 \to \Sigma'_2$.

Proof. Observe that $\Delta \cap f(\Delta) \cap f^{\circ 2}(\Delta)$ consists of four thin rectangles denoted by

$$\Delta_{i,j} = \Delta_i \cap f(\Delta_j), \quad \text{for } i, j \in \{1, 2\}.$$

Continuing inductively, for each $n \geqslant 1$, the intersection

$$\bigcap_{i=0}^{n-1} f^{\circ i}(\Delta) = \bigcap_{i=0}^{n-1} f^{\circ i}(\Delta_1 \cup \Delta_2)$$

consists of 2^n thin and disjoint horizontal rectangles. For $w_0, w_1, \ldots, w_{n-1} \in \{1, 2\}$, let

$$\Delta_{w_0 w_1, \ldots, w_{n-1}} = \bigcap_{i=0}^{n-1} f^{\circ i}(\Delta_{w_i}).$$

On the other hand, for each $n \geqslant 1$, $\bigcap_{i=0}^{n-1} f^{-i}(\Delta)$ consists of 2^n thin and disjoint vertical rectangles. For every finite sequence $w_{-(n-1)}, \ldots, w_{-1}, w_0$ in $\{1, 2\}$, we let $\Delta^{w_0 w_1, \ldots, w_{n-1}} = \bigcap_{i=0}^{n-1} f^{-i}(\Delta^{w_{-i}})$.

For $w = (w_n)_{n \in \mathbb{Z}}$, we define $\pi(w)$ as

$$\pi(w) = \left(\bigcap_{n=0}^{\infty} \Delta_{w_0 w_1 \cdots w_{n-1}} \right) \cap \left(\bigcap_{n=0}^{\infty} \Delta^{w_0 w_{-1} \cdots w_{-n+1}} \right).$$

The image of a cylinder

$$\{(w'_i)_{i \in \mathbb{Z}} \mid w'_i = w_i, \forall i \text{ with } -(n-1) \leqslant i \leqslant n-1\}$$

under π is a square $\Delta_{w_0 \cdots w_{n-1}} \cap \Delta^{w_0 \cdots w_{-(n-1)}}$ of size bounded by $1/2^{n-1} \times 1/2^{n-1}$. It follows that

(i) π is continuous;

(ii) π is invertible (and a homeromorphism);

(iii) Λ is a Cantor set (that is, Λ is compact, totally disconnected, and every point in Λ is a limit of a sequence of points in Λ);

(iv) $\pi \circ \sigma = f \circ \pi$. $\square$

By Proposition 4.11, $f : \Lambda \to \Lambda$ inherits some dynamical features of $\sigma : \Sigma'_2 \to \Sigma'_2$.

Exercise 4.23. *Let $f : \Delta \to \mathbb{R}^2$ be a linear horseshoe map. We have,*

(i) *The periodic points of $f : \Lambda \to \Lambda$ are dense in Λ;*

(ii) *The number of periodic points of $f : \Lambda \to \Lambda$ of period n is 2^n;*

(iii) *$f : \Lambda \to \Lambda$ is topologically mixing.*

4.5   Topological Entropy

We have seen qualitative indications of chaos: transitivity, density of periodic orbits, sensitive dependence on initial conditions. Now, we would like to quantify the complexity of f, to obtain a finer invariant under topological conjugacy.

Let X be a compact set equipped with a metric d, and $f : X \to X$ be a continuous map. For each $n \geqslant 1$, we define a metric

$$d_n(x,y) = \max\{d(f^{\circ i}(x), f^{\circ i}(y)) : 0 \leqslant i \leqslant n-1\}. \qquad (4.3)$$

Then, define

$$B(x,n,\epsilon) = \{y \in X : d_n(x,y) < \epsilon\}.$$

A finite set $E \subseteq X$ is called an (n, ϵ)-dense set if $X \subseteq \bigcup_{x \in E} B(x,n,\epsilon)$. This is also called (n, ϵ)-spanning set. Note that since X is compact and f is continuous, there is always an (n, ϵ)-dense set with a finite number of elements.

Let $S(n, \epsilon)$ be the minimum cardinality of all (n, ϵ)-dense sets. In other words, this is the list of information needed to keep track of all orbits up to time n and ϵ-error. One can ask how fast does the sequence $S(n,\epsilon)$ grow as n tends to infinity. It turns out that it is suitable to look at the exponential growth rate

$$h(f,\epsilon) = \overline{\lim_{n\to\infty}} \frac{1}{n} \log S(n,\epsilon).$$

By definition, if $\epsilon < \epsilon'$, then $S(n,\epsilon) \geqslant S(n,\epsilon')$, and hence,

$$h(f,\epsilon) \geqslant h(f,\epsilon').$$

This implies that as ϵ tends to 0 from the right-hand side, the sequence $h(f,\epsilon)$ increases. Recall that any increasing sequence has a limit (potentially infinite). Thus, we define

$$h(f) = \lim_{\epsilon\to 0} h(f,\epsilon) \geqslant 0.$$

The above quantity is called the *topological entropy* of f. This notion is, to some extent, independent of the choice of the metric on X. Two metrics d and d' on a space X are called equivalent, if the convergence with respect to any of these metrics implies the convergence with respect to the other one.

Lemma 4.4. *Let d and d' be two equivalent metrics on X that make it a compact space, and let $f : X \to X$ be a continuous map. The topological entropy of f with respect to the metrics d and d' are the same.*

Proof. Consider the identity map $I : (X,d) \to (X,d')$. By the equivalence of the metrics d and d', I is a homeomorphisms (it is one-to-one, onto, continuous, and its inverse is also continuous). Moreover, since X is compact, I is indeed uniformly continuous. This implies that, given $\epsilon > 0$, there is $\delta > 0$ such that $d(x,y) < \delta$ implies $d'(x,y) < \epsilon$. In particular, $d_n(x,y) < \delta$ implies that $d'_n(x,y) < \epsilon$. Therefore, any (n,δ)-dense set with respect to d is also a (n,ϵ)-dense set with respect to d'. Hence,

$$S_d(n,\delta) \geqslant S_{d'}(n,\epsilon), \quad \forall n \geqslant 1.$$

Taking limits as n tends to infinity, we obtain $h_d(f,\delta) \geqslant h_{d'}(f,\epsilon)$. Therefore,

$$h_{d'}(f) = \lim_{\epsilon\to 0} h_{d'}(f,\epsilon) \leqslant \lim_{\delta\to 0} h_d(f,\delta) = h_d(f).$$

Repeating the above argument for the map $I : (X,d') \to (X,d)$, we also obtain $h_d(f) \leqslant h_{d'}(f)$. Therefore, the two quantities must be equal. $\qquad\square$

Corollary 4.2. *Topologically conjugate maps have the same topological entropy.*

Proof. Let $\pi : X \to Y$ be a conjugacy between $f : X \to X$ and $g : Y \to Y$, i.e., $\pi \circ f = g \circ \pi$. If d_X is a metric on X, then define d_Y on Y by

$$d_Y(y, y') = d_X(\pi^{-1}(y), \pi^{-1}(y')).$$

Thus, $\pi : (X, d_X) \to (Y, d_Y)$ is an isometry. This implies that $h_{d_X}(f) = h_{d_Y}(g)$, and then by Lemma 4.4, $h(f) = h(g)$. $\square$

Example 4.6. Consider the expanding map $f : S^1 \to S^1$, $f(x) = dx$ (mod 1), $d \geqslant 2$. Observe that for any $n \geqslant 1$, a d_n-ball $B(x, n, \epsilon)$ has diameter $(2\epsilon)/d^n$. Thus, we need at least $d^n/(2\epsilon)$ balls to cover $[0, 1)$, and $d^n/(2\epsilon) + 1$ balls is enough to cover this set. That is,

$$S(n, \epsilon) \leqslant \left(\frac{d^n}{2\epsilon}\right) + 1, \quad \text{and} \quad S(n, \epsilon) \geqslant \left(\frac{d^n}{2\epsilon}\right).$$

In particular,

$$h(T) = \lim_{\epsilon \to 0} \overline{\lim_{n \to \infty}} \frac{1}{n} \log(S(n, \epsilon)) = \log d.$$

By Theorem 4.4, any expanding map $f : S^1 \to S^1$ of degree $d \geqslant 2$ is topologically conjugate to the linear expanding map of degree d. Hence, such maps have the same topological entropy $\log d$.

Proposition 4.12. *If $f : X \to X$ is an isometry, then $h(f) = 0$.*

Proof. By definition, for all $n \geqslant 1$, $d_n(x, y) = d(x, y)$. In particular, $S(n, \epsilon)$ is independent of n, and thus, $h(f) = 0$. $\square$

The rigid rotations $f : S^1 \to S^1$ are isometries, so $h(f) = 0$. Then, by Theorem 4.3, certain homeomorphisms of S^1 are conjugate to a rotation, and must have zero entropy.

Exercise 4.24. *Prove that the topological entropy of any C^1 (continuously differentiable) map of $S^1 \times S^1$ (torus) is finite. Hint: Consider the maximum size of its derivatives.*

Due to the definition of $S(n, \epsilon)$, we often obtain an upper bound on this quantity. That is because, an example of (n, ϵ)-dense set provides an upper bound for $S(n, \epsilon)$. This leads to an upper bound on $h(f)$. Below we give an alternative definition of the topological entropy that is conveniently used to give a lower bound on $h(f)$. The combination of the two methods is often used to calculate $h(f)$.

Given a compact metric space (X,d) and a continuous map $f : X \to X$, consider the metrics d_n defined in Eq. (4.3). Let $N(n,\epsilon)$ be the maximal number of points in X whose pairwise d_n distances are at least $\epsilon > 0$. A set of such points is called an (n,ϵ)-separated set.

Lemma 4.5. *We have,*

(i) $N(n,\epsilon) \geqslant S(n,\epsilon)$;
(ii) $S(n,\epsilon) \geqslant N(n,2\epsilon)$.

Proof. (i). Let E_n be an (n,ϵ)-separated set with $N(n,\epsilon)$ elements. Then, E_n must also be an (n,ϵ)-spanning set, since otherwise, we could enlarge the separating set by adding a point not already covered. Thus, $N(n,\epsilon) \geqslant S(n,\epsilon)$.

(ii). Let E_n be an arbitrary $(n,2\epsilon)$-separated set, and F_n be an arbitrary (n,ϵ)-dense set. We define a map $\phi_n : E_n \to F_n$ as follows. By the definition of (n,ϵ)-dense set, the set $\bigcup_{x \in F_n} B(x,n,\epsilon)$ covers X. Then, for any $x \in E_n$, there is $\phi_n(x) \in F_n$ such that $d_n(x,\phi_n(x)) < \epsilon$. The map ϕ is well defined and one-to-one. That is because, if $\phi(x) = \phi(y)$, then

$$d_n(x,y) \leqslant d_n(x,\phi_n(x)) + d_n(\phi_n(y),y) < \epsilon + \epsilon = 2\epsilon.$$

However, since E_n is $(n,2\epsilon)$-separated, we must have $x = y$.

The injectivity of $\phi_n : E_n \to F_n$ implies that the number of elements in F_n is bigger than the number of elements in E_n. Since E_n and F_n are arbitrary, we must have $S(n,\epsilon) \geqslant N(n,2\epsilon)$. This finishes the proof of the proposition by taking limits as $\epsilon \to 0$. $\qquad\square$

Proposition 4.13.

$$h(f) = \lim_{\epsilon \to 0} \varlimsup_{n \to \infty} \frac{1}{n} \log N(n,\epsilon).$$

Proof. By Part (i) of Lemma 4.5,

$$\varlimsup_{n \to \infty} \frac{1}{n} \log N(n,\epsilon) \geqslant \varlimsup_{n \to \infty} \frac{1}{n} \log S(n,\epsilon) = h(f,\epsilon).$$

By Part (ii) of Lemma 4.5,

$$\varlimsup_{n \to \infty} \frac{1}{n} \log N(n,2\epsilon) \leqslant \varlimsup_{n \to \infty} \frac{1}{n} \log S(n,\epsilon) = h(f,\epsilon).$$

Letting $\epsilon \to 0$, we obtain the desired formula in the proposition. $\square$

Proposition 4.14. *Let $g : Y \to Y$ be a factor of $f : X \to X$, that is, there is a continuous surjective map $\pi : X \to Y$ with $g \circ \pi = \pi \circ f$. Then $h(g) \leqslant h(f)$.*

Proof. Let d^X and d^Y denote the metrics on X and Y, respectively. For $\epsilon > 0$, choose $\delta > 0$ such that if $d^X(x_1, x_2) < \delta$ then $d^Y(\pi(x_1), \pi(x_2)) < \epsilon$. Thus, a δ-ball with respect to d_n^X, $B(x, n, \delta)$, is mapped under π into $B(\pi(x), n, \epsilon)$. In particular,

$$S_{d^X}(n, \delta) \geqslant S_{d^Y}(n, \epsilon).$$

This implies the inequality in the proposition. $\square$

Exercise 4.25. *Let $f : S^1 \times S^1 \times S^1 \to S^1 \times S^1 \times S^1$ be defined as,*

$$f(x, y, z) = (x, x + y, y + z) \quad (\mathrm{mod}\ 1).$$

Find $h_{\mathrm{top}}(f)$.

Exercise 4.26. *Let $D = \{z \in \mathbb{C} : |z| \leqslant 1\}$ and for each $\lambda \in [0, 1]$ define $f_\lambda : D \to D$ as $f_\lambda(z) = \lambda z^2$.*

(i) *Show that $h_{top}(f_\lambda) \geqslant \log 2$, when $\lambda = 1$.*
(ii) *Show that $h_{top}(f_\lambda) = 0$, when $0 \leqslant \lambda < 1$.*

Therefore, topological entropy does not depend continuously on the map.

4.6   Ergodic Theory

One of the main questions the theory of dynamical systems is concerned with is the long-time behaviour of dynamical systems. We have seen in the previous section that even very simple low-dimensional dynamical systems can have rather complicated behaviour. Unfortunately, this kind of complicated structure is actually typical and the complications will accentuate for more realistic systems. However, if one restricts the consideration to the so-called statistical quantities (explained below) then these systems show a much simpler and regular behaviour. In contrast, integrable systems are much more fragile. This explains the advantage of statistical approaches. Note that the statistical approach does not require any source of randomness. The dynamics are purely deterministic.

In order to simplify our consideration we will often make the following assumptions. Please keep in mind that these assumptions are only there to make the consideration easier and transparent, but do not restrict the phenomena we want to look at. Features of the special systems which are not typical for more realistic systems will not be mentioned.

First, we assume mostly that the system is time-homogeneous and second, we assume that the time is often in $\mathbb{Z}$ or $\mathbb{N}$. Clearly, any continuous time dynamics can be related to a discrete time dynamics by restricting the attention to purely integer time values. Then the flow property is reduced to just the following $\Phi_2 = \Phi_1 \circ \Phi_1$ and so in general

$$\Phi_t = \underbrace{\Phi_1 \circ \cdots \circ \Phi_1}_{t\text{-times}}. \tag{4.4}$$

So, it is enough to consider Φ_1 which we will denote in the following for short by Φ.

What is meant by long-time behaviour is in general difficult to understand. Two types of long-time behaviour can be described very easily. The system can converge to a *fixed point* x', that is for all x we have that $\lim_{t\to\infty} \Phi_t(x) = x'$. The system can be periodic, that is there exits a T such that $\Phi_{T+t} = \Phi_T$ or equivalently for all x holds $\Phi_T(x) = x$. However, we have seen a lot of different more irregular behaviours in the previous sections which cannot be captured in this way.

Hence, we consider, as before, a set Ω and a map $\Phi : \Omega \to \Omega$.[2] One way of describing the long-time behaviour is to use a probabilistic description. We can, for example, ask the following question: given an measurable subset $A \subset \Omega$. With which frequency the dynamics hits the set A in the time $t = 0, \ldots, n-1$?[3] In other words, we are interested in the numbers

$$\frac{\#\{k \in \{0, \ldots, n-1\} \; : \; \Phi_k(x) \in A\}}{n}, \tag{4.5}$$

where $\#B$ denotes the number of elements in the set B. This is the fraction of times the dynamics is in A in a length of time n. For what follows, it is convenient to reformulate the above expression as

$$\frac{1}{n} \sum_{k=0}^{n-1} \mathbb{1}_A(\Phi_k(x)), \tag{4.6}$$

---

[2]In the following, we need just the following regularity: Ω is a topological space (for example $\Omega \subset \mathbb{R}^d$) and Φ is continuous.

[3]$n-1$ is just for convenience.

where $\mathbb{1}_A$ denotes the characteristic function, that is $\mathbb{1}_A(y) = 1$ if and only if $y \in A$ otherwise it is zero. One can rewrite the above expression also in terms of measures. Define a measure μ_n to be

$$\mu_n(A) := \frac{1}{n} \sum_{k=0}^{n-1} \mathbb{1}_A(\Phi_k(x)). \tag{4.7}$$

That is $\mu_{x,n}$ associates to each set A the frequency with which the dynamic is in this set in a time period of length n (starting from x).[4] One cannot hope to obtain any simple results for fixed n, but one may hope to obtain an instructive result in the limit $n \to \infty$. One may hope that

$$A \mapsto \lim_{n \to \infty} \frac{1}{n} \sum_{k=0}^{n-1} \mathbb{1}_A(\Phi_k(x)) \tag{4.8}$$

gives rise to a measure $\mu_{x,\infty}$. This measure is called the *ergodic mean*. Unfortunately, one cannot expect convergence of this expression in general. The right type of convergence allows us to develop a general theory. A very common choice is the so-called *weak convergence*, that is a sequence of measures ν_n converges to a ν_∞ in the sense of weak convergence if and only if for *any* continuous bounded function with bounded support $\varphi : \Omega \to \mathbb{R}$ the following limit exists and fulfils

$$\lim_{n \to \infty} \int \varphi(x)\nu_n(dx) = \int \varphi(x)\nu_\infty(dx). \tag{4.9}$$

That one does not consider instead the limit for any measurable function or for any observable of the form $\mathbb{1}_A$, for $A \subset \Omega$ measurable (that is one would consider $\lim_{n \to \infty} \mu_{x,n}(A)$ directly) is due to the effect that even in quite elementary examples the latter convergence does not hold. The reason for that is not a mathematical oddity, but that in the large-time limit often systems show some degree of organisation which appears on all scales. In this case, the latter type of convergence is just not true. However, the weak convergence is equivalent to the following version of convergence on sets: let Ω be a reasonable topological space,[5] then weak convergence is equivalent

---

[4] If one uses the Dirac measure one can also write $\mu_{x,n} := \frac{1}{n} \sum_{k=0}^{n-1} \delta_{T^k(x)}$.

[5] A complete metric space with a countable dense subset.

that for any measurable bounded $A \subset \Omega$ with $\nu_\infty(\partial A) = 0$ holds that

$$\lim_{n\to\infty} \nu_n(A) = \nu_\infty(A), \tag{4.10}$$

where ∂A denotes the boundary of A.[6] This result shows that the problem appears for sets where there is mass in the limiting measure concentrated on the boundary. In this case, the behaviour of the limit will be unstable. The functions φ are "smoothened" versions of $\mathbb{1}_A$ and hence this effect does not appear.

Summarising, in the following, we will study the limit of expressions of the form $\int \varphi(y)\mu_{x,n}(dy)$, note the latter can be expressed explicit by extending Eq. (4.7)

$$\int \varphi(y)\mu_{x,n}(dy) = \frac{1}{n}\sum_{k=0}^{n-1} \varphi(\Phi_k(x)). \tag{4.11}$$

The main question posed is: does there exists an x for which this limit converges and, if yes, in which way the limit will depend on x. The surprising and, at the beginning, even shocking phenomena is that we will show that there exists a lot of such x, but we are not able to find a single one of these explicitly.

There is a nice characterisation of measures which may appear as limits of the type in Eq. (4.11).

Exercise 4.27. *Show that by the definition of $\mu_{x,n}$ we have that*

$$\Phi_{\#}\mu_n = \frac{n+1}{n}\mu_{x,n+1} - \frac{1}{n}\mu_{x,1}. \tag{4.12}$$

Hence, for any function φ it holds that

$$\int \varphi \circ \Phi(y)\mu_{x,n}(dy) = \frac{n+1}{n}\int \varphi(y)\mu_{x,n+1}(dy) - \frac{1}{n}\int \varphi(x)\mu_1(dx). \tag{4.13}$$

As $\varphi \circ \Phi$ is a continuous bounded function we get by taking the limit of both sides that for all x s.t. the limits in Eq. (4.13) exist then

$$\int \varphi \circ \Phi(y)\mu_{x,\infty}(dy) = \int \varphi(y)\mu_{x,\infty}(dy) \tag{4.14}$$

for all continuous bounded functions φ, or for short

$$\Phi_{\#}\mu_{x,\infty} = \mu_{x,\infty}. \tag{4.15}$$

---

[6]The topological boundary that is the closure of A with out its interior.

Definition 4.11. We call a measure μ-*invariant* with respect to the dynamics Φ if and only if[7]

$$\Phi_{\#}\mu = \mu. \tag{4.16}$$

It is not clear that such invariant measures actually exist. The following general result by Krylov–Bogolubov guarantees the following.

Theorem 4.5. *Let Ω be compact*[8] *and suppose there exists a countable dense subset in Ω.*[9] *Let $\Phi : \Omega \to \Omega$ be continuous. Then there exists at least one invariant measure. More specifically, for each $x \in \Omega$ there exists a measure $\mu_{x,\infty}$ on Ω and there exists a subsequence of times $(n_k)_k$ (which may depend on x) such that for all continuous functions φ*

$$\lim_{k \to \infty} \int \varphi(y)\mu_{x,n_k}(dy) \tag{4.17}$$

converges and hence there exists an associated finite invariant measure $\mu_{x,\infty}$ with

$$\lim_{k \to \infty} \int \varphi(y)\mu_{x,n_k}(dy) = \int \varphi(y)\mu_{x,\infty}(dy). \tag{4.18}$$

The problem with this theorem is that we can only establish the convergence for a subsequence and the limiting measure may depend on the choice of the point x and, what is even more unfortunate, it may depend on the choice of the subsequence. It can be quite easily seen in examples, that the latter actually may happen. Hence, the theorem effectively only shows the existence of an invariant measure, but does not give a practical way to construct one. Below, we will consider a much stronger theorem which will achieve convergence for a large class of dynamics.

Proof. The idea is to consider a countable collection $(\varphi_n)_n$ of continuous functions on Ω which is dense in the supremums norm in the space of all continuous functions and use the boundedness of $|\varphi_n(y)|$ for $y \in \Omega$. The details are left as an exercise or alternatively consult Theorem 4.1.2 in Katok and Hasselblatt (1995). $\qquad\square$

---

[7]$\Phi_{\#}\nu = \mu$ means that for all $A \subset \Omega$ measurable holds $\nu(\Phi^{-1}(A)) = \mu(A)$ or equivalently that for all measurable bounded functions $\varphi : \Omega \to \mathbb{R}$ holds that $\int \varphi \circ \Phi(y)\nu(dy) = \int \varphi(y)\mu(dy)$.

[8]For example, $\Omega \subset \mathbb{R}^d$ and Ω is closed and bounded.

[9]A subset $B \subset \Omega$ is called dense, if to every $x \in \Omega$ there exists a sequence $(x_n)_n$ in B with $\lim_{n\to\infty} x_n = x$. For example, $\mathbb{Q}$ is dense in $\mathbb{R}$.

Often in concrete examples one is able to identify some of the invariant measures explicitly.

Example 4.7. Let us consider the unit circle S^1 represented as $[0,1[$ mod 1, that is we consider the description via angles, that is

$$\alpha \mapsto \begin{pmatrix} \cos(2\pi\alpha) \\ \sin(2\pi\alpha) \end{pmatrix}. \tag{4.19}$$

Obviously, we want to consider the angles α and $\alpha + 1$ as identical. In mathematical language, we want to consider α mod 1. The set S^1 has also a natural group structure. For the angle representation one may introduce the following distance

$$d(\alpha,\beta) = \min\{|\alpha - \beta + k| \; : k \in \mathbb{Z}\}. \tag{4.20}$$

Actually, it is sufficient to consider $k = -1, 0, 1$.

We reconsider the two paradigmatic examples considered in Secs. 4.2 and 4.3 with a quite similar definition but an antithetic behaviour.

Example 4.8. For $\alpha \in \mathbb{R}$ the rotation $R_\alpha : S^1 \to S^1$ is defined as follows in the mod 1 representation.

$$R_\alpha(\beta) = \alpha + \beta \text{ mod } 1. \tag{4.21}$$

The second example is the doubling map E_2 defined on S^1 in the mod 1 representation as follows:

$$E_2(\beta) = 2\beta \text{ mod } 1. \tag{4.22}$$

Periodic orbits give immediately rise to an invariant measure.

Definition 4.12. x is called a *periodic point of minimal period p* if p is the smallest positive natural number such that $\Phi_p(x) = x$.

Example 4.9. If y is a periodic point of period p then,

$$\mu_{y,\infty}(dx) := \frac{1}{p}\sum_{k=0}^{p-1} \delta_{\Phi_k(y)}(dx) \tag{4.23}$$

is an invariant measure.[10]

---

[10]Here, δ_x denotes the so-called Dirac measure that is for any measurable subset $A \subset \Omega$ holds that $\delta_x(A)$ is either one or zero and it is one if and only if $x \in A$. Then, for any measurable function $f : \Omega \to \mathbb{R}$ holds that $\int f(x)\delta_y(dx) = f(y)$.

Example 4.10. The restriction of the Lebesgue measure to $[0,1[$ is an invariant measure for the doubling map. Let us denote this measure by m. Let us compute $(E_2)_{\#}m$, that is we have to compute

$$(E_2)_{\#}m(A) = m\left(x \in [0,1[: 2x \bmod 1 \in A\right) \qquad (4.24)$$
$$= m\left(x \in [0,1/2[: 2x \in A\right) + m\left(x \in [1/2,1[: 2x - 1 \in A\right). \qquad (4.25)$$

Let us consider the two pieces separately. Indeed, the map $f : x \mapsto 2x$ as a map from $[0,1/2[\to [0,1[$ is differentiable and

$$m\left(x \in [0,1/2[: 2x \in A\right) = f_{\#}m(A). \qquad (4.26)$$

Note that $Df(x) = 2$ and $f^{-1}(y) = y/2$. Hence by using the transformation formula for the Lebesgue measure, we get that

$$f_{\#}m(A) = \int_A \frac{1}{|\det(Df(f^{-1}(y)))|} dy = \frac{1}{2}m(A). \qquad (4.27)$$

The second case can be treated analogously.

Exercise 4.28. *Does there exists other invariant measure for E_2?*

Exercise 4.29. *Show that for any α the rotation R_α has the Lebesgue measure restricted to $[0,1[$ as invariant measure.*

Example 4.11. The torus can be described in two ways. First, is can be seen as $S^1 \times S^1$. Second, consider $\mathbb{R}^2$ and the unit square $[0,1[\times[0,1[$. Then one can partition $\mathbb{R}^2$ in all the translates of the unit cube by integer vectors $[p,p+1[\times[q,q+1[$. One considers the corresponding points in the different cubes as identically. That is $x,y \in \mathbb{R}^2$ are considered identical if and only if $x - y \in \mathbb{Z}^2$. This is the same as saying that one identifies the opposite sides of the unit square to obtain a torus.

Let M be a 2×2 matrix with integer entries. Then if x,y are considered identical in the above sense then Mx and My are identical. Indeed $Mx - My = M(x - y)$, which is in $\mathbb{Z}^2$ because $x - y \in \mathbb{Z}^2$ and all entries of M are in $\mathbb{Z}$. Then one interprets $M : [0,1[\times[0,1[\mapsto [0,1[\times[0,1[$ if instead of Mx one takes the unique point $z \in [0,1[\times[0,1[$ such that $z - Mx \in \mathbb{Z}^2$. One calls M a *toral endomorphism*.

Exercise 4.30. *Consider $x \mapsto x+1$ on $\mathbb{R}$. Show that there exists no finite invariant measure with respect to this flow. Explain why this does not contradict the statement of Theorem 4.5. Show that the Lebesgue measure is an invariant measure.*

Example 4.12. Consider on $\mathbb{R}^{dN} \times \mathbb{R}^{dN}$ a real valued function H called the Hamiltonian. Denote the variables in this space by $(q_1, \ldots, q_N, p_1, \ldots, p_N)$ and consider an ODE of the form

$$\frac{d}{dt} p_i(t) = -\partial_{q_i} H(q_1(t), \ldots, q_N(t), p_1(t), \ldots, p_N(t)) \qquad (4.28)$$

$$\frac{d}{dt} q_i(t) = \partial_{p_i} H(q_1(t), \ldots, q_N(t), p_1(t), \ldots, p_N(t)). \qquad (4.29)$$

A typical form is

$$H(q_1, \ldots, q_N, p_1, \ldots, p_N) = \sum_{k=1}^{N} \frac{|p_k|^2}{2m} + \sum_{k \neq l = 1}^{N} V(x_k - x_l). \qquad (4.30)$$

Then the Lebesgue measure on $\mathbb{R}^{dN} \times \mathbb{R}^{dN}$ is an invariant measure.

Such a dynamic is called the Hamiltonian dynamics and any mechanical system which conserves energy can be written in this way. The above result is called the Liouville theorem and is at the foundation of statistical physics. The Lebesgue measure restricted to the subset of a given energy $H^{-1}(E)$, $E \in \mathbb{R}$ is called the micro-canonical ensemble. The long-time behaviour of any mechanical system in thermal isolation will be described by this distribution.

Exercise 4.31.

(a) *Show that the Lebesgue measure is an invariant measure for the system in Example* 4.12.
(b) *Show that for a solution $\frac{d}{dt} H(q_1(t), \ldots, q_N(t), p_1(t), \ldots, p_N(t)) = 0$.*
(c) *Hence, one can restrict the dynamics to $\Omega = H^{-1}(E)$. Show that one can define on Ω a dynamical system and that the Lebesgue measure restricted to Ω is an invariant measure. For hints see Sec. 3.11 in Gallavotti* (1983).

4.6.1 *Recurrence*

Let (Ω, Φ) be a dynamical system, that is a set Ω and a map $\Phi : \Omega \mapsto \Omega$. Obviously, not all points in Ω are finitely periodic in general, but there exists the following general related phenomena called recurrence. Recurrence means that a point x will return after some time t almost to x. The following theorem by Poincaré ignited an enormous amount of discussion in physics. It implies for example that if you break an egg on your kitchen floor, if you wait long enough then the egg will jump up from the floor and return almost perfectly restored to your hand. Obviously, an absurd statement, however the statement cannot be dismissed out of hand. One does not see such a restoration in practice because the times considered are too large to be observed. However, the phenomena of recurrence is one indication that a statistic description may be feasible.

Theorem 4.6. *Let $\Phi : \Omega \to \Omega$ be a measurable function and μ a Φ-invariant measure with $\mu(\Omega) < \infty$. Let $A \subset \Omega$ be measurable. Then for μ-a.e. point $x \in A$, there exists a $t_x \in \mathbb{N}$ such that $\Phi_{t_x}(x) \in A$. Moreover, there exists infinitely many such times.*

If the invariant measure is, for example, the Lebesgue measure restricted to a bounded set, then for any ball, however small its radius, the dynamics will return to it. Note that in practice the return-time can be very long, cf. the next section.

Proof. The idea of the proof is to consider the A_{nret} set of all points that never return to A and show that $\Phi_k^{-1}(A_{\mathrm{nret}})$ forms a disjoint family in k. Details are left as an exercise or can be found in the proof of Theorem 3.1 in Krengel (1985) or Theorem 8.1 in Barreira and Valls (2013a). □

Definition 4.13. Let Ω be a metric space.[11] The support of a measure μ, denoted by $\mathrm{supp}(\mu)$ is the set of all x such that for every open ball U around x holds that $\mu(U) > 0$.[12]

By definition the support is closed. Note $\mu(\Omega \setminus \mathrm{supp}(\mu)) = 0$ if Ω has a countable dense subset.[13]

Definition 4.14. Let Ω be a metric space. We say that a point x is positively recurrent if and only if for any ball U containing x there exists an $n_U > 1$ such that $\Phi_{n_U}(x) \in U$. Equivalently,[14] there exists a sequence $(n_k)_k$ with $\lim_{k\to\infty} n_k = \infty$ and $\lim_{k\to\infty} \Phi_{n_k}(x) = x$. Denote by $R_+(\Phi)$ the closure of the set of all positive recurrent points.

Exercise 4.32. *If Ω is a metric space then $\mathrm{supp}(\mu) \subset R_+(\Phi)$.*

---

[11] For simplicity, any subset of $\mathbb{R}^d$ is a metric space.
[12] Actually, neighbourhood is enough, no metric structure is needed.
[13] What one really needs is that one has a countable basis of the topology.
[14] This is only true for metric spaces.

Exercise 4.33. *Consider* $\Omega = \mathbb{R}$, μ *the Lebesgue measure and* $\Phi(x) = x + 1$. *Show that there exists a set of positive measure A to which no element ever returns.*

4.6.2 Kac Lemma

Let us try to investigate how long it will take until the orbit returns. Define the following function called the *first return time*: for $x \in A$ we define

$$R_A(x) := \min\{n \in \mathbb{N} \ : \Phi_n(x) \in A\}, \tag{4.31}$$

where we set $R_A(x) = \infty$ if the set on the right hand side is empty. Note that $R_A(x) \geqslant 1$. R_A is a measurable function. Denote by

$$A^* := \{\omega \in \Omega \setminus A \ : \text{ there exists } n \in \mathbb{N}_0 \text{ with } \Phi_n(\omega) \in A\}. \tag{4.32}$$

That is the set of all points x which are not in A, but enter at least once to A.

One says that a dynamic Φ is *recurrent* with respect to a measure if and only if for any measurable subset $A \subset \Omega$ holds $\mu(A \setminus A_{\text{ret}}) = 0$, where

$$A_{\text{ret}} := \{x \in \Omega \ : \ x \in A \text{ and there exists } n \geqslant 1 \text{ such that } \Phi_n(x) \in A\}. \tag{4.33}$$

In other words the dynamics is recurrent if and only if for any measurable subset A and for μ-a.e. $x \in A$ one has that $R_A(x) < \infty$. Theorem 4.6 shows that for a finite invariant measure the dynamic is always recurrent and hence $\mu(A \setminus A_{\text{ret}}) = 0$.

Theorem 4.7 (Kac's Lemma). *Let* $\Phi : \Omega \to \Omega$ *be a measurable function and* μ *be an invariant measure. Assume that Φ is recurrent with respect to Φ or assume even that $\mu(\Omega) < \infty$. Then*

$$\int_A R_A(\omega)\mu(d\omega) = \mu(A^*). \tag{4.34}$$

Proof. The idea is to show that for $n \geqslant 2$ the following holds

$$\{x \in A \ : \ R_A(x) = n\} \cup \{x \in \Omega \setminus A \ : \ R_A(x) = n\} \tag{4.35}$$

$$= \Phi^{-1}(\{y \in \Omega \setminus A \ : \ R_A(y) = n - 1\}) \tag{4.36}$$

and defining $A_n := \{x \in \Omega \setminus A \ : \ R_A(x) = n\}$ note that $A^* = \bigcup_{n=1}^{\infty} A_n$ to show that $\mu(A^*) = \sum_{k=1}^{\infty} k\mu(\{x \in A \ : \ R_A(x) = k\})$. Details are left as an exercise or can be found in the proof of Theorem 3.6 in Krengel (1985). $\square$

If the dynamics is well spreading in the space, for example is ergodic, cf. Sec. 4.6.3 and μ is a probability measure then $\mu(A^*) = 1$. Hence, the expectation of R_A on A is given by

$$\frac{1}{\mu(A)} \int_A R_A(\omega)\mu(d\omega) = 1. \tag{4.37}$$

Hence, this means that $R_A(\omega)$ should be of the order $1/\mu(A)$, that is if the probability $\mu(A)$ is very small then return time has typically to be very large.

Exercise 4.34. *Consider the following dynamics on* $[0,1[$*. Write* x *in its decimal expansion* $(x_1, x_2, x_3, \ldots)$*. Then* $\Phi((x_1, x_2, x_3, \ldots)) = (x_2, x_3, \ldots)$ *is the shift-map.*

(a) *Show that the Lebesgue measure is an invariant measure for this dynamics.*
(b) *Show that with probability one a real number has seven infinitely often in its decimal representation.*
(c) *Show that any finite motive of digits, like* 4711 *appear infinitely often.*
(d) *What is the frequency with which a word reappears.*

4.6.3   Birkhoff's Ergodic Theorem

One could be also interested in the frequency with which the system returns to A. In the time T, the number of times the orbit return to A can be written as

$$\sum_{t=0}^{T-1} \mathbb{1}_A(\Phi_t(x)), \tag{4.38}$$

where $\mathbb{1}_A$ denotes the characteristic functions of A, that is the function which is one if and only if $\Phi_t(x) \in A$. Hence, the frequency of a recurrence is given by the expression

$$\frac{1}{T} \sum_{t=0}^{T-1} \mathbb{1}_A(\Phi_t(x)). \tag{4.39}$$

As for the convergence of measure, it is more suitable instead of measurable functions, like $\mathbb{1}_A$, to consider continuous functions φ. Intuitively, one can think about φ as a smoothened version of $\mathbb{1}_A$. Then, we define the following *ergodic average* or *Cesaro mean*

$$\overline{\varphi}_T(x) := \frac{1}{T} \sum_{t=0}^{T-1} \varphi(\Phi_t(x)). \tag{4.40}$$

Note that in the existence theorem for the invariant measure, we were only able to establish the existence of limits along carefully chosen subsequences. This is of limited practical use, because in a numerical experiment we do not know which subsequence to choose. Furthermore, one can also consider the ergodic average as a regularised version of the usual limit.

Exercise 4.35. *Let a_n be a sequence which converges to a. Then*

$$\frac{1}{N} \sum_{n=1}^{N} a_n \to a.$$

However, there exist sequences for which the latter expression converges but the sequence itself does not converge.

Hence, it is very surprising that Birkhoff could prove the following very strong convergence result (Breiman, 1973).

Theorem 4.8. *Let $\Phi : \Omega \to \Omega$ be measurable and μ an invariant probability measure.[15] Let $\varphi : \Omega \to \mathbb{C}$ be an integrable function.[16] Then for μ-almost every $x \in \Omega$*

$$\lim_{T \to \infty} \varphi_T(x) = \lim_{T \to \infty} \frac{1}{T} \sum_{t=1}^{T} \varphi(\Phi_t(x)) =: \overline{\varphi}(x) \tag{4.41}$$

exists and we denote the limit by $\overline{\varphi}$.

Note that we only assume that μ is invariant and finite. We do *not* assume that μ is ergodic.

This effect can be also observed numerically. Namely that the sequence $\Phi_t(x)$ may fluctuate but the sequence $\frac{1}{T} \sum_{t=1}^{T} \Phi_t(x)$ converges nicely.

---

[15]It is sufficient to assume that μ is σ-finite, that is, there is a sequence of measurable sets $A_n \subset \Omega$ such that $\mu(A_n) < \infty$ and $\Omega = \bigcup_{n=1}^{\infty} A_n$. Any reasonable measure has this property, in particular all finite measures. The Lebesgue measure in $\mathbb{R}^d$ is σ-finite, for example for A_n the ball around zero with radius n.

[16]That is φ is a measurable function and $\int_{\Omega} |\varphi(x)| \mu(dx) < \infty$. All bounded functions which have a support of finite measure are integrable.

First, we have to prove two auxiliary results. Let us begin with an algebraic one. Consider a finite sequence of real numbers $a_0, a_1, \ldots, a_n$. Let $m \leqslant n$. We say that a_k *recovers in at most m steps* if there exists a $0 \leqslant p \leqslant m$ with $k + p - 1 \leqslant n$ such that $a_k + a_{k+1} + \cdots a_{k+p-1} \geqslant 0$. So if a_k is non-negative then it recovers in one step and hence also in at most m-steps for any $1 \leqslant m \leqslant n$. Conversely, any a_k that recovers in one step has to be non-negative.

Lemma 4.6. *The sum of all a_k that recover in at most m steps is non-negative.*

Proof. The idea of the proof is to consider the first element in the sequence that recovers in at most m steps and then argue that the following $p - 1$ also recover in at most m steps, p as above. Details are left as an exercise or can be found in the proof of the Lemma on page 18 in Halmos (1974). □

The idea to prove Theorem 4.8 is to apply the above lemma to the sequence $a_k = \varphi(\Phi_k(x))$ for each x. Let m be given and n much larger. Define the $D_k^{(m)}$ as follows: $x \in D_k^{(m)}$ if $\varphi(\Phi_k(x))$ recovers in m steps. Hence by the previous lemma

$$\sum_{k=0}^{n} \mathbb{1}_{D_k^{(m)}}(x)\varphi(\Phi_k(x)) \geqslant 0. \tag{4.42}$$

Note that $D_k^{(m)} = \Phi_k^{-1}(D_0)$. Hence using the invariance of μ, we get for the integral of the function in Eq. (4.42) that

$$0 \leqslant \int \sum_{k=0}^{n-m} \mathbb{1}_{\Phi_k^{-1}(D_0^{(m)})}(x)\varphi(\Phi_k(x))\mu(dx)$$

$$+ \sum_{k=n-m+1}^{n} \int \mathbb{1}_{D_k^{(m)}}(x)\varphi(\Phi_k(x))\mu(dx)$$

$$\leqslant (n-m)\int_{D_0^{(m)}} \phi(x)\mu(dx) + m\int_{\Omega} |\phi(x)|\mu dx. \tag{4.43}$$

Dividing by $(n-m)$ and taking the limit $n \to \infty$ yields $0 \leqslant \int_{D_0^{(m)}} \phi(x)\mu(dx)$. Hence by σ-additivity, we get the following.

Theorem 4.9 (Maximal Ergodic Theorem). *Let Ω be a measurable space and $\Phi : \Omega \to \Omega$ be measurable. Let μ be Φ-invariant. Let $\varphi : \Omega \to \mathbb{R}$ be* *an integrable function.*

$$E := \left\{ x \in \Omega \; : \; exists \; n \; with \; \frac{1}{n} \sum_{k=0}^{n-1} \varphi(\Phi_k(x)) \geqslant 0 \right\}. \tag{4.44}$$

Then

$$\int_E \varphi(x)\mu(dx) \geqslant 0. \tag{4.45}$$

Corollary 4.3 (Maximal Ergodic inequality). *Under the same assumption as in Theorem 4.9. Let* $\alpha > 0$ *then*

$$\mu \left(\left\{ x \in \Omega : \; exist \; an \; n \; such \; that \; \frac{1}{n} \sum_{k=0}^{n-1} \varphi(\Phi_k(x)) \geqslant \alpha \right\} \right)$$
$$\leqslant \frac{1}{\alpha} \int |\varphi(x)|\mu(dx). \tag{4.46}$$

Proof. The idea of the proof is to consider Theorem 4.9 for $\psi := (\varphi - \alpha)\mathbb{1}_A$, where A is a subset of the set in the measure μ with $\mu(A) < \infty$. Details are left as an exercise or can be found on page 19 in Halmos (1974). $\qquad\square$

Proof of Theorem 4.8. The idea of the proof is to show that μ-a.s. there exists an $\alpha > 0$ such that $(\frac{1}{n} \sum_{k=0}^{n-1} \varphi(\Phi_k(x)))_n \leqslant \alpha$ for all n using Corollary 4.3. Let $a < b$ with $a, b \in \mathbb{Q}$ denote by $A_{a,b}$ the set of all x such that the limes superior is larger than b and the limit inferior is smaller than an a. Use Theorem 4.9 with $\psi = (\varphi - b)\mathbb{1}_{A_{a,b}}$. Details are left as an exercise or can be found on page 19 in Halmos (1974). $\qquad\square$

Let us denote by

$$\overline{\varphi} := \lim_{n \to \infty} \frac{1}{n} \sum_{k=0}^{n-1} \varphi \circ \Phi_k. \tag{4.47}$$

What can we say about the limit. It will in general depend on x but has the following properties.

Corollary 4.4.

(a) $\overline{\varphi}$ *is measurable and integrable, moreover*

$$\int_\Omega |\overline{\varphi}(x)|\mu(dx) \leqslant \int_\Omega |\varphi(x)|\mu(dx) \tag{4.48}$$

(b) $\overline{\varphi}$ is μ-a.s. Φ invariant, that is $\overline{\varphi}(\Phi(x)) = \overline{\varphi}(x)$ for μ-a.e. x.

(c) For any measurable $A \subset \Omega$ with $\mu(A) < \infty$ which is μ-a.s. invariant, that is A and $\Phi^{-1}(A)$ differ by a set of μ-measure zero, it holds that $\int_A \overline{\varphi}(x)\mu(dx) = \int_A \varphi(x)$. In particular, if $\mu(\Omega) < \infty$ then $\int_\Omega \overline{\varphi}(x)\mu(dx) = \int_\Omega \varphi(x)\mu(dx)$ and furthermore

$$\int_\Omega \left| \frac{1}{n} \sum_{k=0}^{n-1} \varphi(\Phi_k(x)) - \varphi(x) \right| \mu(dx) \to 0. \tag{4.49}$$

Proof. The following ideas are used in the proof: (a) use triangle inequality for integrals and invariance of μ; (b) prove the result first for bounded φ; (c) Use Theorem 4.9 for $\psi(x) = \varphi(x)\mathbb{1}_A(x) - (a - \varepsilon)\mathbb{1}_A(x)$ to show that $a\mu(B_a) \leqslant \int \varphi(x)\mathbb{1}_A(x)\mu(dx)$ where $B_a := \{x \in A : \overline{\varphi}(x) \geqslant a\}$. The analog holds for the opposite inequality. Partition A into $B_{k2^{-n}}$. The L^1-convergence follows directly. Details are left as an exercise or can be found on page 20 in Halmos (1974). □

If μ is a probability measure then this means that $\overline{\varphi} = \mathbb{E}_\mu[\varphi|\mathcal{F}_{\text{inv}}]$, where $\mathcal{F}_{\text{inv}}$ is the σ-algebra of all μ-a.s. invariant measurable subsets of Ω.

4.6.4   *Ergodicity and mixing*

Using the convergence results for arithmetic means in Theorem 4.8, we can strengthen the existence result in Theorem 4.5

Corollary 4.5. *If Ω is additionally compact and metrizable, then for μ-almost every $x \in \Omega$ there exists an invariant measure $\mu_{\infty,x}$ on Ω such that for all continuous φ it holds that*

$$\lim_{T \to \infty} \varphi_T(x) = \lim_{T \to \infty} \frac{1}{T} \sum_{t=0}^{T-1} \varphi(\Phi_t(x)) = \int \varphi(y)\mu_{\infty,x}(dy). \tag{4.50}$$

Note that for different x the measure $\mu_{\infty,x}$ may differ. That is clearly not desirable. Therefore, let us introduce the following very weak notion of uniqueness.

Definition 4.15. Let μ be a probability measure on Ω. A measurable subset $A \subset \Omega$ is called μ-a.s. invariant if and only if $\mu(A \setminus \Phi^{-1}(A)) = 0$ and $\mu(\Phi^{-1}(A) \setminus A) = 0$.

A probability measure μ on Ω is called *ergodic* if and only if for any μ-a.s. invariant set either $\mu(A) = 0$ or $\mu(\Omega \setminus A) = 0$.

This is exactly what one needs to obtain the following:

Corollary 4.6. *If μ is an ergodic probability measure. Then for μ-a.e.
$\omega \in \Omega$ holds that*

$$\lim_{T \to \infty} \frac{1}{T} \sum_{t=0}^{T-1} \varphi(\Phi_t(x)) = \int \varphi(y)\mu(dy). \tag{4.51}$$

Remark 4.2. It is important to note that the ergodicity assumption is not
used to establish the convergence of the arithmetic mean in Theorem 4.8.
It is only used to identify the limit.

The terminology "ergodicity" comes from this latest theorem because
it means that the Cesaro time average is equal to the average over Ω with
respect to μ. In some applications, Ω is considered to describe physical
space and one says for short that space average is equal to time average.

It is important to note that the above theorem does *not* imply that there
exists a unique invariant ergodic measure. That is actually typically not the
case at all. By the choice of the ergodic measure one decides which parts of
the system are considered as important. The set of all invariant measures is
a convex set and one can show that the ergodic measures are the extremal
ones. In other words, this means that any invariant measure can be build
up from ergodic measures. The above result also implies that two ergodic
measures for the same dynamic Φ have to be singular with respect to each
other, that means that the support of one measure has measure zero with
respect to the other measure. Hence, if one has two ergodic measures one
can split Ω in three parts. In one part, the φ_T converges to one measure,
one the other part, it converges to the other measure and finally there is a
part on which φ_T behave in a different way. The latter part has measure
zero for the particular measures under consideration but does not need to
be small at all.

Often there is a natural choice for the ergodic measure to be considered.

Definition 4.16. We call an invariant ergodic measure a physical measure
if and only if there exists a measurable set A with positive Lebesgue measure
and for Lebesgue-a.e. $x \in A$

$$\lim_{T \to \infty} \frac{1}{T} \sum_{t=0}^{T-1} \varphi(\Phi_t(x)) = \int \varphi(y)\mu(dy). \tag{4.52}$$

The notion of ergodicity is often too weak. Furthermore, it is frequently
useful to check stronger conditions which provide additional information,

for example a *mixing*-condition. Let us introduce one of them here as an example.

Definition 4.17. Let μ be an invariant probability measure. The *auto-correlation function* is defined as the following measure c on $\Omega \times \Omega$.

$$c_t(A, B) = \mu(\Phi_t^{-1}(A) \cap B) - \mu(A)\mu(B). \tag{4.53}$$

In order to see better the relation with the ordinary auto-correlation function, let us reformulate the right-hand side in Eq. (4.53) in terms of functions

$$\int \mathbb{1}_A(\Phi_t(x))\mathbb{1}_B(x)\mu(dx) - \int \mathbb{1}_A(x)\mu(dx) \int \mathbb{1}_B(y)\mu(dy). \tag{4.54}$$

Hence, for bounded measurable functions[17] we get the expression

$$\int_{\Omega \times \Omega} \varphi(x)\psi(y)c_t(dx, dy) := \int \varphi(\Phi_t(x))\psi(x)\mu(dx)$$
$$- \int \varphi(x)\mu(dx) \int \psi(y)\mu(dy). \tag{4.55}$$

Definition 4.18. We say that an invariant probability measure is *weakly-mixing* if and only if

$$\lim_{t \to \infty} c_t(A, B) = 0. \tag{4.56}$$

The relation with ergodicity is the following.

Lemma 4.7. *Let μ be an invariant probability measure.*

(a) μ *is ergodic if and only if*

$$\lim_{T \to \infty} \frac{1}{T} \sum_{t=0}^{T-1} c_t(A, B) = 0. \tag{4.57}$$

(b) *If μ is weakly mixing, then it is ergodic.*

Exercise 4.36. *Prove the lemma.*

Exercise 4.37. *Show that if α is irrational the rotation R_α is ergodic.*

---

[17]Square-integrable is sufficient.

4.7 Transfer Operator Techniques

We have given an overview about a basic result of what one can show if one is interested in the time development of probabilities. If one wants to have more precise information which is of vital importance in applications, like speed of convergence, loss of memory and uniqueness of the invariant measure, one needs to work for a slightly restricted class of initial distribution as well as to use more mathematical technology. The aim of this section is to give a showcase of what can be done. We choose to present the technically easiest case for illustration purposes.

We will be quite special in the following by considering the d-dimensional torus $\mathbb{T}^d$, recall for $d = 1$ we have $\mathbb{T}^1 = \mathcal{S}^1$. In analogy with the relation between $\mathcal{S}^1$ and $\mathbb{R}$ modulo 1, we can consider the analogous relation between $[0,1[^d$ as a subset of $\mathbb{R}^d$ if one considers any of the coordinates modulo 1, cf. Example 4.11. In the following $\Omega = \mathbb{T}^d$. The methods can be applied to much wider classes of spaces, but this requires a much higher level of technical sophistication.

Before we can start, we have to consider the following improved version of the transformation formula.

Theorem 4.10. *Let $\Omega' \subset \mathbb{R}^d$ be a measurable subset. Assume that $\Phi : \Omega' \to \mathbb{R}^d$ be differentiable and $\det(D\Phi(x)) > 0$ for all x. If $\mu(dx) = \rho(x)dx$ then $\Phi_\# \mu$ has density ρ' given by*

$$\rho'(y) = \sum_{x \in \Omega' \, : \, \Phi(x) = y} \frac{\rho(x)}{|\det(D\Phi(x))|}. \tag{4.58}$$

This version of the theorem can treat also functions Φ which are not injective but only locally injective; the doubling map is an example.

Exercise 4.38. *Prove the following slightly weaker version from the transformation formula. Assume additionally that Ω' is open and Φ is continuously differentiable.*

Definition 4.19. Let $\Phi : \mathbb{T}^d \to \mathbb{T}^d$ be a differentiable function. We say that Φ is *expanding* if there exists a constant $\sigma > 1$ such that for $x \in \mathbb{T}^d$ and for each $v \in \mathbb{R}^d$ holds that $|(D\Phi)v| \geq \sigma|v|$. Note that we identified $\mathbb{T}^d$ with $[0,1[^d \bmod 1$. Hence, $D\Phi$ is just the matrix of derivative at the point x and $|v|$ denotes the usual Euclidean norm in $\mathbb{R}^d$.

Let us present the overall strategy. Consider a measure of the form $\mu = \rho m$ where $\rho : \mathbb{R}^d \to [0, \infty[$. Then $\mathcal{T}(\rho)$ is the following functions

$$\mathcal{T}(\rho)(y) := \sum_{x \in \Phi^{-1}(y)} \frac{\rho(x)}{|\det(Df(x))|}.$$

We can see that $\mathcal{T}$ is a linear function on the vector space of all continuous functions $\mathbb{T}^d \to \mathbb{R}$. We will denote this space by $\mathcal{C}(M \to \mathbb{R})$. Hence, we consider a linear function

$$\mathcal{T} : \mathcal{C}(M \to \mathbb{R}) \to \mathcal{C}(M \to \mathbb{R}).$$

A linear function on an infinite dimensional vector space is called a (linear) operator.

An invariant measure for the dynamics f is given by a fixed point that is $\mathcal{T}(\rho) = \rho$, cf. Definition 4.11. Hence, to get the existence of an invariant measure, we have to show that $\mathcal{T}$ has a fixed point. We want to define a metric d on the space of all densities ρ such that $\mathcal{T}$ is contracting and then by the following general technique of Banach's contraction principle, we get a good control about the invariant measure. Additionally, this will give us a rate of convergence, ideas about the spectra of the operator and properties of the density of the invariant measure including uniqueness.

Definition 4.20. Let (M, d) be a metric space and $T : M \to M$ be a (not necessarily linear) mapping. If there exists a $\delta < 1$ such that for all $x, y \in M$ it holds that $d(T(x), T(y)) \leqslant \delta d(x, y)$, then we say that T is a (strict) contraction.

Exercise 4.39. *Let (M, d) be a complete*[18] *metric space. Let T be a (strict) contraction. Show that T has a unique fixed point and for every $x \in M$ it holds that the sequence $(T^n(x))_{n \in \mathbb{N}}$ converges to that fixed point. Can you give a bound for the rate of convergence?*

This means that we have to find an appropriate metric and a space of densities on which the transfer operator is a contraction. For this, we introduce a particular class of metric spaces.

---

[18]A metric space is complete if and only if any sequence $(x_n)_n$ in M with $d(x_n, x_m) \leqslant 2^{-n}$ for all $n \leqslant m$ converges to a point $x \in M$.

4.7.1   Metrics on cones

Let V be a vector space. A subset $C \subset V \setminus \{0\}$ is called a *convex cone* if with all $v, w \in C$ and all $\alpha, \beta > 0$ also $\alpha v + \beta w \in C$.

Exercise 4.40. *Show that all the following C sets are convex cones.*

(a) *Let $V = \mathbb{R}^2$ and $C := \{(x, y) : y > |x|\}$.*
(b) *Let $V = \mathcal{C}^0$ the space of all continuous functions on $\mathbb{T}^d$.*
   Define $C := \{\rho \in \mathcal{C}^0 : \rho > 0\}$.
(c) *Let $V = \mathcal{C}(M \to \mathbb{R})$, where M is a metric space. Let $C := \mathcal{C}(a, \nu)$ be the cone of all non-negative functions ρ such that $\ln \rho$ is (a, ν)-Hölder continuous, that is for all $x, y \in \mathbb{T}^d$ holds that*

$$e^{-ad(x,y)^\nu} \rho(y) < \rho(x) < e^{ad(x,y)^\nu} \rho(y).$$

Note that with each vector $v \in C$ also all its rays $\{\alpha v : \alpha > 0\}$ will be in C. The metric we want to introduce only depends on the angle between rays generated by vectors not on the vectors themselves.

Given $v, w \in C$, we can introduce

$$\alpha(v, w) := \sup\{t \geq 0 : w - tv \in C\}.$$

Note that $\alpha(v, w) \geq 0$.

Exercise 4.41. *Show that $\{t \geq 0 : w - tv \in C\}$ is an interval.*

If $\alpha(v, w) = \infty$, that means that there exists a sequence $(t_n)_n$ with $t_n \to \infty$ such that $w - t_n v \in C$ or using the cone property $t_n^{-1} w - v \in t_n^{-1} C \subset C$. We would like to exclude this case. For this, we introduce the following particular type of closure, which is not the closure in the sense of topology, the so-called *relative closure*

$$\overline{C}^r := \{w \in V : \text{ exists } v \in C \text{ and sequence } (t_n)_{n \in \mathbb{R}} \downarrow 0 \text{ s.t.}$$
$$\text{for all } n \in \mathbb{N} \text{ holds } w + t_n v \in C\}. \tag{4.59}$$

In words, these are all the points which can be approximated from within the cone along a line, that is why the set is called the relative closure. In finite dimensions, this is just the closure of the usual cone.

We assume additionally that the relative closure of C is proper, that is

(CP) $\overline{C}^r \cap \overline{-C}^r = \{0\}$ or equivalently $v = 0$ whenever $v \in \overline{C}^r$ and $-v \in \overline{C}^r$.

Then from our previous consideration, we get that if $\alpha(v,w) = \infty$ then $-v \in \overline{C}^r$, that is $v \in C \cap \overline{-C}^r \subset \{0\}$ which by the definition of the convex cone at the beginning of Sec. 4.7.1 is impossible. Hence, we proved the following lemma.

Lemma 4.8. *For all $v, w \in V$ is $\alpha(v,w) < \infty$*

The expression for α is not symmetric in v, w. But as $\alpha(v,w) \in [0, \infty[$ we can define the following metric.

Definition 4.21.

$$\Theta_C(v,w) = -\ln\left(\alpha(v,w)\alpha(w,v)\right).$$

Note that $\Theta_C(v,w) = \infty$ if $\alpha(v,w) = 0$ or $\alpha(w,v) = 0$.

By definition $\Theta_C(v,w) = \Theta_C(w,v)$.

Exercise 4.42. *Show the following properties of α*

(a) $\alpha(v,v) = 1$.
(b) *For $u, v, w \in V$ holds that*

$$\alpha(v,w) \geqslant \alpha(v,u)\alpha(u,w).$$

(c) *Let $v, w \in C$ and $t, s > 0$ then*

$$\alpha(rv, sw) = \tfrac{s}{r}\alpha(v,w).$$

This allows us to prove the following properties of Θ_C

Lemma 4.9. *On C the function Θ_C has the following properties for $v, w, u \in C$*

(a) $\Theta_C(v,w) = \Theta_C(w,v)$,
(b) $\Theta_C(v,v) = 0$,
(c) $\Theta_C(v,w) \leqslant \Theta_C(v,u) + \Theta_C(u,w)$,
(d) $\Theta_C(v,w) \in [0, \infty]$,
(e) *for all $r, s > 0$ holds $\Theta_C(rv, sw) = \Theta_C(v,w)$,*
(f) $\Theta_C(v,w) = 0$ *if and only if there exists a $t > 0$ with $v = tw$.*

Proof. The proof is left as an exercise. Hint: For (f) show that $w - \alpha(v,w)v \in \overline{C}^r \cap \overline{-C}^r$.   □

Remark 4.3. Lemma 4.9 means that Θ_C is a metric on the space of all rays which are in C. The metric between two points is zero if and only if they are on the same ray. If one considers two different rays then the metric does not depend on which point one chooses on each of these rays. So this is a metric on the direction and the distance from the origin does not play any role. Hence, the metric is called *projective metric*.

Formally, one can express that as follows: let $v \in V$, the we can construct the half ray $[v] := \{\alpha v : \alpha > 0\}$ associated to v. We say v, w are equivalent if they lie on the same ray, that is $[v] = [w]$ and we write in symbols $v \sim w$. By $C/\sim$ we denote the collection of all different half rays contained in C. Then d is a metric on this space.

Exercise 4.43.

(a) *Let $V = \mathbb{R}^2$ and $C := \{(x,y) : y > |x|\}$. Show that*

$$\Theta_C\left(\begin{pmatrix} x^{(1)} \\ y^{(1)} \end{pmatrix}, \begin{pmatrix} x^{(2)} \\ y^{(2)} \end{pmatrix}\right) = \left|\ln\left(\frac{y^{(2)} - x^{(2)}}{y^{(1)} - x^{(1)}} \frac{y^{(1)} + x^{(1)}}{y^{(2)} + x^{(2)}}\right)\right|.$$

(b) *Let $V = C(M \to \mathbb{R})$. Define $C := C_+ := \{\rho \in C : \rho > 0\}$. Show that*

$$\Theta_{C_+}(\rho_1, \rho_2) = \ln\sup\left\{\frac{\rho_1(x)\rho_2(y)}{\rho_1(y)\rho_2(x)} : x, y \in M\right\}.$$

(c) *Let $V = C(M \to \mathbb{R})$ be as before. Let $C := C(a, \nu)$ be the cone of all function $\rho > 0$ such that $\ln\rho$ is (a, ν) Hölder continuous. Show that*

$$a(\rho_1, \rho_2) = \inf\left\{\min\left[\frac{\rho_1(x)}{\rho_2(x)} ; \frac{e^{ad(x,y)^\nu}\rho_2(x) - \rho_2(y)}{e^{ad(x,y)^\nu}\rho_1(x) - \rho_1(y)}\right] : x \neq y \in M\right\}.$$

Hint: Do (b) first.

Our final aim is to use Kolomogorov's contraction principle to show existence and convergence to an invariant measure.

Definition 4.22. Let V_1, V_2 be vector spaces and $C_i \subset V_i \setminus \{0\}$ be convex cones. Then we say that a linear operator $T : V_1 \to V_2$ *preserves the cones* if and only if $T(C_1) \subset C_2$.

The surprising property of the metric is that it already implies that T is a contraction.

Proposition 4.15. *T as above fulfils $\Theta_{C_2}(Tv, Tw) \leqslant \Theta_{C_1}(v, w)$.*

Proof. The proof is essentially based on the structure of the metric.   □

This however is not sufficient to apply Banach's contraction principle because we need a strict contraction. The following proposition gives an easy criteria to check.

Proposition 4.16. *Define*

$$\mathrm{diam}_{\Theta_{C_2}}(T(C_1)) := \sup\{\Theta_{C_2}(Tv, Tw) : v, w \in C_1\}.$$

If $\mathrm{diam}_{\Theta_{C_2}}(T(C_1)) < \infty$, *then for all* $v, w \in C_1$ *it holds that*

$$\Theta_{C_2}(Tv, Tw) \leqslant (1 - e^{-\mathrm{diam}(T(C_1))})\Theta_{C_1}(v, w).$$

For a proof see Sec. 1.2 in Baladi (2000).

4.7.2 *Transfer operator*

In the following, let $\Phi : \mathbb{T}^d \to \mathbb{T}^d$ be an expansive map with rate σ which is differentiable and the derivative is ν_0 Hölder continuous. We rescale the distance such that the diameter of the space $\mathbb{T}^d$ is 1, where with diameter we denote $\sup_{x,y \in \mathbb{T}^d} d(x, y)$. Denote by m the Riemannian volume, that is the normal "surface area" on $\mathbb{T}^d$, that is the measure that locally looks like the usual measure on $\mathbb{R}^d$ up to a constant. Choose this constant such that $m(\mathbb{T}^d) = 1$.

The expansivity of Φ implies that the Jacobi matrix of derivative $D\Phi$ is invertible. Then by a classical theorem of analysis there exists for each point $x \in \mathbb{T}^d$ a ball $B_r(x)$ around x of radius r such that Φ restricted to $B_r(x)$ is injective, the image will be an open set on which there exists a differentiable function which is the inverse function of Φ with an image contained in $B_r(x)$. It is a consequence of the compactness of $\mathbb{T}^d$ that one can find an $r > 0$, independent of x, which works for any x.

Each point $y \in \mathbb{T}^d$ can have several pre-images. As $\mathbb{T}^d$ is connected, we get that the number of pre-images has to be the same for all points. Denote this number by k and it is called the degree of the mapping Φ. Because of compactness, we can find a radius r_0 such that the inverse exists in each ball of radius r_0, that is for each ball $B_{r_0}(y)$ there exists $x_1, \ldots, x_k$ with $\Phi(x_i) = y$ and the inverse has exactly k-branches on $B_{r_0}(y)$, which we denote by $\Phi_1^{-1}, \ldots, \Phi_k^{-1}$. As Φ is an expansion, we have that the Φ_i^{-1} are contractions with rate σ^{-1}. In particular if $y^{(1)}$ and $y^{(2)}$ have distance less than r_0, then there exists unique order of the labelling of the pre-images $\{x_1^{(1)}, \ldots, x_k^{(1)}\}$ of $y^{(1)}$ such that $d(x_i^{(1)}, x_i^{(2)}) \leqslant \sigma^{-1} r_0$. That is we can allocate locally each point uniquely to a particular branch.

Let r_0 as above and $0 < \nu < \nu_0$ where ν_0 was the Hölder continuity index of $D\Phi$. We define $\mathcal{C}(a, \nu)$ as the convex cone of all functions $\rho : \mathbb{T}^d \to \mathbb{R}$ which have the following properties

(a) ρ is continuous.

(b) $\rho \geqslant 0$.

(c) $\ln \rho$ is (a, ν)-Hölder continuous on a r_0 neighbourhood, that is

$$\text{if } d(x, y) \leqslant r_0 \quad \text{then } \rho(x) \leqslant e^{ad(x,y)^\nu} \rho(y).$$

As $A \mapsto \det(A)$ is infinitly differentiable and for each x_0 the matrix $Df(x_0)$ is invertible and hence $\det(D\Phi(x_0)) \neq 0$. Hence also the function $x \mapsto \ln|\det(D\Phi(x))|$ has the same smoothness as Φ, that is it $\det(D\Phi) \in \mathcal{C}(a_0, \nu_0)$ for some a_0.

Proposition 4.17. *If one chooses a sufficiently large a, then there exists an $a' < a$ such that $T(\mathcal{C}(a, \nu)) \subset \mathcal{C}(a', \nu) \subset \mathcal{C}(a, \nu)$.*

Proof. Let $y^{(1)}, y^{(2)}$ be near enough and denote by $\{x_1^{(j)}, \ldots, x_k^{(j)}\}$ the pre-images under Φ. The idea of the proof is to use that by log-Hölder continuity $\rho(x_j^{(1)}) \leqslant e^{ad(x_j^{(1)}, x_j^{(2)})^\nu} \rho(x_j^{(2)})$ and the analogous inequality for $\det(D\Phi)$ to estimate $\Theta_{\mathcal{C}(a', \nu)}$ using that $d(x_j^{(1)}, x_j^{(2)})^{\nu_0} \leqslant \sigma^{-\nu} d(y^{(1)}, y^{(2)})^\nu$. Details are left as an exercise or can be found in the proof of Lemma 2.1 in Baladi (2000). $\square$

So after we showed that T preserves a particular choice of cones it remains to be shown that the diameter of $T(\mathcal{C}(a, \nu))$ is finite in $\mathcal{C}(a', \nu)$. So the following is sufficient by Theorem 4.16.

Proposition 4.18. *For $a' < a$*

$$\text{diam}_{\Theta_{\mathcal{C}(a, \nu)}}(\mathcal{C}(a', \nu)) < \infty.$$

Proof. The idea of the proof is to consider x, y sufficiently close and to show that

$$\frac{e^{ad(x,y)^\nu} \rho_2(x) - \rho_2(y)}{e^{ad(x,y)^\nu} \rho_1(x) - \rho_1(y)} \geqslant C \frac{\rho_2(x)}{\rho_1(x)}.$$

then extend to general points x, y using intermediate points. Details are left as an exercise or can be found in the proof of Proposition 2.5 in Viana (1998). □

Corollary 4.7. *Let* $\Phi : \mathbb{T}^d \to \mathbb{T}^d$ *be a differentiable expansive map with derivative* $D\Phi$ *that is* ν_0-*Hölder continuous. Then there exists a large enough* a *such that the transfer operator is a mapping* $\mathcal{T} : \mathcal{C}(a, \nu) \to \mathcal{C}(a, \nu)$ *which is a strict contraction for* $\Theta_{\mathcal{C}(a,\nu)}$ *with* $\delta = 1 - e^{-D_1}$ *with* $D_1 = \text{diam}_{\Theta_{\mathcal{C}(a,\nu)}}(\mathcal{C}(a', \nu)).$

4.7.3 *Existence of invariant measure*

In order to use Banach's contraction principle, we have to establish the following.

Lemma 4.10. *The cone* $\mathcal{C}(a, \nu)$ *is complete with respect to the metric* $\Theta_{\mathcal{C}(a,\nu)}.$

Proof. The idea of the proof is to use that a Cauchy sequence with respect to $\Theta_{\mathcal{C}(a,\nu)}$ is also a Cauchy sequence with respect to $\Theta_{\mathcal{C}_+}$. Details are left as an exercise or can be found in the proof of Proposition 2.6 in Viana (1998). □

Lemma 4.11. *The cone* $\mathcal{C}_+$ *is complete with respect to* $\Theta_{\mathcal{C}_+}$. *If* $(\rho_n)_n$ *is a sequence in* $\mathcal{C}_+$ *with* $\int \rho_n dm = 1$, *then* $(\rho_n)_n$ *converges also uniformly.*

Proof. Let $(\rho_n)_n$ be a Cauchy-sequence. Without loss of generality $\int \rho_n dm = 1$. The basic ideas are to estimate

$$|\rho_n(x) - \rho_m(x)| \leqslant \left| 1 - \frac{\rho_m(x)}{\rho_n(x)} \right| \sup\{\rho_n(z) : z \in \mathbb{T}^d\}.$$

Then one shows that

$$\rho_n(x) \leqslant e^{\sup\{\Theta_{\mathcal{C}_+}(\rho_n, \rho_1) : n \in \mathbb{N}\}} \sup \left\{ \frac{\rho_1(z_1)}{\rho_1(z_2)} : z_1, z_2 \in \mathbb{T}^d \right\} \inf\{\rho_n(y); y \in \mathbb{T}^d\}$$

$$< \infty$$

and $1 \leqslant \sup\{\rho_n(y); y \in \mathbb{T}^d\}$ to conclude that $\rho_n(x)$ is uniformly bounded in x and n. Finally use a similar approach for $\rho_n(x)/\rho_m(x)$ to obtain the estimate

$$\sup\left\{\frac{\rho_n(x)}{\rho_m(x)} : x \in \mathbb{T}^d\right\} \leqslant e^{\Theta_{\mathcal{C}_+}(\rho_n,\rho_m)}.$$

Details are left as an exercise or can be found in the proof of Proposition 2.6 in Viana (1998). □

Corollary 4.8. *Let* Φ *and* D_1 *be as in Corollary 4.7. Then,* $\mathcal{T}$ *has a unique fixed point* $\rho_\infty \in \mathcal{C}(a,\nu)$ *and for all* $\rho \in \mathcal{C}(a,\nu)$ *we get the following convergence*[19]:

$$\Theta_{\mathcal{C}(a,\nu)}(\mathcal{T}^n\rho,\rho_\infty) \leqslant (1-e^{-D_1})^n D_1 \tag{4.60}$$

and

$$\sup\{|\mathcal{T}^n\rho(x)-\rho_\infty(x)| : x \in \mathbb{T}^d\} \leqslant D_1 \sup\left\{\left|\frac{\mathcal{T}^n\rho(x)}{\rho_\infty(x)}-1\right| : x \in \mathbb{T}^d\right\},$$

where

$$\sup\left\{\left|\frac{\mathcal{T}^n\rho(x)}{\rho_\infty(x)}-1\right| : x \in \mathbb{T}^d\right\} \leqslant e^{D_1}(1-e^{-D_1})^{n+1}.$$

We have seen in Sec. 4.6 that for $\varphi \in L^1(\mathbb{T}^d \to \mathbb{R})$ it holds that

$$\int \varphi \circ \Phi \, \rho dm = \int \varphi \, \mathcal{T}(\rho)dm.$$

That shows that ρ_∞ in the previous corollary as a fixed point of $\mathcal{T}$ gives rise to an invariant measure for $\mu_\infty := \rho_\infty \, m$. Indeed,

$$\int \varphi \circ \Phi \, d\mu_\infty = \int \varphi \circ \Phi \, \rho_\infty dm = \int \varphi \, \mathcal{T}(\rho_\infty)dm = \int \varphi \, \rho_\infty dm = \int \varphi \, d\mu_\infty.$$

It is an invariant measure with several nice additional properties, which we will investigate in the following.

4.7.4 *Decay of correlation*

The autocorrelation was already considered in Sec. 4.6.4. In this section, we want to show that the autocorrelation decays exponentialy, in n and hence

---

[19] D_1 is such that $\sup\{\rho(x) : x \in \mathbb{T}^d\} \leqslant D_1$ for all $\rho \in \mathcal{C}(a,\nu)$.

for large n the random variables $\varphi \circ \Phi^n$ and ρ are essentially uncorrelated, that is

$$\int \varphi \circ \Phi^n \, \rho dm \sim \int \varphi \rho_\infty dm \int \rho dm.$$

This means that under the time development of f one loses memory of the initial distribution and the limit distribution of f^n is given by the invariant measure $\rho_\infty dm$. Let us be more precise.

Let $\varphi \in L^1(\mathbb{T}^d, m)$ and $\rho \in \mathcal{C}(a, \nu)$. Dividing both sides by $\int \rho dm$, we can assume that $\int \rho dm = 1$. Then by Corollary 4.8

$$\left| \int \varphi \circ \Phi^n \, \rho dm - \int \varphi \rho_\infty dm \right| \leqslant \int |\varphi| \, |\mathcal{T}^n \rho - \rho_\infty| \, dm$$

$$= \int \varphi \left| \frac{\mathcal{T}^n \rho}{\rho_\infty} - 1 \right| \rho_\infty dm$$

$$\leqslant \int \varphi \rho_\infty dm \, e^{D_1} (1 - e^{-D_1})^{n+1} \downarrow 0,$$

because $1 - e^{-D_1} < 1$. We summarise the result in the following lemma.

Lemma 4.12. *Let $\varphi \in L^1(\mathbb{T}^d, m)$ and ρ be Hölder continuous with index ν. Then there exists a constant $c > 0$ such that*

$$\left| \int \varphi \circ \Phi^n \, \rho dm - \int \varphi d\mu_\infty \int \rho dm \right| \leqslant c \int \rho dm (1 - e^{-D_1})^n.$$

Proof. The idea is to use the previous calculation for $\rho_\pm + B$ as for B large enough $\rho_\pm + B \in \mathcal{C}(A, \nu)$. Details are left as an exercise, cf. also Proposition 2.7 in Viana (1998). □

Corollary 4.9. *Let $\varphi_1 \in L^1(m)$ and φ_2 and ρ a Hölder continuous function of index ν. Then there exists a constant $c > 0$ such that we have exponential decay of correlation in the following sense.*

$$\left| \int \varphi_1 \circ \Phi_n \, \varphi_2 \rho dm - \int \varphi_1 d\mu_\infty \int \varphi_2 \rho dm \right| \leqslant c(1 - e^{-D_1})^n.$$

Proof. Use Lemma 4.12 $\varphi := \varphi_1$ and $\rho := \varphi_2 \rho'$. □

Proposition 4.19. *μ_∞ is ergodic and it is the only Φ-invariant measure which is absolutely continuous with respect to Lebesgue measure. This measure is called the Sinai–Ruelle–Bowen (SRB) measure associated to f.*

Proof. The idea of the proof is to consider the function $\varphi := \mathbb{1}_A - \mu_\infty(A)$ for A an a.s. invariant set. Apply decay of correlation to obtain that $\int \varphi \psi \, dm = 0$. For the final statement use Theorem 4.8. Details are left as an exercise or can be found in the proof of Corollary 2.9 in Viana (1998).

$\square$

4.7.5 Central Limit Theorem

Using decay of correlation one can show more than the ergodic theorem, that is that one can get an idea about the fluctuation of the ergodic sum. More precisely, we know that μ_∞-a.e.

$$\frac{1}{N} \sum_{n=0}^{N-1} \varphi \circ \Phi_n \to \int \varphi \, d\mu_\infty.$$

This means that for any $\varepsilon > 0$ for N large enough

$$\int \varphi \, d\mu_\infty - \varepsilon \leqslant \frac{1}{N} \sum_{n=0}^{N-1} \varphi \circ \Phi_n \leqslant \int \varphi \, d\mu_\infty + \varepsilon$$

or in other words

$$N \int \varphi \, d\mu_\infty - \varepsilon N \leqslant \sum_{n=0}^{N-1} \varphi \circ \Phi_n \leqslant N \int \varphi \, d\mu_\infty + \varepsilon N.$$

We see the first fluctuation if we find ε_N such that

$$N \int \varphi \, d\mu_\infty - \varepsilon_N \leqslant \sum_{n=0}^{N-1} \varphi \circ \Phi_n \leqslant N \int \varphi \, d\mu_\infty + \varepsilon_N$$

does not hold μ_∞-a.e. The next theorem shows that ε_N has to be of order $\sqrt{N}$.

Theorem 4.11. *Let φ be a ν-Hölder continuous function. Define $f := \varphi - \int \varphi \, d\mu_\infty$. Define* [20]

$$\sigma^2 = \int f^2(x) \mu_\infty(dx) + 2 \sum_{n=1}^{\infty} \int f(x) \, f(\Phi_n(x)) \, \mu_\infty(dx).$$

---

[20]Note that $\sigma = 0$ if and only if one can write $f = u \circ \Phi - u$ for some $u \in L^2(\mu_\infty)$ and hence all probability concentrate in zero.

If $\sigma > 0$ then for all interval A.

$$\mu_\infty\left(\left\{x \in \mathbb{T}^d \; : \; \frac{1}{\sqrt{N}}\sum_{n=0}^{N-1}\left(\varphi \circ \Phi_n - \int \varphi d\mu_\infty\right) \in A\right\}\right) \to \int_A \frac{e^{-x^2/2\sigma^2}}{\sqrt{2\pi}\sigma}dx.$$

The proof is essentially based on the exponential decay of correlations and a general Central Limit Theorem suitable for this situation, cf. Sec. 2.5 in Viana (1998).

4.7.6   *Linear response*

Often, it is interesting to understand the change of the system if the dynamics contains a parameter, that is one has not one function Φ describing the dynamics, but a family Φ_α with α taking values in an interval around zero and Φ_α depends smoothly on α. As you have seen in Secs. 4.2–4.4, one cannot expect any regularity for most of the objects under consideration but the SRB measure behaves surprisingly regularly.

Theorem 4.12. *Denote by μ_α the density of an SRB measure associated to Φ_α. Then for all three times continuous differentiable functions φ it holds that*

$$\frac{d}{d\alpha}\bigg|_{\alpha=0}\int\varphi(x)\mu_\alpha(dx) = \int\varphi(x)\sum_{k=0}^{\infty}\mathcal{T}^k\nabla\left(\rho\,\frac{d}{d\alpha}\bigg|_{\alpha=0}\Phi_\alpha\right)\mu_0(dx) \quad (4.61)$$

where $\mathcal{T}$ denotes the transfer operator associated to Φ_0.

Note that the theorem expresses a derivative with respect to α at $\alpha = 0$, that is a response of the system to the change of the parameter α via an expression of the right-hand side that only contains the invariant measure for $\alpha = 0$.

Chapter 5

Numerical Methods

Colin Cotter* and Hilary Weller[†]

*Imperial College London, UK
[†]University of Reading, UK

5.1 Introduction and the Navier–Stokes Equations

Weather prediction models solve the Navier–Stokes equations numerically starting from initial conditions based on the latest observed state of the atmosphere. Model domains typically extend from the ground up to around 40–80 km. In addition, climate model domains also include the global ocean. Figure 5.1 shows the surface pressure and winds calculated by numerically solving the Navier–Stokes equations using the European Centre for Medium Range Weather Forecasts (ECMWF) prediction model. Fast and accurate numerical methods suitable for massively parallel super-computers are therefore needed to solve the Navier–Stokes equations. This chapter provides an introduction to numerical modelling for solving simplified versions of the Navier–Stokes equations with an emphasis on the requirements of weather prediction models.

In the atmosphere and ocean, viscous stresses are very small in comparison to other terms of the Navier–Stokes equations and so, when developing numerical methods, attention is focused on their inviscid counterpart, the Euler equations. We recall from Chapter 2 that the Euler equations are coupled, nonlinear equations which predict the evolution of winds ($\mathbf{u}$), density (ρ) and temperature (T) with time, t. For an ideal gas, the Euler equations consist of the momentum equation, the continuity equation and an energy equation,

$$\frac{\partial \mathbf{u}}{\partial t} + \mathbf{u} \cdot \nabla \mathbf{u} = -2\Omega \times \mathbf{u} - \frac{\nabla p}{\rho} + \mathbf{g}, \qquad (5.1)$$

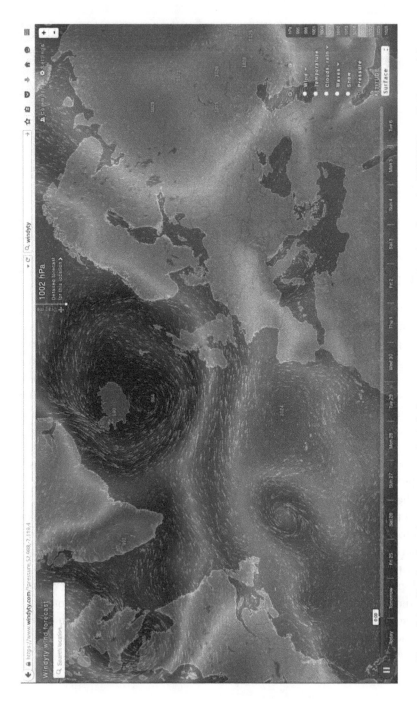

Figure 5.1.   Simulated surface pressure and winds from www.windyty.com using data from the ECMWF prediction model.

$$\frac{\partial \rho}{\partial t} + \mathbf{u} \cdot \nabla \rho = -\rho \nabla \cdot \mathbf{u}, \tag{5.2}$$

$$\frac{\partial T}{\partial t} + \mathbf{u} \cdot \nabla T = \frac{1}{\rho c_v} (Q - p \nabla \cdot \mathbf{u}), \tag{5.3}$$

in a frame of reference rotating with angular velocity Ω (e.g., the rotation of the planet). For a perfect gas, pressure (p) and density are related via the perfect gas law

$$p = \rho R T, \tag{5.4}$$

where R is the gas constant of air. In addition, we will need to know the gravitational acceleration vector, $\mathbf{g}$, the heat capacity of air at constant volume, c_v, and sources of heat, Q due to, for example, solar radiation (sunshine), condensation of water droplets to form clouds and longwave cooling of the atmosphere to space. Complete weather prediction models also predict the transport of water vapour and other constituents of the atmosphere such as pollutants (with concentrations ϕ) using transport equations such as

$$\frac{\partial \phi}{\partial t} + \mathbf{u} \cdot \nabla \phi = \text{Source} - \text{Sink}. \tag{5.5}$$

All of the Navier–Stokes and Euler equations have advection terms (or transport terms, the $\mathbf{u} \cdot \nabla X$ part) and transport by the atmospheric winds is a crucially important part of weather and climate prediction. Some numerical methods for advection are described in Sec. 5.2 and numerical analysis is presented which predicts the stability and accuracy of these methods.

Depth-averaging and other simplifying approximations of the Euler equations leads to the shallow water equations in a rotating frame (recall from Chapter 2):

$$\frac{\partial \mathbf{u}}{\partial t} + \mathbf{u} \cdot \nabla \mathbf{u} = -2\Omega \times \mathbf{u} - g \nabla (h + h_0), \tag{5.6}$$

$$\frac{\partial h}{\partial t} + \mathbf{u} \cdot \nabla h = -h \nabla \cdot \mathbf{u}, \tag{5.7}$$

where h is the fluid depth and h_0 is the height of the underlying topography. These equations represent the evolution of gravity and Rossby waves which play a very large role in the weather. We therefore need to develop numerical methods that can solve these equations accurately. Some simple techniques for solving wave equations are presented in Sec. 5.3 and numerical analysis is presented which highlights some of the problems that can occur when naive techniques are used.

Algorithms for solving large linear systems of equations lie at the heart of many numerical models in the geosciences. From Sec. 5.4 onwards, we shall discuss some examples that motivate the introduction of general approaches for solving these systems in the MPE context. With the advent of high performance parallel linear solver libraries such as PETSc, most researchers working in the geosciences are unlikely to develop their own linear solver codes. However, it is important to understand which of these methods should be selected for a given application problem and what we expect out of them. In the sections that follow Sec. 5.4, we shall start by giving an example of where challenging linear systems of equations arise in geoscientific models. We then introduce the workhorse of high performance linear solver algorithms, the conjugate gradient method, and use error estimates for it to motivate the development of preconditioners. Whilst homemade linear solver code should not be considered when so much effort has been put into PETSc and other open source libraries, we believe that implementing mathematical algorithms as code is an important part of understanding how they work, and we encourage the reader to write their own code to implement the (unpreconditioned) conjugate gradient algorithm. Moving on, selection of preconditioners is one way that researchers can really influence the performance of their simulation codes, and in this chapter, we aim to provide some signposts to the main issues surrounding this. We then recall some classical iterative methods, and reinterpret them as preconditioners. We examine their convergence properties and link them to their performance as preconditioners. We hope that this collection of material will equip Mathematics of Planet Earth researchers with some theoretical tools that they can use to select the right tools for their application problem.

5.2   Modelling Advection

Advection is the process by which the wind moves things around. It is also sometimes called transport or atmospheric dispersion, although we will use the word dispersion to have a more specific, mathematical meaning in this chapter. Figure 5.2 shows a Saharan dust storm in which sand is advected by the winds from the Sahara over the ocean.

Figure 5.2. A Saharan dust storm observed by SeaWiFS `http://oceancolor.gsfc.nasa.gov/SeaWiFS/HTML/dust.html` showing dust advected by the winds.

All of the Navier–Stokes equations include an advection term: density, temperature as well as the winds themselves are advected by the wind. Advection equations are also solved in atmospheric models to predict the transport of pollutants, water vapour and volcanic ash. Nonlinear advection is the process by which the wind advects itself (the advection term in Eq. (5.1)). In this chapter, we will focus on numerical methods for solving the linear advection equation in one spatial dimension, x, without sources or sinks of the advected (dependent) variable, ϕ, recalled from Chapter 2,

$$\frac{\partial \phi}{\partial t} + u\frac{\partial \phi}{\partial x} = 0, \tag{5.8}$$

for a prescribed, constant and uniform wind, u, and with given initial conditions, $\phi(x,0) = \phi_0$. We saw in Chapter 2 that this equation has an analytic solution

$$\phi(x,t) = \phi_0(x - ut), \tag{5.9}$$

enabling calculations of errors of implementations of numerical methods.

5.2.1 *Linear finite difference schemes*

To solve Eq. (5.8) using a finite difference scheme, space and time can be divided into equal intervals of size Δx and Δt so that so that point x_j is at position $j\Delta x$ for $j = 0, 1, \ldots, n_x$ and time t_n is at time $n\Delta t$ for $n = 0, 1, 2, \ldots$. Then we define the dependent variable, ϕ, at these points in space and time: $\phi_j^{(n)} = \phi(x_j, t_n)$. Discrete values of ϕ are represented in Fig. 5.3. If we wish to represent Eq. (5.8) discretely at (x_j, t_n), we will need to use values from other points in space and at other times. If we go *backward* in space we use values of ϕ from points in space to the left of x_j, i.e., we use $\phi_{j-1}, \phi_{j-2}, \ldots$

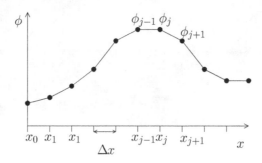

Figure 5.3. Dependent variable ϕ at grid points x_j.

If we go *forward* in time we use values of ϕ from times in the future, i.e., we use $\phi^{(n+1)}$. So a forward in time, backward in space (FTBS) approximation for Eq. (5.8) at (x_j, t_n) would be

$$\frac{\phi_j^{(n+1)} - \phi_j^{(n)}}{\Delta t} + u\frac{\phi_j^{(n)} - \phi_{j-1}^{(n)}}{\Delta x} = 0. \tag{5.10}$$

This can be rearranged to get the value at the unknown, future time, $\phi_j^{(n+1)}$, on the left-hand side and all other terms on the right-hand side so that we can calculate ϕ at the new time step at all locations based on values at previous time-steps. Also in this equation, we can remove u, Δt and Δx by substituting in the Courant number, $c = u\Delta t/\Delta x$ to obtain

$$\phi_j^{(n+1)} = \phi_j^{(n)} - c\big(\phi_j^{(n)} - \phi_{j-1}^{(n)}\big). \tag{5.11}$$

This looks like a useful numerical approximation for solving the advection equation; the new time-step values are evaluated entirely from known values from the previous time-step and, if $u > 0$ and consequently $c > 0$, the additional values of ϕ are taken from upstream of the point we are evaluating which seems sensible. However, there are many other possibilities for similar, linear finite difference schemes as shown in Table 5.1.

Numerical solutions of the linear advection equation using CTCS (as defined in Table 5.1) and FTBS on a periodic domain for 100 time-steps using 40 independent grid points and a Courant number of 0.2 are shown in Fig. 5.4 in comparison to the analytic solution (which remains the same shape as the initial conditions but moves to the right). CTCS is more accurate than FTBS which is consistent with the second order of accuracy of CTCS and 1st order accuracy of FTBS. Both schemes produce solutions which mostly propagate at the correct speed although FTBS is very damping and CTCS produces unbounded results. The analysis in the following sections will explain some of this behaviour.

Exercise 5.1. *Verify the formulae in Table 5.1.*

Exercise 5.2. *The code in Fig. 5.5 is a implementation in Python 2.7 of one of the schemes in Table 5.1.*

(a) *Which scheme is the code in Fig. 5.5 an implementation of?*
(b) *What are the initial conditions written in mathematical notation?*
(c) *What are the boundary conditions at $x = 0$ and $x = 1$?*
(d) *Type this code into a file and run it. Put in comments based on your answers above.*

Table 5.1.   Some linear finite difference for solving the advection equation, their stability limit (in terms of the Courant number, c), order of accuracy and classification into implicit or explicit. In the acronyms for the schemes, "F" is "forward", "B" is "backward", "C" is "centred", "T" is "time" and "S" is space. "CN" means "Crank–Nicolson".

Name	Formula	Stability limit	Order of accuracy		Implicit/ Explicit
			Space	Time	
FTBS	$\phi_j^{(n+1)} = \phi_j^{(n)} - c\left(\phi_j^{(n)} - \phi_{j-1}^{(n)}\right)$	$c \in [0,1]$	1	1	Explicit
FTFS	$\phi_j^{(n+1)} = \phi_j^{(n)} - c\left(\phi_{j+1}^{(n)} - \phi_j^{(n)}\right)$	$c \in [-1,0]$	1	1	Explicit
FTCS	$\phi_j^{(n+1)} = \phi_j^{(n)} - \frac{c}{2}\left(\phi_{j+1}^{(n)} - \phi_{j-1}^{(n)}\right)$	$c = 0$	1	2	Explicit
CTBS	$\phi_j^{(n+1)} = \phi_j^{(n-1)} - 2c\left(\phi_j^{(n)} - \phi_{j-1}^{(n)}\right)$	$c = 0$	2	1	Explicit
CTFS	$\phi_j^{(n+1)} = \phi_j^{(n-1)} - 2c\left(\phi_{j+1}^{(n)} - \phi_j^{(n)}\right)$	$c = 0$	2	1	Explicit
CTCS	$\phi_j^{(n+1)} = \phi_j^{(n-1)} - c\left(\phi_{j+1}^{(n)} - \phi_{j-1}^{(n)}\right)$	$c \in [-1,1]$	2	2	Explicit
BTBS	$\phi_j^{(n+1)} = \phi_j^{(n)} - c\left(\phi_j^{(n+1)} - \phi_{j-1}^{(n+1)}\right)$	$c \geq 0$	1	1	Implicit
BTFS	$\phi_j^{(n+1)} = \phi_j^{(n)} - c\left(\phi_{j+1}^{(n+1)} - \phi_j^{(n+1)}\right)$	$c \leq 0$	1	1	Implicit
BTCS	$\phi_j^{(n+1)} = \phi_j^{(n)} - \frac{c}{2}\left(\phi_{j+1}^{(n+1)} - \phi_{j-1}^{(n+1)}\right)$	$c \in \mathbb{R}$	1	2	Implicit
CNCS	$\phi_j^{(n+1)} = \phi_j^{(n)} - \frac{c}{4}\left(\phi_{j+1}^{(n+1)} - \phi_{j-1}^{(n+1)} + \phi_{j+1}^{(n)} - \phi_{j-1}^{(n)}\right)$	$c \in \mathbb{R}$	2	2	Implicit

(e) *If we assume that the domain is 1 meter wide and that $u = 1m\ s^{-1}$, what are Δx, Δt and the total duration of the simulation?*

(f) *Choose another explicit scheme from Table 5.1 and write a code implementing this scheme. How do the results differ from those of the given implementation? Can this be explained in terms of the properties in Table 5.1?*

Table 5.1 brings up a number of questions which motivate the rest of this section:

(a) How do we know which scheme to chose?

(b) What does the order of accuracy mean? How is this calculated?

(c) How can we predict which schemes will be stable and for what Courant numbers they will be stable?

(d) Some of the schemes in Table 5.1 are *implicit*. This means that values at the next time level are needed in order to calculate values at the next time level. How can this be done?

(e) What are the advantages of implicit and explicit schemes?

We will start with the order of accuracy.

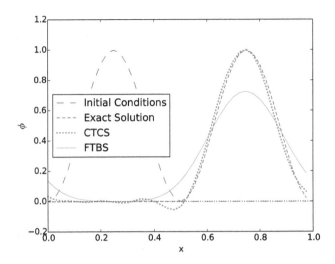

Figure 5.4. Numerical solutions of the linear advection equation using the CTCS and FTBS schemes on a periodic domain for 100 time-steps using 40 independent grid points and a Courant number of 0.2 in comparison to the analytic solution. The initial conditions are $\phi = \frac{1}{2}(1 - \cos 4\pi x)$ for $x \leq 1/2$ and $\phi = 0$ elsewhere.

5.2.2 *Order of accuracy and order of convergence*

If we increase the spatial or temporal resolution (decrease Δx or Δt), we would expect the solution of our numerical method to improve. The rate at which it improves is related to the order of accuracy.

Definition 5.1. The order of accuracy of a discretisation scheme is the maximum value of n such that the error ε satisfies

$$|\varepsilon| \leq A\Delta x^{n}, \tag{5.12}$$

for some constant, A. We say that the scheme is nth order accurate.

If we then calculate the errors of a numerical implementation of the scheme for different values of Δx we will be able to calculate the order of convergence, n, which satisfies Eq. (5.12). If the scheme is implemented correctly and the mathematical analysis is correct, then for small enough Δx, the order of accuracy should be equal to the order of convergence. This numerical experiment is an important part of testing the analysis and implementation of numerical methods.

```
from pylab import *
# Numerical solution of the 1D linear advection equation
# using a finite difference scheme
nx = 61                  # Number of points from x=0
# to x=1 (inclusive)
nt = 50                  # Number of time-steps
c = 0.4                  # Courant number
x = linspace(0,1,nx)     # Points in the x direction
# Initial conditions for dependent variable phi
phi = where(x<0.5, 0.5*(1-cos(4*pi*x)), 0)
phiOld = phi.copy()

# Plot the initial conditions
plt.clf()
plt.ion()
plot(x, phi, label='Initial conditions')
legend(loc='best')
axhline(0, linestyle=':', color='black')
plt.ylim([-0.2,1.2])
show()
raw_input("press return to start simulation")

# Loop over all time-steps
for it in range(nt):
    # Loop over space (excluding end points)
    for ix in range(1,nx-1):
        phi[ix] = phiOld[ix] - 0.5*c*(phiOld[ix+1] -
                                      phiOld[ix-1])
    # Compute boundary values
    phi[0] = phiOld[0] - 0.5*c*(phiOld[1] - phiOld[nx-2])
    phiOld[nx-1] = phiOld[0]
    # Update old time value
    phiOld = phi.copy()
    # Plot results
    plt.clf()
    plot(x, phi, label="Results at time-step "+str(it+1))
    axhline(0, linestyle=':', color='black')
    plt.ylim([-0.2,1.2])
    legend(loc='best')
    draw()
raw_input("press return to end")
```

Figure 5.5.   Python 2.7 code to solve the linear advection equation using a finite difference scheme.

Exercise 5.3. *The errors of integrating a curve using the trapezium rule are 0.02645 for $\Delta x = 0.4$ and 0.00666 for $\Delta x = 0.2$. Calculate the order of convergence.*

Finite difference formulae can be derived and their order of accuracy found using Taylor series. For example, if we wish to find a finite difference approximation for $f'_j = \partial f/\partial x$ at $x = x_j = j\Delta x$ given f_{j-1}, f_j and f_{j+1} (values of f at x_{j-1}, x_j and x_{j+1}) then we start by writing down Taylor series for f_{j-1} and f_{j+1} centred about x_j,

$$f_{j+1} = f_j + \Delta x f'_j + \frac{\Delta x^2}{2!} f''_j + \frac{\Delta x^3}{3!} f'''_j + O(\Delta x^4), \qquad (5.13)$$

$$f_{j-1} = f_j - \Delta x f'_j + \frac{\Delta x^2}{2!} f''_j - \frac{\Delta x^3}{3!} f'''_j + O(\Delta x^4). \qquad (5.14)$$

This gives us two equations for our unknown, f'_j, in terms of our knowns, f_{j-1}, f_j and f_{j+1}. There are some other unknowns in these equations, f''_j, f'''_j,.... Therefore, we can eliminate the unknown which has the largest coefficient from Eqs. (5.13) and (5.14) and leave the other as an error. The largest unknown is $\Delta x^2 f''_j$ which can be eliminated by subtracting one equation from the other which, after rearranging, give us a formula for our unknown, f'_j, in terms of our knowns,

$$f'_j = \frac{f_{j+1} - f_{j-1}}{2\Delta x} + O\left(\Delta x^2\right). \qquad (5.15)$$

The error terms is proportional to Δx^2 because the leading error term is $\Delta x^3 f'''_j/(3\Delta x)$ and so the centred in space finite difference formula for f' is second-order accurate.

Exercise 5.4. *Use a Taylor series to find the order of accuracy of the forward in time finite difference formula. Compare this to the order given in Table 5.1.*

5.2.3 *Implicit and explicit numerical methods*

We saw in Table 5.1 that some numerical methods use values at just the current (n) and previous $(n-i)$ time-steps in order calculate values at the next time-step $(n+1)$. These are called explicit schemes. Implicit schemes also use values at time-step $n+1$ to calculate values at time-step $n+1$. We will see in the following subsections that explicit schemes usually have time-step restrictions, meaning that they are only stable for a sufficiently small time-step. In contrast, implicit schemes, if correctly designed, can be stable for arbitrarily large time-step, although they will not necessarily give high accuracy in time and they can be computationally expensive per time-step. In this section, we will describe how to use an implicit scheme to update a variable from one time-step to the next.

The backward in time, centred in spaced (BTCS) scheme is

$$\phi_j^{(n+1)} = \phi_j^{(n)} - \frac{c}{2}(\phi_{j+1}^{(n+1)} - \phi_{j-1}^{(n+1)}). \tag{5.16}$$

In order to evaluate all the values of $\phi_j^{(n+1)}$ from the values of $\phi_j^{(n)}$, we represent all of the values, $\phi_j^{(n)}$, as a vector, $\underline{\phi}^{(n)} = (\phi_0^{(n)},\ \phi_1^{(n)},\dots,\phi_j^{(n)},$ $\dots,\phi_{N-1}^{(n)})^T$ and all the values $\phi_j^{(n+1)}$ as a vector $\underline{\phi}^{(n+1)}$ and define BTCS as a matrix M such that $M\underline{\phi}^{(n+1)} = \underline{\phi}^{(n)}$. This matrix will depend on the domain and boundary conditions. We will assume $x : 0 \to 1$ with periodic boundary conditions so that $\phi_0 = \phi_N$. Then, the formulae for BTCS at (or near) the boundary points are

$$\phi_0^{(n+1)} = \phi_0^{(n)} - \frac{c}{2}(\phi_1^{(n+1)} - \phi_{N-1}^{(n+1)}), \tag{5.17}$$

$$\phi_{N-1}^{(n+1)} = \phi_{N-1}^{(n)} - \frac{c}{2}(\phi_0^{(n+1)} - \phi_{N-2}^{(n+1)}). \tag{5.18}$$

Equations (5.16)–(5.18) can be rearranged so that $\phi_j^{(n+1)}$ is on the left-hand side and $\phi_j^{(n)}$ is on the right-hand side. This can then be written as a matrix equation:

$$\begin{pmatrix} 1 & c/2 & 0 & 0 & 0 & & & -c/2 \\ -c/2 & 1 & c/2 & 0 & 0 & & & 0 \\ & -c/2 & 1 & c/2 & 0 & & & 0 \\ & & \vdots & \vdots & \vdots & & & \\ & & & \vdots & \vdots & & & \\ 0 & & & & & -c/2 & 1 & c/2 \\ c/2 & 0 & & & & 0 & -c/2 & 1 \end{pmatrix} \begin{pmatrix} \phi_0^{(n+1)} \\ \phi_1^{(n+1)} \\ \phi_2^{(n+1)} \\ \vdots \\ \phi_j^{(n+1)} \\ \vdots \\ \phi_{N-1}^{(n+1)} \end{pmatrix} = \begin{pmatrix} \phi_0^{(n)} \\ \phi_1^{(n)} \\ \phi_2^{(n)} \\ \vdots \\ \phi_j^{(n)} \\ \vdots \\ \phi_{N-1}^{(n)} \end{pmatrix}.$$

$$\tag{5.19}$$

Linear numerical methods to solve linear equations can always be written like this. Nonlinear equations require linearisation so that only the linear terms are treated implicitly. There are numerous computational tools to solve sets of linear, simultaneous equations, such as Eq. (5.19) in order to find $\underline{\phi}^{(n+1)} = M^{-1}\underline{\phi}^{(n)}$ such as, for example Gaussian elimination, the Jacobi method, Gauss–Seidel and conjugate gradient which will be discussed in Sec. 5.4.

Exercise 5.5. *Write the matrix equation for solving a numerical method that is Crank–Nicolson in time (see Table 5.1) and backward in space.*

Exercise 5.6. *Choose an implicit scheme from Table 5.1 and write a code implementing this scheme. You can use Python package* scipy *to find the solution of a matrix equation, importing it as:*
import scipy.linalg as la
If you are using package numpy *or* pylab *to provide function zeros then you can create the matrix using*
nx=50
M = zeros([nx,nx])
and then set the elements of M. *If your dependent variable is an array* phi, *then* phi *can be updated to the next time-step using*
phi = la.solve(M,phi)
Experiment to see if you can use very large time-steps. How does this affect the accuracy?

5.2.4 *Some definitions of numerical properties*

Before presenting the numerical analysis which predicts the stability of numerical methods, we will provide some definitions.

Definition 5.2.

(a) **Convergence:** A finite difference scheme is convergent if the solutions of the scheme converge to the solutions of the PDE as Δx and Δt tend to zero.

(b) **Consistency:** A finite difference scheme is consistent with a PDE if the errors in approximating all of the terms tend to zero as Δx and/or Δt tend to zero. (Terms of the finite difference scheme are typically analysed using Taylor series.)

(c) **Order of accuracy:** Error $\propto \Delta x^n$ (error is $O(\Delta x^n)$) means scheme is nth-order accurate. Errors of an nth-order scheme converge to zero with order n.

(d) **Stability:** Errors do not tend to infinity for any number of time-steps. Stability can be proved using von Neumann stability analysis. A scheme is

  (i) Conditionally stable — if stable only for a sufficiently small time-step,

  (ii) Unconditionally stable — if stable for any time-step,

  (iii) Unconditionally unstable — if unstable for any time-step.

(e) **Conservation:** If, e.g., mass, energy, potential vorticity are conserved by the PDEs, then a numerical method may or may not conserve some of the same properties.

(f) **Boundedness:** If the initial conditions are bounded between values a and b then a bounded solution will remain bounded between a and b for all time.

(g) **Monotonicity:** Monotone schemes do not generate new extrema or amplify existing extrema. If the initial conditions are monotonic then they will remain monotonoic after the action of a monotonic numerical method.

These definitions will be useful in the following sections and useful when describing results of numerical methods.

Exercise 5.7. *Based on the results in Fig. 5.4, what numerical properties might CTCS and FTBS have and what numerical properties can we be sure that they do not have?*

5.2.5 The Lax Equivalence Theorem

The Lax Equivalence Theorem relates stability, consistency and accuracy.

Theorem 5.1 (Lax Equivalence Theorem). *For a consistent finite difference method for a well-posed linear initial value problem, the method is convergent if and only if it is stable. Further, if a finite difference scheme is linear, stable, and accurate of order (p,q), then it is convergent of order (p,q).*

See, e.g., Durran (1999) for discussion of the proof. It is therefore necessary to study order of accuracy and stability of a linear numerical method. We do not independently need to prove convergence.

5.2.6 The domain of dependence and the Courant–Friedrichs–Lewy (CFL) criterion

The domain of dependence of the solution of a PDE at position $\mathbf{x}$ and at time t is the set of points at a previous time that influence the solution at position $\mathbf{x}$ and at time t. So for the one-dimensional linear advection equation in Eq. (5.8), the domain of dependence of (x,t) is point $(x - ut)$ at time $t = 0$. Numerical methods for solving PDEs also have domains of

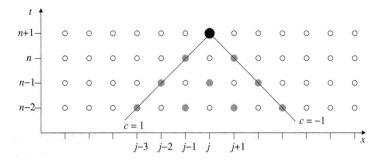

Figure 5.6. The grey circles show the domain of dependence of point (x_j, t_{n+1}) for the CTCS advection scheme. The black lines indicate the relationship between space and time for a Courant numbers, $c = \pm 1$.

dependence which consists of sets of points in space–time. For example, the points in (x,t) that are defined for a grid with time-step Δt and spatial resolution Δx can be marked on a diagram such as in Fig. 5.6. The grey filled circles show the domain of dependence of point (x_j, t_{n+1}) for the CTCS advection scheme (as defined in Table 5.1).

The Courant–Friedrichs–Lewy (CFL) criterion (Durran, 1999) states that for stability, the domain of dependence of the numerical solution should contain the domain of dependence of the original PDE. For any Courant number, $c = u\Delta t / \Delta x$, a line can be drawn on Fig. 5.6 which represents the true domain of dependence on the grid. If this line is not surrounded by points in the numerical domain of dependence on the CFL criterion will not be satisfied and the numerical method will be unstable. The CFL criterion is necessary but not sufficient so if the real domain of dependence does lie within the numerical domain of dependence, the scheme could still be unstable. Figure 5.6 shows that CTCS will be unstable for $c > 1$ and $c < -1$.

Exercise 5.8. *Draw the domain of dependence of the FTBS advection scheme. Based on this, for what values of the Courant number will FTBS be unstable? Is this consistent with the stability reported in Table 5.1?*

5.2.7 *Von Neumann stability analysis*

For linear finite difference schemes, we can prove stability using von Neumann stability analysis. Von Neumann stability analysis uses Fourier analysis. We assume that, in one spatial dimension on a uniform grid, the numerical solution of a PDE can be expressed as a sum of Fourier modes,

$$\phi = \sum_{k=-\infty}^{\infty} A_k e^{ikx}, \tag{5.20}$$

each with wavenumber k and we will consider the stability of an individual wavenumber, e^{ikx}. If we can prove that a linear finite difference scheme is stable for all $k \in \mathbb{I}$, then we will know that it is stable since the initial conditions and solutions are linear combinations of wavenumbers and the linear scheme cannot create any interactions between wavenumbers. Hence, for each $k \in \mathbb{I}$ we search for an amplification factor, A (which may be a function of k, u, Δt and Δx) such that

$$\phi_k^{(n+1)} = A\phi_k^{(n)}, \tag{5.21}$$

where $\phi_k = e^{ikx}$ is the part of the solution with wavenumber k. The amplification factor will tell us the following about the scheme.

- If $|A| < 1$ for all k then the scheme is stable and damping.
- If $|A| = 1$ for all k then the scheme is neutrally stable (waves do not amplify or decay).
- If $|A| > 1$ for any k then the scheme is unstable (at least one wavenumber is amplified).

Von Neumann stability analysis is by far easiest to compute if we have a uniform, one-dimensional grid so that we can use $x = j\Delta x$ for grid-points numbered by j. Then we can substitute $x = j\Delta x$ and $\phi_k = e^{ikx}$ into Eq. (5.21) and simplify and rearrange to find A. For numerical methods with solutions that propagate in space (which would be expected for linear advection), A will be complex and $|A|^2 = AA^*$ where A^* is the complex conjugate.

Example 5.1 (Von Neumann stability analysis of FTBS). As an example, we will find the stability limits of FTBS,

$$\phi_j^{(n+1)} = \phi_j^{(n)} - c\left(\phi_j^{(n)} - \phi_{j-1}^{(n)}\right). \tag{5.22}$$

Substituting in $\phi_j^{(n)} = A^n e^{ikx}$ and $x = j\Delta x$ gives

$$A^{n+1} e^{ikj\Delta x} = A^n e^{ikj\Delta x} - cA^n\left(e^{ikj\Delta x} - e^{ik(j-1)\Delta x}\right), \tag{5.23}$$

cancelling powers of $A^n e^{ikj\Delta x}$ and rearranging to find A in terms of c and $k\Delta x$ gives

$$A = 1 - c\left(1 - e^{-ik\Delta x}\right). \tag{5.24}$$

We need to find the magnitude of A so we need to write it down in real and imaginary form. So substitute $e^{-ik\Delta x} = \cos k\Delta x - i\sin k\Delta x$,

$$A = 1 - c(1 - \cos k\Delta x) - ic\sin k\Delta x, \qquad (5.25)$$

and calculate $|A|^2 = AA^*$,

$$|A|^2 = 1 - 2c(1 - \cos k\Delta x) + c^2(1 - 2\cos k\Delta x + \cos^2 k\Delta x)$$
$$+ c^2 \sin^2 k\Delta x,$$
$$\implies |A|^2 = 1 - 2c(1-c)(1 - \cos k\Delta x).$$

We need to find for what value of Δt or c is $|A| \leq 1$ in order to find when FTBS is stable. We see that

$$|A| \leq 1 \Leftrightarrow |A|^2 - 1 \leq 0$$
$$\Leftrightarrow -2c(1-c)(1 - \cos k\Delta x) \leq 0$$
$$\Leftrightarrow c(1-c)(1 - \cos k\Delta x) \geq 0.$$

We know that $1 - \cos k\Delta x \geq 0$ for all $k\Delta x$ so FTBS is stable when $c(1-c) \geq 0 \Leftrightarrow 0 \leq c \leq 1$. So we have proved that FTBS is unstable if $u < 0$ or if $\frac{u\Delta t}{\Delta x} > 1$. We will now define the upwind scheme given by

$$\phi^{(n+1)} = \begin{cases} \text{FTBS} & \text{when } u \geq 0, \\ \text{FTFS} & \text{when } u < 0. \end{cases} \qquad (5.26)$$

The upwind scheme is first order accurate in space and time, conditionally stable and damping. The damping of FTBS is clear in the solutions in Fig. 5.4.

Exercise 5.9. *Verify the stability limits given in Table 5.1 for FTFS, FTCS, BTBS, BTFS, BTCS and CTCS. The centred in time schemes are more difficult but CTCS is manageable.*

Exercise 5.10. *Forward in time, finite volume advection schemes can be defined generically as:*

$$\frac{\phi_j^{(n+1)} - \phi_j^{(n)}}{\Delta t} = -u\frac{\phi_{j+\frac{1}{2}} - \phi_{j-\frac{1}{2}}}{\Delta x}.$$

Two specific finite volume schemes can be defined by:

$$\text{Lax–Wendroff} \quad \phi_{j+\frac{1}{2}} = \tfrac{1}{2}(1+c)\phi_j + \tfrac{1}{2}(1-c)\phi_{j+1},$$

$$\text{Warming and Beam} \quad \phi_{j+\frac{1}{2}} = \tfrac{1}{2}(3-c)\phi_j - \tfrac{1}{2}(1-c)\phi_{j-1}.$$

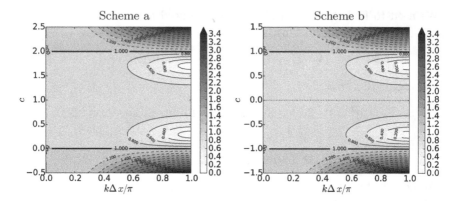

Figure 5.7. Magnitudes of the amplification factors for Lax–Wendroff and Warming and Beam against c and $k\Delta x$. (Identifying which scheme is which left as an exercise.)

Draw the domains of dependence of Lax–Wendroff and Warming and Beam. The magnitudes of the amplification factors for Lax–Wendroff and Warming and Beam are plotted against c and $k\Delta x$ in Fig. 5.7. Based only on the domains of dependence and the CFL criterion and without doing any calculations, say which amplification factor plot is associated with which scheme and why.

5.2.8   Dispersion errors

Dispersion occurs when waves of different wavenumbers propagate at different speeds. For example, the wake created by a ship is dispersive. A ship's wake consists of wave of different wavenumbers (i.e., different wavelengths). These propagate at different speeds and hence the over-all shape of the wake changes as it propagates away from the ship. In contrast, sound waves have low dispersion. If we assume that sound waves of all wavelengths propagate at the same speed (an approximation) then high and low notes (sound waves of different wavelengths) will propagate at the same speed and a piccolo and a bassoon will stay in time even if we are at the back of a concert hall.

   Linear advection is not dispersive; waves of all wavelengths propagate with speed u and the analytic solution confirms that an initial profile moves without changing shape, consistent with all waves that make up the initial condition propagating at the same speed. However, Fig. 5.4 shows that the numerical solutions do change shape and so the numerical solutions are dispersive. That is not all wavelengths propagate at the same speed. We can use numerical analysis to predict what dispersion errors a numerical method will have. We will start by considering how the complex amplification factor relates to the wave propagation speed for the analytic solution of linear advection.

For initial conditions consisting of a single Fourier mode: $\phi(x,0) = A_k e^{ikx}$, the solution at time t is

$$\phi(x,t) = A_k e^{ik(x-ut)}.$$

This can be represented as an (amplification) factor times e^{ikx},

$$\phi(x,t) = A_k e^{-ikut}\, e^{ikx}.$$

So if $t = n\Delta t$ we have $A^n = e^{-ikut} = e^{-ikun\Delta t}$. Therefore, for real linear advection, the amplification factor is $A = e^{-iku\Delta t}$ which has $|A| = 1$ and so, under the influence of the linear advection equation, waves should not amplify or decay. We can also see that the argument of the complex amplification factor tells us the propagation speed. If the amplification factor is $A = ae^{i\theta}$ then the propagation (or phase) speed is $-\theta/(k\Delta t)$. If this phase speed is dependent on k for a numerical method, then the numerical method is dispersive.

Von Neumann analysis (Sec. 5.2.7) can give us the amplification factor for CTCS,

$$A = -ic\sin k\Delta x \pm \sqrt{1 - c^2 \sin^2 k\Delta x}. \tag{5.27}$$

The argument of A tells us the numerical phase speed,

$$u_n = \frac{1}{k\Delta t} \tan^{-1}\left(\frac{c\sin k\Delta x}{\pm\sqrt{1 - c^2\sin^2 k\Delta x}}\right). \tag{5.28}$$

This can be simplified by substituting in $u\Delta t = c\Delta x$ and $\sin\alpha = c\sin k\Delta x$ to obtain

$$\frac{u_n}{u} = \pm\frac{\alpha}{ck\Delta x}.$$

So there are two possible phase speeds for each mode. These can be plotted against $k\Delta x$ to find out how waves propagate when advected by CTCS. A plot of u_n/u against $k\Delta x$ is called a dispersion relation. The dispersion relation for CTCS for $c = 0.4$ is shown in Fig. 5.8. Dispersion relations can also be given as wave frequency, $\omega = ku$, against $k\Delta x$ as in the right-hand side of Fig. 5.8.

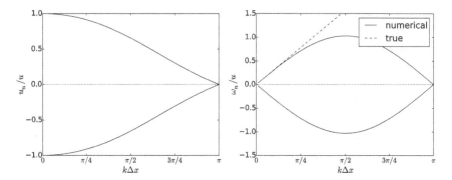

Figure 5.8.   The dispersion relation for CTCS plotted as u_n/u against $k\Delta x$ and as ω_n/u against $k\Delta x$. The true solution for linear advection has wave frequency $\omega = ku$ and so the dipsersion relation is a straight line.

Interpreting dispersion relations requires us to understand the significance of $k\Delta x$. The wavenumber is related to the wavelength, λ, by $\lambda = 2\pi/k$. If $k\Delta x$ is small (approaching zero) then waves have long wavelengths in comparison to Δx which means that waves are well resolved. If $k\Delta x = \pi$ then the wavelength is $\lambda = 2\Delta x$. Representing a wave over two grid intervals is the lowest resolution that a wave can have and so these are very poorly resolved, grid-scale waves.

Figure 5.8 shows us that well-resolved waves propagate at the correct speed (looking at the positive branch), poorly resolved waves propagate too slowly and grid-scale waves are stationary. These dispersion errors help to explain the results in Fig. 5.4 in which CTCS produces an under-shoot to the left of the cosine bell shape. This undershoot is caused by poorly resolved wavenumbers propagating too slowly.

The dispersion relation for CTCS in Fig. 5.8 also tells us that CTCS can give two possible solutions, a physical mode with $u_n/u \geq 0$ and a computational mode with $u_n/u < 0$ and that the computational mode propagates in the wrong direction. This is an undesirable feature of CTCS which is caused by the fact that three time-levels are used to update ϕ leading to a quadratic equation for the amplification factor with two possible solutions. Not all values within the domain of dependence at time-level n effect the solution at time-level $n+1$. In particular, the value of $\phi_j^{(n)}$ has no influence on $\phi_j^{(n+1)}$. The domain of dependence for CTCS in Fig. 5.6 shows that only half of the points within the range of the domain of dependence for CTCS are actually included in the domain of dependence.

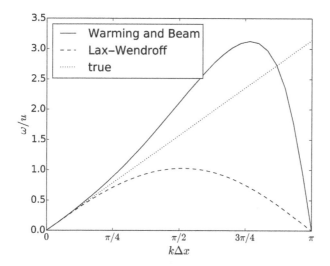

Figure 5.9. The dispersion relations for Lax–Wendroff and Warming and Beam for $c = 0.2$.

There could be two completely independent solutions evolving on alternating grid points. This is a feature of the centred (or leapfrog) time-stepping scheme. However, leap-frog time-stepping is widely used in atmospheric science due to its accuracy and efficiency and the computational mode is controlled by a filter which blends values from consecutive time-levels (Williams, 2009).

Exercise 5.11. *The dispersion relations for Lax–Wendroff and Warming and Beam are given in Fig. 5.9 and numerical solutions of an initial square wave advecting to the right using $c = 0.2$ and 100 points for 100 time-steps are shown in Fig. 5.10 and are compared with the analytic solution. Based on the dispersion relation, which scheme in Fig. 5.10 is Lax–Wendroff and which is Warming and Beam.*

5.2.9 *Conservation*

Linear advection conserves the integrated sum of the dependent variable (i.e., the mass). If we define the mass of dependent variable ϕ on the domain $[0, 1)$ to be

$$M = \int_0^1 \phi \, dx, \qquad (5.29)$$

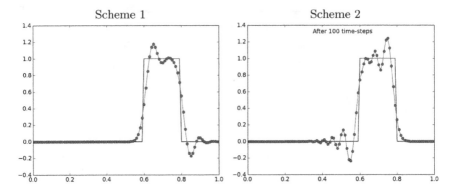

Figure 5.10.   Results of advecting an initial square wave for 100 time-steps using Lax–Wendroff and Warming and Beam with a Courant number of 0.2 on a grid of 100 points in comparison to the analytic solution.

then we can show that, assuming periodic boundary conditions, $\phi(0,t) = \phi(1,t)$, the mass of ϕ is conserved under linear advection,

$$\frac{dM}{dt} = \frac{d}{dt}\left(\int_0^1 \phi dx\right) = \int_0^1 \frac{\partial \phi}{\partial t}dx. \tag{5.30}$$

We can now use the linear advection equation to substitute $\frac{d\phi}{dt} = -u\frac{d\phi}{dx}$,

$$\implies \frac{dM}{dt} = -\int_0^1 u\frac{\partial \phi}{\partial x}dx = -u\int_0^1 d\phi = -u\,[\phi]_0^1 = 0, \tag{5.31}$$

since $\phi(0,t) = \phi(1,t)$. Therefore, M is conserved (it does not change in time). This proof assumes that u is constant in x which is not necessary but makes the proof simpler.

We can also find if total mass is conserved by a numerical scheme. For example, consider FTBS,

$$\phi_j^{(n+1)} = \phi_j^{(n)} - c\big(\phi_j^{(n)} - \phi_{j-1}^{(n)}\big),$$

and define the integrated mass at time n as

$$M^{(n)} = \sum_{j=0}^{n_x-1} \phi_j^{(n)}. \tag{5.32}$$

From this, we can calculate $M^{(n+1)}$ as a function of $M^{(n)}$,

$$M^{(n+1)} = \sum_{j=0}^{n_x-1} \Delta x \phi_j^{(n+1)} = \Delta x \sum_{j=0}^{n_x-1} (\phi_j^{(n)} - c(\phi_j^{(n)} - \phi_{j-1}^{(n)})), \qquad (5.33)$$

$$= M^{(n)} - c\Delta x \left(\sum_{j=0}^{n_x-1} \phi_j^{(n)} - \sum_{j=0}^{n_x-1} \phi_{j-1}^{(n)} \right) \qquad (5.34)$$

$$= M^{(n)} - c\Delta x \left(\sum_{j=1}^{n_x} \phi_j^{(n)} - \sum_{j=0}^{n_x-1} \phi_j^{(n)} \right) \qquad (5.35)$$

$$= M^{(n)} - c\Delta x (\phi_{n_x}^{(n)} - \phi_0^{(n)}) = M^{(n)}, \qquad (5.36)$$

which requires use of the periodic boundary conditions, $\phi_{n_x}^{(n)} = \phi_0^{(n)}$. Therefore $M^{(n+1)} = M^{(n)}$ which means that mass is conserved.

Exercise 5.12. *Find the mass conservation properties of CTCS.*

5.3   The Shallow Water Equations

Second-order wave equations take the form

$$\frac{\partial^2 h}{\partial t^2} = c^2 \nabla^2 h, \qquad (5.37)$$

for dependent variable h and wave speed c. For example, if h is the pressure and c is the speed of sound then Eq. (5.37) represents the propagation of acoustic waves. This is a form of Helmholtz equation. We will focus on numerical methods for solving the linearised shallow water equations because so many important atmospheric processes are governed by the shallow water equations. Ignoring topography, the nonlinear shallow water equations in Eq. (5.6) and Eq. (5.7) can be linearised about a state of rest to give

$$\frac{\partial \mathbf{u}}{\partial t} = -2\Omega \times \mathbf{u} - g\nabla h, \qquad (5.38)$$

$$\frac{\partial h}{\partial t} = -H\nabla \cdot \mathbf{u} \qquad (5.39)$$

(where H is the mean fluid depth, h is the wave height above or below H, g is the acceleration due to gravity and Ω is the rotation of the domain).

These can be expressed in a form like Eq. (5.37) by taking the divergence of Eq. (5.38) and taking a second time derivative of Eq. (5.39) and combining to eliminate $\nabla \cdot \mathbf{u}$. For the rotating shallow water equations, the Coriolis term, $2\Omega \times \mathbf{u}$, remains when they are expressed as a Helmholtz equation,

$$\frac{\partial^2 h}{\partial t^2} = gH\nabla^2 h + 2H\nabla \cdot (\Omega \times \mathbf{u}). \tag{5.40}$$

From Eq. (5.40), we can see that the gravity wave speed is $\sqrt{gH}$. The linearised shallow water equations also represent inertial oscillations and, if Coriolis varies with latitude (i.e., if the angle that Ω makes with $\mathbf{u}$ varies in space as on the surface of the sphere) then these equations also represent Rossby waves. In order to visualise how solutions of the shallow water equations mimic the weather, Fig. 5.11 shows the velocity and height from a solution of the nonlinear shallow water equations in Eqs. (5.6) and (5.7) on the sphere with westerly flow over a mid-latitude mountain. The height behaves in a similar manner to the surface pressure, showing a low pressure and cyclonic winds to the east of the mountain and a Rossby wave train (alternating high and low pressure) to the northeast.

5.3.1   *Finite difference schemes for the one-dimensional, linearised shallow water equations*

The one-dimensional linearised shallow water equations without rotation are

$$\frac{\partial u}{\partial t} = -g\frac{\partial h}{\partial x}, \tag{5.41}$$

$$\frac{\partial h}{\partial t} = -H\frac{\partial u}{\partial x}. \tag{5.42}$$

As there are two equations that depend on each other, it is quite natural to solve them using forward–backward time-stepping — forward for u and backward for h. (This is not the same as implicit time-stepping.) We will also start by assuming that h and u are defined at the same spatial positions (this is called co-located, unstaggered or A-grid) and we will use centred spatial discretisation,

$$\frac{u_j^{(n+1)} - u_j^{(n)}}{\Delta t} = -g\frac{h_{j+1}^{(n)} - h_{j-1}^{(n)}}{2\Delta x}, \tag{5.43}$$

$$\frac{h_j^{(n+1)} - h_j^{(n)}}{\Delta t} = -H\frac{u_{j+1}^{(n+1)} - u_{j-1}^{(n+1)}}{2\Delta x}, \tag{5.44}$$

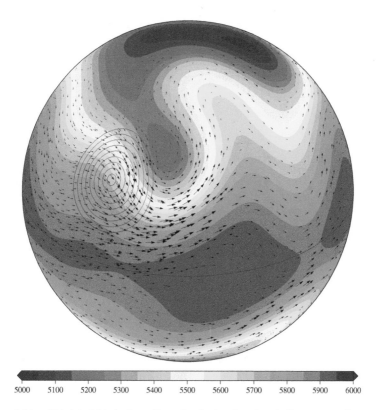

5000 5100 5200 5300 5400 5500 5600 5700 5800 5900 6000

Figure 5.11. Total height (coloured) and velocity for the shallow water flow over a mid-latitude mountain (contoured) after 15 days, test case 5 of Williamson *et al.* (1992).

where $x_j = j\Delta x$, $t^{(n)} = n\Delta t$, $h_j^{(n)} = h(x_j, t^{(n)})$ and $u_j^{(n)} = u(x_j, t^{(n)})$. We can find the stability limits and dispersion properties of this scheme using von Neumann stability analysis. However, first we will look at a numerical solution using this scheme. In order to trigger the spurious behaviour of this co-located scheme, we will add a forcing term, f, to the equation for the height where

$$f = \begin{cases} \sqrt{gH} \sin \omega t, & \text{at } x = 0.5, \\ 0, & \text{elsewhere.} \end{cases} \tag{5.45}$$

Numerical solutions at time $t = \frac{1}{2}$ for the co-located scheme are shown on the left-hand side of Fig. 5.12 for $g = H = 1$ on a periodic domain of size 1 with 60 points in space and using a Courant number, $\sqrt{gH}\Delta t/\Delta x = 0.1$ and

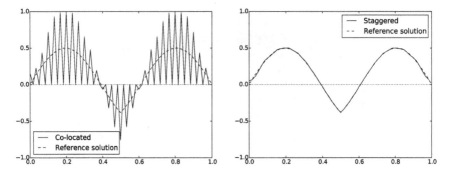

Figure 5.12.   Numerical solutions at time $t = \frac{1}{2}$ for the co-located forward–backward scheme and the staggered forward-backward scheme. The height equation has a forcing term at the mid-point of the domain which oscillates with frequency ω. The simulations use $g = H = 1$ on a periodic domain of size 1 with 60 points in space and using a Courant number, $\sqrt{gH}\Delta t/\Delta x = 0.1$ and with $\omega = 8$. (Test case courtesy of John Thuburn.) These are compared with a reference solution using the staggered scheme and 5 times the resolution.

with $\omega = 8$. These are compared with a reference solution using 5 times the resolution and a more accurate numerical method (to follow). Rather than the forcing term creating a gentle wave, the co-located scheme generates grid-scale oscillations in response to the low frequency, point forcing.

Exercise 5.13. *Write a code to simulate the shallow water equations using the co-located scheme defined by Eqs. (5.43) and (5.44). Use a setup the same as that described for Fig. 5.12, but without the forcing term and with 20 grid points. Initialise the model with $h = 0$ and $u = 0$ everywhere but $h = 1$ at one point near the centre of the domain and run for 50 time-steps. Visualise the results. What is particularly unrealistic about the results?*

We can find the stability limits and dispersion relation for the co-located scheme in Eqs. (5.43) and (5.44) using von Neumann stability analysis. To calculate an amplification factor, A, for each wavenumber, k, we assume wave-like solutions for h and u,

$$h_j^{(n)} = \mathbb{H}\, A^n\, e^{ikj\Delta x}, \qquad (5.46)$$

$$u_j^{(n)} = \mathbb{U}\, A^n\, e^{ikj\Delta x}, \qquad (5.47)$$

for some constants $\mathbb{H}$ and $\mathbb{U}$. Substituting these into Eqs. (5.43) and (5.44) and defining the Courant number $c = \frac{\sqrt{gH}\Delta t}{\Delta x}$ gives

$$A = 1 - \frac{c^2}{2}\sin^2 k\Delta x \pm \frac{ic}{2}\sin k\Delta x \sqrt{4 - c^2 \sin^2 k\Delta x}. \qquad (5.48)$$

There are two solutions for A but this is correct because there are also two analytic solutions because these are second-order wave equations with waves travelling in both directions. For $|c| \leq 2$, Eq. (5.48) gives $|A|^2 = 1$ so the scheme is stable and undamping for sufficiently small time-steps. However, for $|c| > 2$ we have

$$|A|^2 = \left(1 - \frac{c^2}{2}\sin^2 k\Delta x \pm \frac{c}{2}\sin k\Delta x \sqrt{c^2 \sin^2 k\Delta x - 4}\right)^2,$$

which can be greater than 1 and so the scheme is unstable for $|c| > 2$. So this scheme is conditionally stable; stable for $c \leq 2$.

The argument of the amplification factor (Eq. (5.48)) gives us the wave frequency as a function of wavenumber for the co-located scheme (the dispersion relation). For $|c| \leq 2$, the frequency of the numerical method is

$$\omega_n = \pm\frac{1}{\Delta t}\tan^{-1}\frac{\frac{c}{2}\sin k\Delta x \sqrt{4 - c^2 \sin^2 k\Delta x}}{1 - \frac{c^2}{2}\sin^2 k\Delta x}. \qquad (5.49)$$

This can be simplified by assuming that $\frac{c}{2}\sin k\Delta x = \sin\alpha$ to give

$$\omega_n \Delta x = \pm\frac{2\alpha}{c} = \pm\frac{2}{c}\sin^{-1}\left(\frac{c}{2}\sin k\Delta x\right). \qquad (5.50)$$

The positive branch of the dispersion relation is shown in Fig. 5.13 (labelled "A-grid") and compared with the dispersion relation of the analytic solution, $\omega = k\sqrt{gH}$. The analytic solution is non-dispersive (all waves propagate with speed $\sqrt{gH}$) whereas the numerical scheme is dispersive with waves of all wavelengths propagating too slowly. The worse the resolution of the waves, the slower they propagate with grid scale waves ($k\Delta x = \pi$) being stationary. We can see some of these stationary grid-scale waves in the numerical solution in Fig. 5.12. The dispersion relation can also tell us why these waves are generated. The forcing of the central point in Fig. 5.12 has frequency $\omega = 8$ and $\Delta x = 1/60$. This frequency is shown in Fig. 5.13. This frequency should be associated with waves of wavenumber $k = \omega/\sqrt{gH} = 8$ so that $k\Delta x = 8/60$ so these wave have a long wavelength relative to the grid size. However, for the co-located method (A-grid) the frequency $\omega\Delta x = 8/60$ will be associated with two wavelengths, $k\Delta x \approx 0.1$ (a well resolved wave)

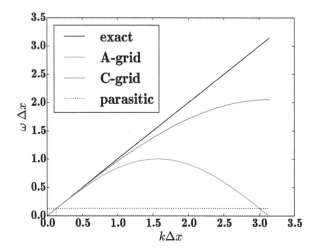

Figure 5.13. Dispersion relation for the co-located (A-grid) and staggered (C-grid) forward–backward finite difference schemes for the one-dimensional linearised shallow water equations in comparison to the analytic solution for $\sqrt{gH} = 1$, $c = \sqrt{gH}\Delta t/\Delta x = 0.4$. The frequency of the parasitic mode of Fig. 5.12 is also shown.

and $k\Delta x \approx 3$ (a near grid-scale wave). Therefore, forcing at this frequency will spuriously excite grid-scale waves as well as realistically exciting long waves. This is known as a parasitic mode. The grid scale waves are parasites, taking energy out of the large scales.

Exercise 5.14. *Use von Neumann stability analysis to find the amplification factor and stability limits for the implicit, co-located scheme for the shallow water equations,*

$$\frac{u_j^{(n+1)} - u_j^{(n)}}{\Delta t} = -g\frac{h_{j+1}^{(n+1)} - h_{j-1}^{(n+1)}}{2\Delta x}, \tag{5.51}$$

$$\frac{h_j^{(n+1)} - h_j^{(n)}}{\Delta t} = -H\frac{u_{j+1}^{(n+1)} - u_{j-1}^{(n+1)}}{2\Delta x}. \tag{5.52}$$

Can you write this scheme as a matrix equation for h to be solved implicitly?

In order to avoid the computational and parasitic modes of the co-located method, we can use a staggered method with h and u being defined

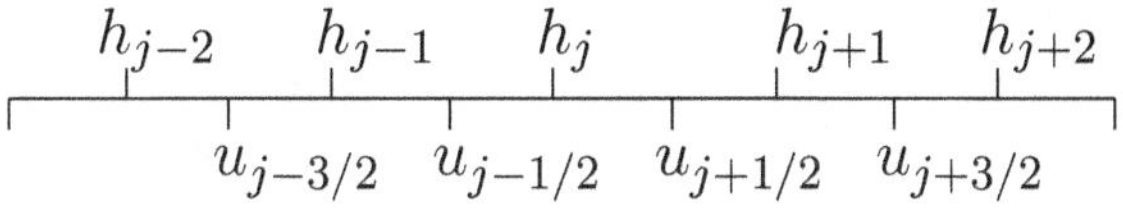

$$h_{j-2} \quad h_{j-1} \quad h_j \quad h_{j+1} \quad h_{j+2}$$

$$u_{j-3/2} \quad u_{j-1/2} \quad u_{j+1/2} \quad u_{j+3/2}$$

Figure 5.14. Location of height and velocity for a one-dimensional staggered (or C-grid) finite difference scheme.

at alternating points in space, as depicted in Fig. 5.14. Discretising the shallow water equations on this grid leads to a staggered (or C-grid) method,

$$\frac{u_{j+\frac{1}{2}}^{(n+1)} - u_{j+\frac{1}{2}}^{(n)}}{\Delta t} = -g \frac{h_{j+1}^{(n)} - h_j^{(n)}}{\Delta x}, \tag{5.53}$$

$$\frac{h_j^{(n+1)} - h_j^{(n)}}{\Delta t} = -H \frac{u_{j+\frac{1}{2}}^{(n+1)} - u_{j-\frac{1}{2}}^{(n+1)}}{\Delta x}. \tag{5.54}$$

Exercise 5.15. *Calculate the amplification factor and the stability limit of $|c| \leq 1$ for the forward–backward, staggered scheme for the shallow water equations.*

It can be shown that the forward–backward, staggered scheme for the shallow water equations is neutrally stable for $|c| \leq 1$ and unstable for $|c| > 1$. The dispersion relation is shown in Fig. 5.13. This scheme is still dispersive; the higher the wavenumber the slower the propagation. We can see that there will be no computational mode, no parasitic mode and the wave frequency is always increasing with wavenumber, an important property for approximating non-dispersive waves. This results in a very accurate solution of the slowly forced oscillation problem in Fig. 5.12. The two-dimensional extension of this staggered grid is the Arakawa C-grid (Arakawa and Lamb, 1977) which is widely used in weather and climate prediction models due to its low dispersion errors and lack of computational modes.

5.4 Iterative Methods for Linear Systems in the Geosciences

In this second part of the chapter, we consider iterative methods for linear systems of equations arising from the discretisation of PDEs. These systems often arise in geoscientific models, where they frequently dominate the overall run time of the model. Some examples are the Helmholtz equation that is solved for the pressure in semi-implicit compressible atmosphere models, the pressure Poisson equation in ocean models (two-dimensional for hydro-static, three-dimensional for non-hydrostatic), and the Stokes' equation in glacier and ice sheet models. In this part of the chapter, we take an abstract approach, since the same general tools for understanding numerical solver algorithms are used in all cases, even though there are application-specific details. We split our discussion into two parts. First we consider Krylov methods, in particular the conjugate gradient method, where performance is good for well-conditioned matrices. The search for good preconditioners leads to the consideration of splitting methods such as Jacobi and SOR. Even though these methods predate the invention of the digital computer by centuries, they are still in active use today in the form of precondition-ers for Krylov methods. These methods are also getting a lease of life as the smoothers for multigrid methods, which represent one aspect of the state of the art in high performance massively parallel scientific comput-ing, yielding iterative schemes with iteration counts that are (virtually) independent of mesh resolution. In fact, the vast majority of linear solver research today concentrates on finding good splitting methods for specific PDEs from various applications, in order to incorporate them as precondi-tioners and smoothers for Krylov methods and multigrid algorithms. As a *Mathematics of Planet Earth* researcher, you may well be involved in the development and use of high performance numerical models that contain these methods at the heart of their engine, and we hope that this chap-ter provides a starting point for understanding how to select and combine different methods, and how to explain their behaviour. For further read-ing, we refer the reader to Golub and Loan (2012); Trefethen and Bau (1997).

5.5   Time-Stepping Algorithms for the One-Dimensional Shallow Water Equations

The one-dimensional shallow water equations take the form,

$$\frac{\partial u}{\partial t} + u\frac{\partial u}{\partial x} + g\frac{\partial h}{\partial x} = 0, \tag{5.55}$$

$$\frac{\partial h}{\partial t} + \frac{\partial}{\partial x}(uh) = 0. \tag{5.56}$$

These equations describe nonlinear water waves that vary in the x-direction only, with speed u, layer depth h, and acceleration due to gravity g. They can be thought of a toy model that shares some properties with more challenging and realistic ocean or atmosphere models, namely they support waves but also evolve through nonlinear advection terms. To gain some understanding of how these waves behave, we linearise the equations about the state of rest $u = 0$, $h = H$,

$$\frac{\partial u}{\partial t} + g\frac{\partial h}{\partial x} = 0, \tag{5.57}$$

$$\frac{\partial h}{\partial t} + H\frac{\partial u}{\partial x} = 0. \tag{5.58}$$

This is a wave equation with wave propagation speed $c = \sqrt{gH}$, written as a first-order system of PDEs. To characterise the relative timescales due to wave propagation and advection by the velocity u, we recall the Froude number from Eq. (2.58), $Fr = \frac{U}{\sqrt{gH}}$ (where U is a typical velocity value). Low Froude number solutions are governed by the advective timescale $T_U = L/U$. In geophysical applications, we are frequently concerned with low Froude number limits. For example, atmospheric winds are usually much slower than the speed of acoustic waves.

We shall now see that efficiently solving PDEs in low Froude number limits motivates the development of efficient solvers for linear systems. As we have seen from earlier in the course, explicit time integration schemes have stability conditions that depend on the wave Courant number $\lambda = c\Delta t/\Delta x$, where c is the wave propagation speed. If the stability is conditional on the Courant number being below a critical value, $\lambda < \lambda^*$, then the total number of time-steps required to solve until T_U is

$$n = \frac{T_U}{\Delta t} = \frac{L}{U\Delta t} = \frac{cL}{U\lambda\Delta x} = \frac{1}{Fr}\frac{L}{\lambda\Delta x} > \frac{1}{Fr}\frac{L}{\lambda^*\Delta x}. \tag{5.59}$$

Computing low Froude number solutions becomes impractical with explicit time-stepping, due to this inverse scaling with Fr.

A standard approach to addressing this problem is to use implicit time-stepping algorithms instead. For example, the implicit midpoint rule replaces

$$\dot{x} = f(x), \tag{5.60}$$

by

$$x^{n+1} - x^n = \Delta t f\left(\frac{x^{n+1} + x^n}{2}\right), \tag{5.61}$$

and is unconditionally stable. Applying the implicit midpoint rule to our toy problem gives

$$u^{n+1} - u^n + \Delta t u^{n+1/2} \frac{\partial}{\partial x} u^{n+1/2} + \Delta t g \frac{\partial}{\partial x} h^{n+1/2} = 0, \qquad (5.62)$$

$$h^{n+1} - h^n + \Delta t \frac{\partial}{\partial x} (u^{n+1/2} h^{n+1/2}) = 0, \qquad (5.63)$$

where $u^{n+1/2} = (u^{n+1} + u^n)/2$, etc. At this point, we have only discretised in time, not in space. This is a nonlinear PDE that we need to (approximately) solve using iterative methods. We solve for u^{n+1}, h^{n+1} by generating an iterative sequence

$$v^0 (= u^n), v^1, v^2, v^3, \ldots, \quad \phi^0 (= h^n), \phi^1, \phi^2, \phi^3, \ldots, \qquad (5.64)$$

by the following Picard iteration procedure.

(a) Set $\bar{u} = (v^i + u^n)/2$, $\bar{h} = (\phi^i + h^n)/2$.
(b) Solve

$$v^{i+1} + \frac{1}{2} \Delta t g \frac{\partial \phi^{i+1}}{\partial x} = u^n - \frac{1}{2} \Delta t g \frac{\partial h^n}{\partial x} - \Delta t \bar{u} \frac{\partial}{\partial x} \bar{u}, \qquad (5.65)$$

$$\phi^{i+1} + \frac{1}{2} \Delta t H \frac{\partial v^{i+1}}{\partial x} = h^n - \frac{1}{2} \Delta t H \frac{\partial u^n}{\partial x} - \Delta t \frac{\partial}{\partial x} \left(\bar{u} \left(\bar{h} - H \right) \right) \quad (5.66)$$

to obtain v^{i+1} and h^{i+1}.

Exercise 5.16. *Show that if this algorithm has a fixed point, then it is a solution of the implicit midpoint rule algebraic system.*

It can be shown that two iterations are sufficient to obtain a stable time-stepping scheme for small enough Δt (2–4 are used in practice) (Durran, 1999). Convergence is guaranteed by the Banach contraction mapping theorem for small enough Δt.

Elimination of v^{i+1} leads to

$$\phi^{i+1} - \frac{\Delta t^2}{4} \frac{\partial^2 \phi^{i+1}}{\partial x^2} = F, \qquad (5.67)$$

which is a Helmholtz equation. After solving, we reconstruct v^{i+1} from known ϕ^{i+1}. Efficient solutions of elliptic problems similar to this Helmholtz problem are crucial for weather, oceans and climate models, and a substantial fraction of the computational time in many models is taken up by iterative methods for solving them.

In one dimension, the centred difference approximation to the Helmholtz equation

$$\phi - \alpha^2 \frac{\partial^2 \phi}{\partial x^2} = F, \tag{5.68}$$

is

$$\phi_k + \frac{\alpha^2}{\Delta x^2}(-\phi_{k-1} + 2\phi_k - \phi_{k+1}) = F_k. \tag{5.69}$$

This is a matrix vector system of the form

$$\begin{pmatrix} \Delta x^2/\alpha^2 + 2 & -1 & 0 & \cdots & 0 \\ -1 & \Delta x^2/\alpha^2 + 2 & -1 & \cdots & 0 \\ \vdots & \vdots & \vdots & \vdots & \vdots \end{pmatrix} \begin{pmatrix} \phi_1 \\ \phi_2 \\ \vdots \\ \phi_{n-1} \end{pmatrix} = \frac{\Delta x^2}{\alpha^2} \begin{pmatrix} F_1 \\ F_2 \\ \vdots \\ F_{n-1} \end{pmatrix}. \tag{5.70}$$

Definition 5.3. A matrix A is positive-definite if $y^T A y > 0$ for any $y \neq 0$.

Definition 5.4. A matrix A is symmetric if $A^T = A$.

Exercise 5.17. *Using direct calculation of the eigenvalues using Fourier methods (similar to the von Neumann analysis in Sec. 5.2.7), show that this matrix is symmetric and positive-definite.*

5.6   The Conjugate Gradient Algorithm

We have already seen that solving positive-definite symmetric systems of equations efficiently is crucial for weather, ocean and climate models. In this section, we introduce an algorithm that is the standard high performance workhorse for these systems, the conjugate gradient (CG) algorithm.

The aim is to find the solution x^* to

$$Ax = b, \tag{5.71}$$

for A symmetric, positive-definite. For symmetric positive-definite A, it becomes useful to define the vector norm

$$\|x\|_A = \sqrt{x^T A x}, \tag{5.72}$$

and the matrix norm

$$\|B\|_A = \|A^{1/2} B A^{-1/2}\|_2. \tag{5.73}$$

These norms are useful for studying convergence of iterative methods for solving $A\boldsymbol{x} = \boldsymbol{b}$.

We describe the idea behind CG in terms of expanding the solution using a basis. We choose a special (to be determined later) basis of vectors $(\boldsymbol{p}_1, \boldsymbol{p}_2, \ldots, \boldsymbol{p}_n)$ and write

$$\boldsymbol{x}^* = \sum_{i=1}^{n} \alpha_i^* \boldsymbol{p}_i. \tag{5.74}$$

We will try to compute α_i^* one by one, so will consider the recursive formula,

$$\boldsymbol{x}^k = \sum_{i=1}^{k} \alpha_i^* \boldsymbol{p}_i \implies \boldsymbol{x}^{k+1} = \boldsymbol{x}^k + \alpha_k^* \boldsymbol{p}_k. \tag{5.75}$$

The calculation of α_i^* can be simplified by requiring that $\boldsymbol{p}_i^T A \boldsymbol{p}_j = 0$ for $i \neq j$. This is called *A-conjugacy* of the basis. Then we have

$$\boldsymbol{p}_i^T \boldsymbol{b} = \boldsymbol{p}_i^T A \boldsymbol{x}^* \tag{5.76}$$

$$= \boldsymbol{p}_i^T \sum_{j=1}^{n} \alpha_j^* A \boldsymbol{p}_j \tag{5.77}$$

$$= \alpha_i^* \boldsymbol{p}_i^T A \boldsymbol{p}_i \implies \alpha_i^* = \frac{\boldsymbol{p}_i^T \boldsymbol{b}}{\boldsymbol{p}_i^T A \boldsymbol{p}_i} \tag{5.78}$$

providing a direct formula for α_i^*. Under the assumption of A-conjugacy, we have the iterative update,

$$\boldsymbol{x}^{k+1} = \boldsymbol{x}^k + \alpha_k^* \boldsymbol{p}_k = \boldsymbol{x}^k + \frac{\boldsymbol{p}_k^T \boldsymbol{b}}{\boldsymbol{p}_k^T A \boldsymbol{p}_k} \boldsymbol{p}_k. \tag{5.79}$$

This algorithm will terminate in n iterations once we have the full expansion of $\boldsymbol{x}^*$. Hence, it takes the appearance of a direct solver. However, it might be that we can choose $(\boldsymbol{p}_1, \ldots, \boldsymbol{p}_n)$ so that $\boldsymbol{x}^k$ gets close to $\boldsymbol{x}^*$. Then, we can terminate the algorithm to get an approximate solution as quickly as possible.

To determine how to optimally choose the basis, we reinterpret the linear system $A\boldsymbol{x} = \boldsymbol{b}$ as an optimisation problem, seeking $\boldsymbol{x}^*$ that minimises

$$\phi(\boldsymbol{x}) = \tfrac{1}{2} \boldsymbol{x}^T A \boldsymbol{x} - \boldsymbol{x}^T \boldsymbol{b}. \tag{5.80}$$

After completing the square,

$$\phi(x) = \tfrac{1}{2}(Ax - b)^T A^{-1}(Ax - b) - \tfrac{1}{2}b^T A^{-1}b, \qquad (5.81)$$

we see that $Ax^* = b$ as required. Taking the first k basis functions, we write $x = \sum_{i=1}^{k} \alpha_i p_i$, and we have

$$\phi_k(x) = \phi\left(\sum_{i=1}^{k} \alpha_i p_i\right) = \sum_{i=1}^{k}\left(\frac{\alpha_i^2}{2}p_i^T A p_i - \alpha_i b^T p_i\right). \qquad (5.82)$$

To minimise this over $\alpha_1, \alpha_2, \ldots, \alpha_k$ we solve

$$0 = \frac{\partial}{\partial \alpha_j}\sum_{i=1}^{k}\left(\frac{\alpha_i^2}{2}p_i^T A p_i - \alpha_i b^T p_i\right) = \alpha_j p_j^T A p_j - b^T p_j, \qquad (5.83)$$

which gives

$$\alpha_i = \alpha_i^* = \frac{p_i^T b}{p_i^T A p_i}. \qquad (5.84)$$

Now, say we have computed $(p_1, p_2, \ldots, p_{k-1})$, and x_{k-1}, and we want to improve the approximate solution by adding an extra basis vector. To reduce the error as quickly as possible, we choose p_k so that $\phi(x_{k-1}) - \phi(x_k)$ is maximised by searching in the "down-hill" direction, i.e.,

$$p_k = -\nabla\phi(x_{k-1}) = b - A x_{k-1} = r_{k-1}, \qquad (5.85)$$

where r_{k-1} is the *residual* at iteration $k-1$. This proposed basis vector does not satisfy A-conjugacy, so we modify it by projection,

$$p_k = r_{k-1} - \sum_{j=1}^{k-1}\frac{p_j^T A r_{k-1}}{p_j^T A p_j}p_j. \qquad (5.86)$$

This is the conjugate gradient direction.

The process described above can now be formulated as an iterative algorithm.

Definition 5.5 (Conjugate gradient algorithm version 1.0). The CG algorithm is as follows:

- $x_0 = 0$.
- for $k = 1$ to n
- $\quad r_{k-1} = b - A x_{k-1}$. If $|r_{k-1}| < \epsilon$ then quit (for a given tolerance ϵ).
- $\quad$ if $k = 1$ then $p_1 = r_0$

- else $p_k = r_{k-1} - \sum_{j=1}^{k-1} \frac{p_j^T A r_{k-1}}{p_j^T A p_j} p_j.$
- $x_k = x_{k-1} + \frac{b^T p_k}{p_k^T A p_k} p_k.$
- end k.

At this point, we note a few remarks about this algorithm. First of all, note that the algorithm only requires matrix-vector multiplications with A. This means that it can be implemented very efficiently when A is sparse, for example, and so matrix-vector multiplication is very cheap. However, there are a number of these multiplications per iteration. We will address this by showing that this can be reduced to one matrix-vector multiplication per iteration. It is also not practical to store the entire set of p_ks. We will address this by showing that only the p_k for the current iteration needs to be stored.

To address the first point, note that it appears that we need all of $p_1,\ldots,p_{k-1}$ to compute p_k since

$$p_k = r_{k-1} - \sum_{j=1}^{k-1} \beta_j p_j, \quad \beta_j = \frac{p_j^T A r_{k-1}}{p_j^T A p_j}. \tag{5.87}$$

In fact, $p_j^T A r_{k-1} = 0$ for $0 < j < k - 1$. We will show this in the following lemmas.

Lemma 5.1. $p_j^T r_k = 0$ *for* $1 \leqslant j \leqslant k$.

Proof. If $1 \leqslant j \leqslant k$, then

$$p_j^T r_k = p_j^T (A x^k - b) \tag{5.88}$$

$$= p_j^T A \sum_{i=1}^{k} \alpha_i^* p_i - p_j^T b \tag{5.89}$$

$$= 0 \tag{5.90}$$

from the definition of α_j^*. □

Lemma 5.2. $r_j^T r_k = 0$ *for* $1 \leqslant j < k$.

Proof. Rearranging the formula for p_{k+1} gives

$$r_j = p_{j+1} + \sum_{i=1}^{j} \beta_i p_i, \tag{5.91}$$

so if $1 \leqslant j < k$ then

$$r_j^T r_k = p_{j+1}^T r_k + \sum_{i=1}^{j} \beta_i p_i^T r_k = 0. \tag{5.92}$$

$\square$

Lemma 5.3. $p_j^T A r_{k-1} = 0$ *for* $0 < j < k - 1$.

Proof.

$$\alpha_j^* p_j^T A r_{k-1} = (\alpha_j^* A p_j)^T r_{k-1} \tag{5.93}$$

$$= (A x_j - A x_{j-1})^T r_{k-1} \tag{5.94}$$

$$= (r_{j-1} - r_j)^T r_{k-1} = 0, \tag{5.95}$$

because the residuals are orthogonal. So,

$$p_k = r_{k-1} - \sum_{j=1}^{k-1} \frac{p_j^T A r_{k-1}}{p_j^T A p_j} p_j \tag{5.96}$$

$$= r_{k-1} - \frac{p_{k-1}^T A r_{k-1}}{p_{k-1}^T A p_{k-1}} p_{k-1}. \tag{5.97}$$

$\square$

Having shown that we do not need to store the entire list of basis functions, we can turn our attention to reducing the number of multiplications by A per iteration. As written above, the algorithm requires four multiplications by A, as follows:

$$r_{k-1} = b - A x_{k-1}, \tag{5.98}$$

$$p_k = r_{k-1} - \frac{p_{k-1}^T A r_{k-1}}{p_{k-1}^T A p_{k-1}} p_{k-1}, \tag{5.99}$$

$$\alpha_k^* = \frac{b^T p_k}{p_k^T A p_k}. \tag{5.100}$$

We will reduce this to just one, the multiplication $A p_k$. Since

$$(x_{k-1})^T A p_k = \left(\sum_{j=1}^{k-1} \alpha_j^* p_j \right)^T A p_k = 0, \tag{5.101}$$

by A-conjugacy and

$$r_{k-1}^T p_k = r_{k-1}^T \left(r_{k-1} - \sum_{j=1}^{k-1} \beta_j p_j \right) = r_{k-1}^T r_{k-1}, \tag{5.102}$$

we have

$$\alpha_k^* = \frac{b^T p_k}{p_k^T A p_k} = \frac{(r_{k-1} + A x_{k-1})^T p_k}{p^T A p_k} \tag{5.103}$$

$$= \frac{r_{k-1}^T p_k}{p_k^T A p_k} = \frac{r_{k-1}^T r_{k-1}}{p_k^T A p_k}. \tag{5.104}$$

Further, since

$$p_{k-1} = r_{k-2} - \sum_{j=1}^{k-2} \beta_j p_j, \tag{5.105}$$

we have

$$\frac{p_{k-1}^T A r_{k-1}}{p_{k-1}^T A p_{k-1}} = \frac{p_{k-1}^T A r_{k-1}}{p_{k-1}^T A r_{k-2}}$$

$$= \frac{(r_{k-2} - r_{k-1})^T r_{k-1}}{(r_{k-2} - r_{k-1})^T r_{k-2}} = -\frac{r_{k-1}^T r_{k-1}}{r_{k-2}^T r_{k-2}}. \tag{5.106}$$

Finally,

$$x_k = x_{k-1} + \alpha_k^* p_k \implies r_k = r_{k-1} - \alpha_k^* A p_k. \tag{5.107}$$

After these calculations, we can write down an improved version of the conjugate gradient algorithm.

Definition 5.6 (Conjugate gradient algorithm version 1.1). The simplified CG algorithm is

- $x_0 = 0$.
- $r_0 = b$.
- for $k = 1$ to n
- If $r_{k-1} < \epsilon$ then quit.
- if $k = 1$ then $p_1 = r_0$
- else $p_k = r_{k-1} + \dfrac{r_{k-1}^T r_{k-1}}{r_{k-2}^T r_{k-2}} p_{k-1}$.
- $\alpha_k^* = \dfrac{r_{k-1}^T r_{k-1}}{p_k^T A p_k}$.

- $\boldsymbol{x}_k = \boldsymbol{x}_{k-1} + \alpha_k^* \boldsymbol{p}_k.$
- $\boldsymbol{r}_k = \boldsymbol{r}_{k-1} - \alpha_k^* A \boldsymbol{p}_k.$
- end k.

We strongly advise the reader to reinforce their understanding of this algorithm by implementing it as code and exploring the results when applied to simple matrices.

5.7   Convergence of CG

In this section, we study the convergence rates of the CG algorithm, which motivates the need for preconditioners. To do this, we return to the optimisation formulation of the linear system. We start from the completed square form of ϕ,

$$\phi(\boldsymbol{x}) = (\boldsymbol{b} - A\boldsymbol{x})^T A^{-1} (\boldsymbol{b} - A\boldsymbol{x}) - \tfrac{1}{2} \boldsymbol{b}^T A^{-1} \boldsymbol{b}, \qquad (5.108)$$

and substitute $\boldsymbol{b} = A\boldsymbol{x}^*$, to obtain

$$\phi(\boldsymbol{x}) = (\boldsymbol{x} - \boldsymbol{x}^*)^T A (\boldsymbol{x} - \boldsymbol{x}^*) - \tfrac{1}{2} \boldsymbol{b}^T A^{-1} \boldsymbol{b}. \qquad (5.109)$$

Hence, the error $\boldsymbol{x}_k - \boldsymbol{x}^*$ measured in the A-norm is

$$\|\boldsymbol{e}_k\|_A^2 = (\boldsymbol{x}_k - \boldsymbol{x}^*)^T A (\boldsymbol{x}_k - \boldsymbol{x}^*) \qquad (5.110)$$

$$= \min_{\boldsymbol{x} \in V_k} ((\boldsymbol{b} - A\boldsymbol{x})^T A^{-1} (\boldsymbol{b} - A\boldsymbol{x})), \qquad (5.111)$$

where

$$V_k = \left\{ \boldsymbol{x} : \boldsymbol{x} = \sum_{i=1}^{k} \gamma_i \boldsymbol{p}_i \right\}. \qquad (5.112)$$

By inspecting the conjugate gradient algorithm, we see that each basis vector takes the form,

$$\boldsymbol{p}_k = c_{k0} \boldsymbol{b} + c_{k1} A \boldsymbol{b} + \cdots + c_{kk} A^{k-1} \boldsymbol{b}. \qquad (5.113)$$

This means that $\boldsymbol{p}_k$ can be expanded in the first k vectors of the *Krylov* basis

$$\boldsymbol{b}, \, A\boldsymbol{b}, \, A^2\boldsymbol{b}, \, A^3\boldsymbol{b}, \ldots. \qquad (5.114)$$

We conclude that the CG algorithm at iteration n solves the following problem.

Definition 5.7 (CG approximation problem). Let P_k be the set of polynomials p of degree $\leqslant k$ with $p(0) = 1$, and let $e_0 = x_0 - x^*$ be the initial error. Find $p_k \in P_k$ such that

$$\|p_k(A)e_0\|_A = \text{minimum}. \tag{5.115}$$

Note that this is not a practical algorithm since e_0 is not known. We just use it to understand the convergence properties of CG. This reformulation of the algorithm leads immediately to the following error estimate.

Theorem 5.2 (Error estimate for CG). *If the CG iteration has not already converged before step k (i.e., $r_{k-1} \neq 0$), then the CG approximation problem has a unique solution $p_k \in P_k$, with error $e_k = p_k(A)e_0$. We have*

$$\frac{\|e_k\|_A}{\|e_0\|_A} = \min_{p \in P_k} \frac{\|p(A)e_0\|_A}{\|e_0\|_A} \leqslant \inf_{p \in P_k} \max_\lambda |p(\lambda)|, \tag{5.116}$$

where the maximum is taken over the eigenvalues of A.

Proof. For existence and uniqueness, we just note that $\|p_k(A)e_0\|_A$ is a convex function over a finite dimensional set P_k.
Expanding $e_0 = \sum_{j=1}^n \mu_j u_j$ in orthonormal eigenvectors u_j of A (with eigenvalues λ_j), we have

$$\|e_0\|_A^2 = \sum_{j=1}^n \mu_j^2 \lambda_j, \quad \|p(A)e_0\|_A^2 = \sum_{j=1}^n \mu_j^2 \lambda_j (p(\lambda_j))^2. \tag{5.117}$$

Hence,

$$\|p(A)e_0\|_A^2 \leqslant \sum_{j=1}^n \mu_j^2 \lambda_j \max_\lambda (p(\lambda))^2 = \|e_0\|_A^2 \max_\lambda (p(\lambda))^2, \tag{5.118}$$

hence the result.   □

We can make some immediate deductions from this result. If A has only s distinct eigenvalues then we can find a sth order polynomial with $p_s(0) = 1$ and $p_s(\lambda_i) = 0$, $i = 1, \ldots, s$. Then the algorithm converges in s iterations. If A has eigenvalues clustered into s groups, then we can find an sth order polynomial which is small near each of the clusters. The error will be very small after s iterations. Unfortunately, the more usual case when we discretise PDEs is a broad spectrum of eigenvalues with no clusters,

$$\lambda_1 < \lambda_2 < \cdots < \lambda_n. \tag{5.119}$$

In this case, we need to estimate the size of the kth order polynomial with $p(0) = 1$ that is minimised in $\lambda_1 < t < \lambda_k$. It turns out that this can be done by using Chebyshev polynomials, leading to the following theorem.

Theorem 5.3 (Error estimate for CG from condition number). *If the CG iteration has not already converged before step k (i.e., $r_{k-1} \neq 0$), then the error satisfies*

$$\frac{\|e_k\|_A}{\|e_0\|_A} \leqslant \frac{2}{\left(\frac{\sqrt{\kappa}+1}{\sqrt{\kappa}-1}\right)^k + \left(\frac{\sqrt{\kappa}+1}{\sqrt{\kappa}-1}\right)^{-k}} \leqslant 2\left(\frac{\sqrt{\kappa}-1}{\sqrt{\kappa}+1}\right)^k, \tag{5.120}$$

where $\kappa = |\lambda_{\max}/\lambda_{\min}|$ is the condition number of the matrix A.

For a proof of this theorem, see Trefethen and Bau (1997). This is a pessimistic estimate (better but harder to prove estimates exist), but illustrates that need a well-conditioned matrix for fast convergence.

5.8 Preconditioning

To optimise the convergence rate, we try to find a symmetric positive-definite matrix $\hat{A}$ which is cheap to invert, and that approximates A. Since we only require matrix-vector multiplications, we can consider applying the conjugate gradient algorithm to

$$(\hat{A}^{-1}A)x = \hat{A}^{-1}b. \tag{5.121}$$

This is called *left preconditioning*. In the CG algorithm, the matrix-vector products then get replaced by evaluating $y = Ax$, then solving $\hat{A}z = y$, which can be done cheaply. The problem with this idea is that $\hat{A}^{-1}A$ is not symmetric positive-definite even when both A and $\hat{A}$ are. This problem can be solved with symmetric preconditioning.

If $\hat{A}$ is symmmmetric positive-definite, then it has a Cholesky factorisation, $\hat{A} = U^T U$. For efficient preconditioning, it must be cheap to solve $Ux = y$. We then apply the CG algorithm to

$$\tilde{A}y = ((U^T)^{-1}AU^{-1})y = (U^T)^{-1}b, \tag{5.122}$$

before solving $Ux = y$ for x. This is called *symmetric preconditioning*. In this case, matrix-vector multiplications $(U^T)^{-1}AU^{-1})p$ can be cheaply evaluated by sequentially solving

$$Ut = p, \quad U^T q = At. \tag{5.123}$$

This idea leads to the following algorithm.

Definition 5.8 (Preconditioned conjugate gradient algorithm).
The preconditioned conjugate gradient algorithm (PCG) is

- $x_0 = 0$, $r_0 = b$.
- for $k = 1$ to n
- If $r_{k-1} < \epsilon$ then quit.
- Solve $\hat{A} z_{k-1} = r_{k-1}$.
- if $k = 1$ then $p_1 = z_0$
- else $p_k = z_{k-1} - \dfrac{z_{k-1}^T r_{k-1}}{z_{k-2}^T r_{k-2}} p_{k-1}$.
- $\alpha_k^* = \dfrac{z_{k-1}^T r_{k-1}}{p_k^T A p_k}$.
- $x_k = x_{k-1} + \alpha_k^* p_k$, $r_k = r_{k-1} - \alpha_k^* A p_k$.
- end k.

Exercise 5.18. *Show that the preconditioned conjugate gradient algorithm is equivalent to applying the original conjugate gradient algorithm to the preconditioned system*

$$(U^T)^{-1} U^{-1} A U^{-1} y = (U^T)^{-1} b. \tag{5.124}$$

Remarkably, the PCG algorithm does not require any knowledge of U.

CG is the classic Krylov method that builds up the solution from the Krylov basis

$$b, Ab, A^2 b, A^3 b, \ldots. \tag{5.125}$$

Krylov methods all have the strength that they only require matrix-vector multiplication, and the weakness that they require preconditioning. For non-symmetric systems, there are a number of Krylov subspace methods: GMRES, BICG-stab, CGR, etc. (Trefethen and Bau, 1997; Golub and Loan, 2012). All have convergence rates that scale with condition number κ. Non-symmetric systems do not require symmetric preconditioners, and we can simply apply them to

$$\hat{A}^{-1} A x = \hat{A}^{-1} b, \tag{5.126}$$

where $\hat{A} y = z$ is cheap to solve.

The most common (and most immediately parallel-friendly) preconditioners are adapted from classical iterative methods, such as Jacobi
iteration, Gauss–Seidel/SOR (symmetrised as SSOR). For scalable parallel
performance in massive problems these are combined with multigrid methods, domain decomposition methods, and other approaches that represent
the state of the art. For linear systems arising from discretisations of PDE,
preconditioners often derive from specific properties of the PDE in question.

We shall move on to look at classical iterative methods and how they
can be used as preconditioners.

5.9   Classical Iterative Methods as Preconditioners

For the equation $A\boldsymbol{x} = \boldsymbol{b}$, iterative methods provide a way of obtaining a
(hopefully) better approximate solution $\boldsymbol{x}^{k+1}$ from a previous approximate
$\boldsymbol{x}^k$. In this section, we shall concentrate on a particular class of iterative
methods that are suitable for use as preconditioners, based on writing A as
a splitting $A = M + N$, and then defining the iteration $M\boldsymbol{x}^{k+1} = -N\boldsymbol{x}^k + \boldsymbol{b}$,
where M is chosen so that this equation is cheap to solve. These methods
are *stationary methods*, that is the iterative update is defined as an affine
transformation on the previous iterate $\boldsymbol{x}^k$. Stationary methods are suitable
for use as preconditioners since they define approximations $\hat{A}$ of the matrix
A, as we shall see shortly.

In this section, we shall introduce some classical iterative methods,
illustrating them using the finite difference approximation of the two-
dimensional Poisson problem given by

$$-\nabla^2 p(x,y) = f(x,y), \quad 0 \leqslant x,y \leqslant 1,$$
$$p(x,y) = 0, \quad x = 0 \text{ or } x = 1 \quad \text{or} \quad y = 0 \text{ or } y = 1. \tag{5.127}$$

The standard centred finite difference approximation of this equation is

$$4p_{i,j} - p_{i-1,j} - p_{i+1,j} - p_{i,j+1} - p_{i,j-1} = h^2 f_{i,j}, \quad 0 < i,j < N,$$
$$p_{i,j} = 0, \quad i = 0 \text{ or } i = N \quad \text{or} \quad j = 0 \text{ or } j = N, \tag{5.128}$$

where h is the grid spacing in the finite difference mesh. If we write

$$\boldsymbol{p} = (p_{1,1}, \ldots, p_{N-1,1}, p_{1,2}, \ldots, p_{N-1,2}, \ldots, p_{1,N-1}, \ldots, p_{N-1,N-1}), \tag{5.129}$$

then equation can be written in the form

$$
\begin{pmatrix}
D+2I & -I & \cdots & 0 \\
-I & D+2I & \ddots & \vdots \\
\vdots & \ddots & \ddots & -I \\
0 & \cdots & -I & D+2I
\end{pmatrix}
\boldsymbol{p} = \boldsymbol{b},
\tag{5.130}
$$

where

$$
D =
\begin{pmatrix}
2 & -1 & \cdots & 0 \\
-1 & 2 & \ddots & \vdots \\
\vdots & \ddots & \ddots & -1 \\
0 & \cdots & -1 & 2
\end{pmatrix},
\tag{5.131}
$$

and I is the $(N-1) \times (N-1)$ identity matrix.

5.9.1 Richardson's method

We take $M = I$, so $N = A - I$, leading to

$$
\boldsymbol{x}^{k+1} = \boldsymbol{x}^k + \boldsymbol{b} - A\boldsymbol{x}^k.
\tag{5.132}
$$

This will diverge in general if any eigenvalues λ of A satisfy $|1-\lambda| > 1$, so we consider a rescaling $M = I/\omega$, for $\omega > 0$, so $N = A - I/\omega$, leading to

$$
\boldsymbol{x}^{k+1} = \boldsymbol{x}^k + \omega(\boldsymbol{b} - A\boldsymbol{x}^k).
\tag{5.133}
$$

When applied to the two-dimensional Poisson model problem, Richardson iteration becomes

$$
p_{i,j}^{k+1} = p_{i,j}^k - \omega(4p_{i,j}^k - p_{i-1,j}^k - p_{i+1,j}^k - p_{i,j+1}^k - p_{i,j-1}^k + \Delta x^2 f_{i,j}).
\tag{5.134}
$$

Note that each update can be evaluated independently, so the algorithm can be easily executed on a parallel computer.

5.9.2 *Jacobi's method*

Next we consider the splitting $A = L + D + U$ with L strictly lower triangular, D diagonal and U strictly upper triangular, i.e.,

$$L_{ij} = 0, j \geqslant i, \quad D_{ij} = 0, i \neq j, \quad U_{ij} = 0, i \geqslant j. \tag{5.135}$$

This leads to Jacobi's method,

$$D\boldsymbol{x}^{k+1} = \boldsymbol{b} - (L+U)\boldsymbol{x}^k. \tag{5.136}$$

Since D is is diagonal, is very cheap to solve this equation. When applied to the two-dimensional model problem, we get

$$4p_{i,j}^{k+1} = p_{i-1,j}^k + p_{i+1,j}^k + p_{i,j+1}^k + p_{i,j-1}^k + \Delta x^2 f_{i,j}. \tag{5.137}$$

This algorithm is also very easy to implement on a parallel computer.

5.9.3 *Gauss–Seidel method*

We consider the splitting $A = L + D + U$ with L strictly lower triangular, D diagonal and U strictly upper triangular. The forwards and backwards Gauss–Seidel methods are

$$(L+D)\boldsymbol{x}^{k+1} = \boldsymbol{b} - U\boldsymbol{x}^k, \quad \text{or} \quad (U+D)\boldsymbol{x}^{k+1} = \boldsymbol{b} - L\boldsymbol{x}^k, \tag{5.138}$$

respectively. Each iteration requires the solution of a triangular system by forward–backward substitution. The Gauss–Seidel algorithm is an improvement of Jacobi's method using new values as soon as possible within the same iteration.

When applied to the two-dimensional model problem, we get

$$4p_{i,j}^{k+1} = p_{i-1,j}^{k+1} + p_{i+1,j}^{k+1} + p_{i,j+1}^{k+1} + p_{i,j-1}^{k+1} + \Delta x^2 f_{i,j}. \tag{5.139}$$

To make the algorithm more general, we can introduce a scaling/relaxation parameter ω so that

$$\left(\frac{1}{\omega}D + L\right)\boldsymbol{x}^{k+1} = \boldsymbol{b} + \left(\left(\frac{1}{\omega} - 1\right)D - U\right)\boldsymbol{x}^k. \tag{5.140}$$

For $\omega = 1$, we recover Gauss–Seidel, but for $1 < \omega < 2$, we often obtain faster convergence. This is called Successive Over-Relaxation (SOR). For some specific classes of problem the optimal value of ω can be calculated or estimated. Unlike Richardson and Jacobi, the ordering of the vector changes the algorithm; Gauss–Seidel and SOR are *successive displacement methods* where one can overwrite the old vector with the new one element by element.

5.10   Using Splitting Methods as Preconditioners

Recall that we seek an approximation $\hat{A}$ of A to use as a preconditioner. A (non-symmetric) preconditioner $\hat{A}$ can be built from a splitting method by approximating the inverse of A by applying one iteration with the initial guess $z_0 = 0$. Then,

$$\hat{A}^{-1}Ax = z, \tag{5.141}$$

where

$$Mz = -Nz_0 + Ax = Ax, \tag{5.142}$$

i.e., $\hat{A} = M$.

For the CG algorithm, we require $\hat{A}$ to be symmetric. In the case of SOR, $M = D + L$ which is not symmetric. To address this, we consider that a second splitting method can be built from $A = M^T + N^T$. Combining them leads to

$$Mx^{k+\frac{1}{2}} = -Nx^k + b, \quad M^T x^{k+1} = -N^T x^{k+\frac{1}{2}} + b. \tag{5.143}$$

For example, SOR $(L = U^T)$ takes the form

$$\left(L + \frac{1}{\omega}D\right)x^{k+\frac{1}{2}} = \left(\left(\frac{1}{\omega} - 1\right)D - U\right)x^k + b, \tag{5.144}$$

$$\left(U + \frac{1}{\omega}D\right)x^{k+1} = \left(\left(\frac{1}{\omega} - 1\right)D - L\right)x^{k+\frac{1}{2}} + b. \tag{5.145}$$

In the general scaled SOR case, this is called Symmetric Successive Over-Relaxation (SSOR).

To convert a symmetric splitting into a preconditioner, we write the symmetric iteration as a single step with $x^0 = 0$.

$$M^T x^1 = (M - A)^T x^{\frac{1}{2}} + b \tag{5.146}$$

$$= (M - A)^T M^{-1} b + b \tag{5.147}$$

$$= (M^T + M - A)M^{-1} b, \tag{5.148}$$

so $\hat{A}^{-1} = M^{-T}(M^T + M - A)M^{-1}$. For example, the symmetric Gauss–Seidel preconditioning matrix is $\hat{A}^{-1} = (L+D)^{-T}D(L+D)^{-1}$.

5.11 Convergence Properties for Splitting Methods

In this section, we develop tools for studying convergence properties of splitting methods, and show how these properties relate to their suitability as preconditioners.

Recall that a splitting $A = M + N$, defines the iterative method

$$M\boldsymbol{x}^{k+1} = -N\boldsymbol{x}^k + \boldsymbol{b}. \tag{5.149}$$

On the other hand, the solution $\boldsymbol{x}^*$ of $A\boldsymbol{x} = \boldsymbol{b}$ satisfies

$$M\boldsymbol{x}^* = -N\boldsymbol{x}^* + \boldsymbol{b}. \tag{5.150}$$

Subtracting Eq. (5.150) from Eq. (5.149) gives

$$M\boldsymbol{e}_{k+1} = -N\boldsymbol{e}_k, \quad \boldsymbol{e}_k = \boldsymbol{x}^* - \boldsymbol{x}_k, \tag{5.151}$$

and so

$$\boldsymbol{e}_{k+1} = C\boldsymbol{e}_k \implies \boldsymbol{e}_k = C^k\boldsymbol{e}_0, \quad C = -M^{-1}N = I - M^{-1}A. \tag{5.152}$$

C is called the iteration matrix. For the corresponding symmetric iterative method,

$$M\boldsymbol{x}_{k+\frac{1}{2}} = -N\boldsymbol{x}_k + \boldsymbol{b}, \quad M^T\boldsymbol{x}_{k+1} = -N^T\boldsymbol{x}_{k+\frac{1}{2}} + \boldsymbol{b}. \tag{5.153}$$

The symmetric iteration matrix is

$$C = (I - (M^T)^{-1}A)(I - M^{-1}A) \tag{5.154}$$

$$= I - (M^S)^{-1}A, \tag{5.155}$$

where

$$M^s = M(M + M^T - A)^{-1}M^T. \tag{5.156}$$

Definition 5.9. An iterative method based on the splitting $A = M + N$ with iteration matrix $C = -M^{-1}N$ is called convergent if

$$\boldsymbol{y}_k = C^k \boldsymbol{y}_0 \to \boldsymbol{0} \tag{5.157}$$

for any initial vector $\boldsymbol{y}_0$.

This means that $\boldsymbol{e}_k = \boldsymbol{x}^* - \boldsymbol{x}_k \to \boldsymbol{0}$, i.e., $\boldsymbol{x}_k \to \boldsymbol{0}$ as $k \to \infty$.

We now present two convergence criteria for splitting methods.

Proposition 5.1. *If $\|C\|_p < 1$ for $p = 1$, 2 or ∞, then the iterative method converges.*

Proof.

$$\|\boldsymbol{y}_k\|_p = \|C^k \boldsymbol{y}_0\|_p \tag{5.158}$$

$$\leqslant \|C^k\|_p \|\boldsymbol{y}_0\|_p \tag{5.159}$$

$$\leqslant (\|C\|_p)^k \|\boldsymbol{y}_0\|_p \to 0 \quad \text{as } k \to \infty. \tag{5.160}$$

$\square$

This is only a sufficient condition, there may be matrices C with $\|C\|_p > 1$ for some $p = 1, 2, \infty$, but the method is still convergent.

A better convergence criteria makes use of the following definition.

Definition 5.10. The spectral radius $\rho(C)$ of a matrix C is the maximum of the absolute values of all the eigenvalues λ_i of C:

$$\rho(C) = \max_{1 \leqslant i \leqslant n} |\lambda_i|. \tag{5.161}$$

Proposition 5.2. *An iterative method converges if and only if $\rho(C) < 1$.*

Proof. If $\rho(C) \geqslant 1$, then C has an eigenvector $\boldsymbol{v}$ with $\|\boldsymbol{v}\|_2 = 1$ and eigenvalue λ with $|\lambda| > 1$. Then,

$$\|C^k \boldsymbol{v}\|_2 = \|\lambda^k \boldsymbol{v}\|_2 = |\lambda|^k \|\boldsymbol{v}\|_2 \geqslant 1, \tag{5.162}$$

which does not converge to zero. Assuming a complete set of eigenvectors $\boldsymbol{v}_i$, we expand

$$\boldsymbol{z} = \sum_{i=1}^{n} \alpha_i \boldsymbol{v}_i. \tag{5.163}$$

Then,

$$C^k z = \sum_{i=1}^{n} \alpha_i C^k v_i = \sum_{i=1}^{n} \alpha_i \lambda^k v_i \to 0. \qquad (5.164)$$

$\square$

The proof can be extended to more general matrices without a linearly independent eigenvector decomposition, but is more complicated.

We now make some remarks about these two criteria.

(i) For symmetric matrices B, $\rho(B) = \|B\|_2$, so the two criteria are equivalent.

(ii) If $\|C\|_p = c < 1$, then

$$\|e_{k+1}\|_p = \|C e_k\|_p \leqslant \|C\|_p \|e_k\|_p = c \|e_k\|_p. \qquad (5.165)$$

This guarantees that the error will be reduced by a factor of at least c in each iteration.

(iii) If we only have $\rho(C) < 1$, but not $\|C\|_p < 1$ then the error may not converge monotonically.

Exercise 5.19. *In this exercise, we use the iteration matrix to analyse the SOR parameter ω. For SOR, show that the iteration matrix C is*

$$C = \left(\frac{1}{\omega} D + L \right)^{-1} \left(\frac{1-\omega}{\omega} D - U \right) = (D + \omega L)^{-1} ((1-\omega)D - \omega U). \quad (5.166)$$

By computing the determinant of C, show that $\rho(C) < 1$ requires $|1 - \omega| < 1$.

We now relate these criteria to the use of splitting methods as preconditioners. Recall that a splitting $A = M + N$ leads to a preconditioner $\hat{A} = M$.

Proposition 5.3. *Let A be a matrix with splitting $M + N$, such that*

$$\rho(C) < c < 1. \qquad (5.167)$$

Then, the eigenvalues of the left preconditioned matrix $\hat{A}^{-1} A$ are located in a disk of radius c around 1 in the complex plane.

We deduce that good convergence of the CG algorithm occurs when c is small.

Proof.

$$C = -M^{-1}N = M^{-1}(M - A) = I - M^{-1}A. \qquad (5.168)$$

Then,

$$1 > c > \rho(C) = \rho(I - M^{-1}A), \qquad (5.169)$$

and the result follows since I and $M^{-1}A$ have a simultaneous eigendecomposition (i.e., if λ is an eigenvalue of $M^{-1}A$ then $1 - \lambda$ is an eigenvalue of $I - M^{-1}A$. $\qquad \square$

Now we consider the case of symmetric splitting methods,

$$M\boldsymbol{x}^{k+\frac{1}{2}} = -N\boldsymbol{x}^k + \boldsymbol{b}, \quad M^T\boldsymbol{x}^{k+1} = -N^T\boldsymbol{x}^{k+\frac{1}{2}} + \boldsymbol{b}. \qquad (5.170)$$

The iteration matrix is

$$C = (I - (M^T)^{-1}A)(I - M^{-1}A) \qquad (5.171)$$

$$= I - (M^S)^{-1}A, \qquad (5.172)$$

where

$$M^s = M(M + M^T - A)^{-1}M^T. \qquad (5.173)$$

We have

$$C = I - \hat{A}^{-1}A, \qquad (5.174)$$

where $\hat{A}^{-1}$ is the symmetric splitting preconditioner. Hence, we have shown the following.

Proposition 5.4. *Let A be a matrix with splitting $M + N$, such that the symmetric splitting has iteration matrix*

$$\rho(C) < c < 1. \qquad (5.175)$$

Then, the eigenvalues of the symmetric preconditioned matrix $\hat{U}^{-T}A\hat{U}^{-1}$ are contained in the interval $[1 - c, 1 + c]$ (where $U^{-T} = (U^T)^{-1}$).

This means that a convergent symmetric splitting method leads to a preconditioned matrix with condition number

$$\kappa = \frac{1 + c}{1 - c}, \qquad (5.176)$$

which gets closer to 1 as c gets smaller.

5.12 Convergence Analysis for Splitting Methods

In this section, we apply the convergence criteria to the splitting methods that we have already introduced.

5.12.1 *Richardson iteration*

For scaled Richardson iteration we have

$$x^{k+1} = x^k - \omega(Ax^k - b), \ M = \frac{I}{\omega}, N = A - \frac{I}{\omega}, \implies C = I - \omega A. \quad (5.177)$$

If A has eigenvalues $\lambda_1, \lambda_2, \ldots, \lambda_n$ then C has eigenvalues $1 - \omega\lambda_1, 1 - \omega\lambda_2, \ldots, 1 - \omega\lambda_n$. We need $|1 - \omega\lambda_i| < 1$, $i = 1, \ldots, n$ for convergence. If A is symmetric positive-definite, then all the eigenvalues are real and positive. Therefore, all of the eigenvalues of C lie between $1 - \omega\lambda_{\min}$ and $1 - \omega\lambda_{\max}$, and we can minimise $\rho(C)$ by choosing $\omega = 2/(\lambda_{\min} + \lambda_{\max})$. We obtain

$$\rho(C) = 1 - 2\frac{\lambda_{\min}}{\lambda_{\min} + \lambda_{\max}} = \frac{\lambda_{\max} - \lambda_{\min}}{\lambda_{\min} + \lambda_{\max}}. \quad (5.178)$$

The eigenvectors for the model Poisson problem are

$$E_{i,j}^{k,l} = \sin(ik\pi h)\sin(jl\pi), \quad i, j, k, l = 1, \ldots, m-1, \quad (5.179)$$

with eigenvalues

$$\lambda_{k,l} = 4\left(\sin^2\frac{k\pi h}{2} + \sin^2\frac{l\pi h}{2}\right), \quad k, l = 1, \ldots, m-1. \quad (5.180)$$

The min and max eigenvalues are $\lambda_{\min} = 8\sin^2(\pi h/2)$ and $\lambda_{\max} = 8\cos^2(\pi h/2)$, so optimal value of ω is

$$\omega = \frac{2}{\lambda_{\min} + \lambda_{\max}} = \frac{1}{4}. \quad (5.181)$$

This is the same as Jacobi's method for our model problem, because the diagonal values are all the same. The 2-norm of C is

$$\|C\|_2 = \frac{\cot^2\left(\frac{\pi h}{2}\right) - 1}{\cot^2\left(\frac{\pi h}{2}\right) + 1} = \cos\pi h = 1 - \frac{\pi^2 h^2}{2} + \mathcal{O}(h^4), \quad (5.182)$$

as $h \to \infty$, so convergence gets slower and slower as the mesh is refined. This is a generic issue for iterative methods applied to discretisations of PDEs.

5.12.2   *Convergence analysis for symmetric matrices*

For the rest of this section, we restrict our attention to symmetric positive-definite matrices. First, we note that for a symmetric positive-definite matrix A,

$$\lambda_{\max} = \max_{x \neq 0} \frac{x^T A x}{x^T x} \equiv \|A\|_2^2, \quad \lambda_{\min} = \min_{x \neq 0} \frac{x^T A x}{x^T x}, \tag{5.183}$$

and so

$$\lambda_{\min} \|y\|_2^2 \leqslant y^T A y \leqslant \lambda_{\max} \|y\|_2^2 \tag{5.184}$$

for any non-zero vector y.

We also note the following useful form of the iteration matrix,

$$C = -M^{-1} N = -M^{-1}(A - M) = I - M^{-1} A. \tag{5.185}$$

Proposition 5.5. *For a splitting $A = M + N$, if the (symmetric) matrix $M + M^T - A$ is positive-definite then*

$$\|I - M^{-1} A\|_A < 1. \tag{5.186}$$

Proof. If $y = (I - M^{-1} A)x$, $w = M^{-1} A x$, then

$$\|y\|_A^2 = (x - w)^T A(x - w) = x^T A x - 2 w^T M w + w^T A w \tag{5.187}$$

$$= x^T A x - w^T (M + M^T) w + w^T A w \tag{5.188}$$

$$= x^T A x - w^T (M + M^T - A) w \tag{5.189}$$

$$\leqslant \|x\|_A^2 - \mu_{\min} \|w\|_2^2, \tag{5.190}$$

where $\mu_{\min}$ is the minimum eigenvalue of $M^T + M - A$. We have

$$\|w\|_2^2 = x^T A (M^{-1})^T M^{-1} A x \tag{5.191}$$

$$= (A^{1/2} x)^T A^{1/2} (M^{-1})^T M^{-1} A^{1/2} (A^{1/2} x) \tag{5.192}$$

$$\geqslant \hat{\mu}_{\min} \|A^{1/2} x\|_2^2 = \hat{\mu}_{\min} \|x\|_A^2, \tag{5.193}$$

where $\hat{\mu}_{\min}$ is the minimum eigenvalue of $A^{1/2} (M^{-1})^T M^{-1} A^{1/2}$, i.e., the square of the minimum eigenvalue of $M^{-1} A^{1/2}$, which is invertible so $\hat{\mu}_{\min} > 0$.

If $y = (I - M^{-1}A)x$, $w = M^{-1}Ax$, then

$$\|y\|_A^2 \leqslant (1 - \mu_{\min}\hat{\mu}_{\min}) \|x\|_A^2 < \|x\|_A^2. \tag{5.194}$$

$\square$

Proposition 5.6. *For a splitting $A = M + N$, if M is positive-definite, then*

$$\rho(I - M^{-1}A) = \|I - M^{-1}A\|_A = \|I - M^{-1}A\|_M. \tag{5.195}$$

Proof.

$$I - A^{1/2}M^{-1}A^{1/2} = A^{1/2}(I - M^{-1}A)A^{-1/2}, \tag{5.196}$$

$$I - M^{-1/2}AM^{-1/2} = M^{1/2}(I - M^{-1}A)M^{-1/2}, \tag{5.197}$$

so $I - M^{-1}A$, $I - A^{1/2}M^{-1}A^{1/2}$, and $I - M^{-1/2}AM^{-1/2}$ all have the same eigenvalues. Hence,

$$\rho(I - M^{-1}A) = \rho(I - A^{1/2}M^{-1}A^{1/2}) \tag{5.198}$$

$$= \|I - A^{1/2}M^{-1}A^{1/2}\|_2 \tag{5.199}$$

$$= \|I - M^{-1}A\|_A, \tag{5.200}$$

and similarly for $I - M^{-1/2}AM^{-1/2}$. $\square$

If $M + M^T - A$ is symmetric positive-definite then there is guaranteed reduction in the A-norm of the error in each iteration. If M is also symmetric positive-definite the there is guaranteed reduction in the M-norm of the error in each iteration.

Exercise 5.20 (Convergence of Jacobi). *In this exercise we concentrate on scaled Jacobi iteration $(M = D/\omega)$.*

(a) *Show that $M^T + M - A = 2D/\omega - A$.*

(b) *Show that $2D/\omega - A$ and $2I/\omega - D^{-1}A$ have the same eigenvalues.*

(c) *Let λ be the maximum eigenvalue of $D^{-1}A$ (which is symmetric positive-definite). Show that $2/\omega - \lambda$ is the minimum eigenvalue of $2I/\omega - D^{-1}A$ and hence of $2D/\omega - D^{-1}A$, and hence that $2D/\omega - A$.*

(d) *Show that scaled Jacobi converges provided that $\omega < 2/\lambda$.*

Example 5.2 (Convergence of Gauss–Seidel). For Gauss–Seidel,

$$M^T + M - A = (D+L)^T + D + L - A = D + U + D + L - A = D, \quad (5.201)$$

which is symmetric positive-definite, so Gauss–Seidel always converges. For SOR,

$$M^T + M - A = \left(\frac{1}{\omega}D + L\right)^T + \frac{1}{\omega}D + L - A \quad (5.202)$$

$$= \frac{2}{\omega}D + U + L - (L + D + U) = \left(\frac{2}{\omega} - 1\right)D, \quad (5.203)$$

which is symmetric positive-definite provided that $0 < \omega < 2$. For our model problem, it can be shown that the optimal value of ω is

$$\omega = \frac{2}{1 + \sqrt{1 - \rho_J^2}}, \quad (5.204)$$

where ρ_J is the spectral radius of the Jacobi iteration,

$$\rho_J = \rho\left(I - D^{-1}A\right), \quad (5.205)$$

for which the convergence rate is

$$\rho_{\text{SOR}} = \frac{1 - \sqrt{1 - \rho_J^2}}{1 + \sqrt{1 - \rho_J^2}} = \frac{1 - \sin \pi h}{1 + \sin \pi h} = 1 - 2\pi h + \mathcal{O}(h^2). \quad (5.206)$$

Example 5.3 (Symmetric iterative methods). For symmetric iterative methods, the iteration matrix is

$$C = (I - (M^T)^{-1}A)(I - M^{-1}A) \quad (5.207)$$

$$= I - (M^S)^{-1}A, \quad (5.208)$$

where

$$M^s = M(M + M^T - A)^{-1}M^T. \quad (5.209)$$

For convergent methods, $\|I - M^{-1}A\|_A < 1$ and $\|I - (M^T)^{-1}A\|_A < 1$, so

$$\|I - (M^S)^{-1}A\|_A \leqslant \|I - M^{-1}A\|_A \|I - (M^T)^{-1}A\|_A < 1, \quad (5.210)$$

and the method also converges.

See Golub and Loan (2012) for more details.

Bibliography

Ablowitz, M. J. and Fokas, A. S. (2003). *Complex variables: Introduction and Applications* (Cambridge University Press, Cambridge).

Albert, J. (2007). *Bayesian Computation with R* (Springer).

Allen, J. S. and Holm, D. D. (1996). Extended-geostrophic Hamiltonian models for rotating shallow water motion, *Physica D* **98**, pp. 229–248.

Arakawa, A. and Lamb, V. (1977). Computational design of the basic dynamical processes of the UCLA general circulation model, *Methods in Computational Physics* **17**, pp. 173–265.

Baladi, V. (2000). *Positive transfer operators and decay of correlations [Advanced Series in Nonlinear Dynamics, Vol. 16]* (World Scientific Publishing Co., Inc., River Edge, NJ).

Barker, A. A. (1965). Monte Carlo calculations of the radial distribution functions for a proton-electron plasma, *Australian Journal of Physics* **18**, pp. 119–133.

Barreira, L. and Valls, C. (2013a). *Dynamical systems. An Introduction* (Springer, London).

Barreira, L. and Valls, C. (2013b). *Dynamical systems. An Introduction [Translated from the 2012 Portuguese original]* (Springer, London).

Beichl, I. and Sullivan, F. (2000). The Metropolis algorithm, *Computing in Science and Engineering* **2**, pp. 65–69.

Benamou, J. D. and Brenier, Y. (1998). Weak existence for the semigeostrophic equations formulated as a coupled Monge–Ampere/transport problem, *SIAM Journal on Applied Mathematics* **58**, pp. 1450–1461.

Bolstad, W. M. (2010). *Understanding Computational Bayesian Statistics* (Wiley, New York).

Breiman, L. (1973). *Probability* (Addison-Wesley, Reading, MA).

Brenier, Y. (1991). Polar factorization and monotone rearrangement of vector-valued functions, *Communications on Pure and Applied Mathematics* **44**, pp. 375–417.

Browning, G. L., Kasahara, A., and Kreiss, H. O. (1980). Initialization of the primitive equations by the bounded derivative method, *Journal of Atmospheric Science* **37**, pp. 1424–1436.

Browning, G. L. and Kreiss, H. O. (1987). Reduced systems for the shallow water equations, *Journal of Atmospheric Science* **44**, pp. 2813–2822.

Chemin, J. Y., Desjardins, B., Gallagher, I., and Grenier, E. (2006). Basics of mathematical geophysics, *Oxford Lecture Series in Mathematics and its Applications* **32**, (Oxford University Press, New York).

Chen, M. H., Shao, Q. M., and Ibrahim, J. G. (2000). *Monte Carlo Methods in Bayesian Computation* (Springer, New York).

Chib, S. and Greenberg, E. (1996). Understanding the Metropolis-Hastings algorithm, *The American Statistician* **49**, pp. 327–335.

Cotter, C. J., Holm, D. D., and Percival, J. R. (2010). The square root depth wave equations, *Proceedings of the Royal Society of London – Series A: Mathematical and Physical Science* **466**, pp. 3621–3633.

Cullen, M. J. P. (2006). *A Mathematical Theory of Large-scale Atmosphere/ocean Flow* (World Scientific, Singapore).

de Moivre, A. (1967). *The Doctrine of Chances or, a Method of Calculating the Probabilities of Events in Play* (Frank Cass & Co., Ltd., London).

Devroye, L. (1986). *Non-Uniform Random Variate Generation* (Springer-Verlag, New York).

Doob, J. L. (1994). *Measure Theory* (Springer, New York).

Dudley, R. M. (1989). *Real Analysis and Probability* (Chapman & Hall, CRC, London).

Durran, D. R. (1999). *Numerical Methods for Wave Equations in Geophysical Fluid Dynamics* (Springer, New York).

Feller, W. (1966). *An Introduction to Probability Theory and Its Applications* (John Wiley & Sons, Inc., New York).

Feller, W. (1970). *An Introduction to Probability Theory and Its Applications* (John Wiley & Sons, Inc., New York).

Gallavotti, G. (1983). *The Elements of Mechanics* (Springer-Verlag, New York).

Gamerman, D. and Lopes, H. F. (2006). *Markov Chain Monte Carlo: Stochastic simulation for Bayesian inference* (Chapman & Hall, CRC, London).

Gelman, A., Carlin, J. B., Stern, H. S., and Rubin, D. B. (2004). *Bayesian Data Analysis* (Chapman & Hall, CRC, London).

Geyer, C. (1992). Practical Markov chain Monte Carlo, *Statistical Science* **7**, pp. 473–483.

Golub, G. H. and Loan, C. F. V. (2012). *Matrix Computations, Vol. 3* (JHU Press, USA).

Halmos, P. R. (1974). *Measure Theory* (Springer, New York).

Hastings, W. (1970). Monte Carlo sampling methods using Markov chains and their application, *Biometrika* **57**, pp. 97–109.

Held, I. M. (1999). The macroturbulence of the troposphere, *Tellus* **51**, pp. 59–70.

Held, I. M. and Hou, A. Y. (1980). Nonlinear axially symmetric circulations in a nearly inviscid atmosphere, *Journal of Atmospheric Science* **37**, pp. 515–533.

Holm, D. D. (1996). Hamiltonian balance equations, *Physica D* **98**, pp. 379–414.

Holm, D. D. and Kupershmidt, B. A. (1983). Poisson brackets and Clebsch representations for magnetohydrodynamics, multifluid plasmas, and elasticity, *Physica D* **6**, pp. 347–363.

Holm, D. D., Kupershmidt, B. A., and Levermore, C. D. (1983). Canonical maps between Poisson brackets in Eulerian and Langrangian descriptions of continuum mechanics, *Physics Letters A.* **98**, pp. 389–395.

Holm, D. D., Marsden, J. E., and Ratiu, T. (1998). The Euler-Poincaré equations and semidirect products with applications to continuum theories, *Advances in Mathematics* **137**, pp. 1–81.

Holm, D. D., Marsden, J. E., and Ratiu, T. S. (2000). *The Euler–Poincaré equations in geophysical fluid dynamics.* [*Large-Scale Atmosphere-Ocean Dynamics 2: Geometric Methods and Models*] (Cambridge University Press, Cambridge).

Holm, D. D., Marsden, J. E., Ratiu, T. S., and Weinstein, A. (1985). Nonlinear stability of fluid and plasma equilibria, *Physics Reports* **123**, pp. 1–116.

Hoskins, B. J. and Karoly, D. J. (1981). The steady linear response of a spherical atmosphere to thermal and orographic forcing, *Journal of Atmospheric Science* **38**, pp. 1179–1196.

Jacod, J. and Protter, P. E. (2000). *Probability Essentials* (Springer, New York).

Jazwinski, A. H. (1970). *Stochastic Processes and Filtering Theory* (Academic Press, USA).

Kaper, H. and Engler, H. (2013). *Mathematics & Climate* (Society for Industrial and Applied Mathematics, Philadelphia, PA).

Katok, A. and Hasselblatt, B. (1995). *Introduction to the modern theory of dynamical systems* [*Encyclopedia of Mathematics and its Applications*] (Cambridge University Press, Cambridge).

Klein, R. (2010). Scale-dependent models for atmospheric flows, *Annual Review of Fluid Mechanics* **42**, pp. 249–274.

Klenke, A. (2014). *Probability Theory* (Springer, New York).

Koltermann, K. P., Gouretski, V. V., and Jancke, K. (2011). *Hydrographic Atlas of the World Ocean Circulation Experiment (WOCE)* [*Vol. 3: Atlantic Ocean*] (International WOCE Project Office, Southampton, U.K.).

Krengel, U. (1985). *Ergodic Theorems* (Walter de Gruyter & Co., Berlin).

Laplace, P. (1995). *Philosophical Essay on Probabilities* (Springer-Verlag, New York).

Liu, J. S. (2004). *Monte Carlo Strategies in Scientific Computing* (Springer, New York).

Lorenz, E. N. (1969). The predictability of a flow which possesses many scales of motion, *Tellus* **21**, pp. 289–307.

Lynch, P. (1989). The slow equations, *Quarterly Journal of the Royal Meteorological Society* **115**, pp. 201–219.

Majda, A. J. and Bertozzi, A. L. (2001). *Vorticity and Incompressible Flows* (Cambridge University Press, Cambridge).

Marsden, J. E. and Chorin, A. J. (1993). *A Mathematical Introduction to Fluid Dynamics* (Springer, New York).

Metropolis, N., Rosenbluth, A., M. Rosenbluth, A. T., and Teller, E. (1953). Equations of state calculations by fast computing machines, *Journal of Chemical Physics* **21**, pp. 1087–1092.

Olver, P. J. (2014). *Introduction to Partial Differential Equations* (Springer, New York).

Owen, B. (2013). *Monte Carlo Theory, Methods and Examples*.

Pedlosky, J. (1987). *Geophysical Fluid Dynamics, 2nd Edition* (Springer-Verlag, New York).

Peskun, P. H. (1973). Optimum Monte-Carlo sampling using Markov chains, *Biometrika* **60**, pp. 607–612.

Price, J. F. (1981). Upper ocean response to a hurricane, *Journal of Physical Oceanography* **2**, pp. 153–175.

Robert, C. P. and Casella, G. (2004). *Monte Carlo Statistical Methods* (Springer, New York).

Robert, C. P. and Casella, G. (2010). *Introducing Monte Carlo Methods with R* (Springer, New York).

Rubinstein, R. Y. and Kroese, D. P. (2008). *Simulation and the Monte Carlo Method* (Wiley, New York).

Savage, L. J. (1971). Elicitation of personal probabilities and expectation, *Journal of the American Statistical Association* **66**, pp. 783–801.

Schervish, M. J. (1995). *Theory of Statistics [Springer Series in Statistics]* (Springer-Verlag, New York).

Shepherd, T. G. (2003). *Hamiltonian dynamics [Encyclopedia of Atmospheric Sciences, pp. 929–938]* (Academic Press, USA).

Shepherd, T. G. (2014). Atmospheric circulation as a source of uncertainty in climate change projections, *Nature Geoscience* **7**, pp. 703–708.

Simmons, A. J. and Hoskins, B. J. (1978). The life cycles of some nonlinear baroclinic wave, *Journal of Atmospheric Science* **35**, pp. 414–432.

Sivia, D. S. and Skilling, J. (2006). *Data Analysis: A Bayesian Tutorial* (Oxford University Press, London).

Talley, L. D. (2007). *Hydrographic Atlas of the World Ocean Circulation Experiment (WOCE) [Vol. 2: Pacific Ocean]* (International WOCE Project Office, Southampton, U.K.).

Talley, L. D., Pickard, G. L., Emery, W. J., and Swift, J. H. (2011). *Descriptive Physical Oceanography* (Academic Press, USA).

Thorpe, A. J., Hoskins, B. J., and Innocentini, V. (1989). The parcel method in a baroclinic atmosphere, *Journal of Atmospheric Science* **46**, pp. 1274–1284.

Trefethen, L. N. and Bau, D. (1997). *Numerical Linear Algebra, Vol. 50* (SIAM, PA).

Vallis, G. K. (1996a). Approximate geostrophic models for large-scale flow in the ocean and atmosphere, *Physica D* **98**, pp. 647–651.

Vallis, G. K. (1996b). Potential vorticity inversion and balanced equations of motion for rotating and stratified flows, *Quarterly Journal of the Royal Meteorological Society* **122**, pp. 291–322.

Vallis, G. K. (2006). *Atmospheric and Oceanic Fluid Dynamics: Fundamentals and Large-scale Circulation* (Cambridge University Press, Cambridge).

van der Vaart, A. W. (2000). *Asymptotic Statistics* (Cambridge University Press, Cambridge).

Vanneste, J. and Shepherd, T. G. (1999). On wave action and phase in the non-canonical Hamiltonian formulation, *Proceedings of the Royal Society of London* **455**, pp. 3–21.

Viana, M. (1998). Dynamics: A probabilistic and geometric perspective, *Document of Mathematica Journal* **DMV, Extra Volume**, pp. 557–578.

von Neumann, J. (1951). Various techniques used in connection with random digits. Monte Carlo methods, *National of Bureau Standards*, pp. 36–38.

Warn, T., Bokove, O., Shepherd, T. G., and Vallis, G. K. (1995). Rossby number expansions, slaving principles and balance dynamics, *Quarterly Journal of the Royal Meterological Society* **121**, pp. 723–739.

Warneford, E. S. and Dellar, P. J. (2013). The quasi-geostrophic theory of the thermal shallow water equations, *Journal of Fluid Mechanics* **723**, pp. 374–403.

Whitham, G. B. (1974). *Linear and Nonlinear Waves* (John Wiley & Sons Inc.).

Williams, P. D. (2009). A proposed modification to the Robert-Asselin time filter, *Monthly Weather Review* **137**, pp. 2538–2546.

Williamson, D. L., Drake, J. B., Hack, J. J., Jakob, R., and Swarztrauber, P. N. (1992). A standard test set for numerical approximations to the shallow water equations in spherical geometry, **102** *Journal of Computational Physics*, pp. 211–224.

Zhang, C. (2005). Madden-Julian oscillation, *Review of Geophysics* **43**, RG2003.

Index

CPSIA information can be obtained
at www.ICGtesting.com
Printed in the USA
LVHW062359280519
619382LV00007B/19/P

9 781786 343833